Advances and Applications of Block Copolymers II

Advances and Applications of Block Copolymers II

Guest Editors

Nikolaos Politakos
Apostolos Avgeropoulos

MDPI

Basel • Beijing • Wuhan • Barcelona • Belgrade • Novi Sad • Cluj • Manchester

Guest Editors
Nikolaos Politakos
Institute of Chemical Biology
National Hellenic Research
Foundation
Athens
Greece

Apostolos Avgeropoulos
Department of Materials
Science Engineering
University of Ioannina
Ioannina
Greece

Editorial Office
MDPI AG
Grosspeteranlage 5
4052 Basel, Switzerland

This is a reprint of the Special Issue, published open access by the journal *Polymers* (ISSN 2073-4360), freely accessible at: https://www.mdpi.com/journal/polymers/special_issues/2ZEQ235MMY.

For citation purposes, cite each article independently as indicated on the article page online and as indicated below:

Lastname, A.A.; Lastname, B.B. Article Title. *Journal Name* **Year**, *Volume Number*, Page Range.

ISBN 978-3-7258-8023-2 (Hbk)
ISBN 978-3-7258-8024-9 (PDF)
https://doi.org/10.3390/books978-3-7258-8024-9

Contents

About the Editors

Nikolaos Politakos

Nikolaos Politakos is currently a Senior Researcher within the National Hellenic Research Foundation, Institute of Chemical Biology in Athens, Greece. He received his B.Sc. (2005) in Materials Engineering and M.Sc. (2007) in Chemistry and Technology of Materials at the Polytechnic School of the University of Ioannina (UoI), Ioannina, Greece. He obtained his Ph.D. (in 2010) in Materials Engineering from the Polytechnic School of the University of Ioannina (UoI), Ioannina, Greece. He started his postdoctoral research in 2010 at the University of the Basque Country (UPV/EHU) and completed this in 2014, working with chemical modifications of polymers and composites with fullerenes. From 2014 to 2017, he worked in CIC biomaGUNE in San Sebastian, synthesizing and characterizing responsive copolymer brushes from surfaces and particles for biomedical applications. From 2017 to 2020, he worked in POLYMAT, conducting research centered around the preparation of 3D sponge-like graphene oxide materials for CO_2 capture and photocatalysis. In 2020, he joined the Responsive Polymer Therapeutics Group and focused his research on preparing hydrogels for wound healing. Finally, since 2024, he has been a Postdoctoral Fellow HORIZON-WIDERA-2022-TALENTS within the National Hellenic Research Foundation, working with nanogels for pulmonary drug delivery. He has published 47 peer-reviewed publications, earning him an h-index of 18 and an i10-index of 27, as well as 960 total citations (according to Google Scholar) at present. He has participated in over 36 national and international conferences and over 20 national and international projects. His research interests include the synthesis of copolymers, biomaterials, nanogels, chemical modifications, polypeptides, wound healing, smart materials/responsiveness, hybrid materials, hydrophobic materials, and CO_2 adsorption and photocatalysis.

Apostolos Avgeropoulos

Apostolos Avgeropoulos is a Full Professor within the Department of Materials Science Engineering (DMSE), University of Ioannina (UoI), Greece. He obtained his B.Sc. in Chemistry from the Chemistry Department, University of Athens, in 1992 and his Ph.D. in 1997 in Polymer Science from the same Department, under the supervision of Prof. N. Hadjichristidis in collaboration with Prof. E. L. Thomas at DMSE, MIT, USA. He was a postdoctoral researcher at DMSE-MIT for a total of 16 months from 1997 to 2001, and a post-doctoral researcher at the Industrial Chemistry Laboratory, Chemistry Department, University of Athens and the National Center for Scientific Research "DEMOKRITOS" from 1999 to 2002. He was elected as an Adjunct Assistant Professor at DMSE-UoI in 2002. In 2003, he started as a non-tenured Assistant Professor, obtained tenure in 2007, and was promoted to Associate Professor in 2009. In 2013, he became a Full Professor in the same department, where he continues his research in polymers science and engineering. His current research interests involve high-molecular-weight copolymers, amphiphilic functionalized copolymers, nanocrystal/block copolymer composites (structure and structure properties), as well as nano-lithographic and nano-patterning applications of block copolymers with various architectures, among others. He has supervised more than 180 undergraduate diploma theses as a faculty member, as well as 51 MScs and 15 PhDs. He is the Director of the Polymer Science and Engineering Laboratory of DMSE-UoI. His overall funding from various agencies is approximately 7 M €. He has published 3 book chapters, 2 US patents, 1 EU patent, 1 Chinese patent, and 253 peer-reviewed publications (h-index= 49, 8818 citations—Google Scholar) at present. He has given more than 40 invited lectures in domestic and foreign institutions and has participated in 86 national/international conferences with more than 40 invited talks and 120 poster presentations.

Preface

This Reprint, entitled "Advances and Applications of Block Copolymers II", is a collection of works on polymer science and applications of block copolymers. This Reprint covers different aspects of polymer science, including synthesis, properties, and applications. The works presented focus on the full spectrum of materials science and nanotechnology. The aim of this Reprint is to help materials science researchers, especially those in polymers, understand various aspects of copolymers from a synthetic perspective for potential applications. This Reprint also aims to showcase a collection of works on polymeric materials, with an emphasis on applications. The Guest Editors were motivated by the ongoing, applied research on polymeric materials, being a class of materials used in everyday life, from simple applications to more specialized ones.

Nikolaos Politakos and Apostolos Avgeropoulos
Guest Editors

polymers

MDPI

Editorial

Advances and Applications of Block Copolymers II

Nikolaos Politakos [1,*] and Apostolos Avgeropoulos [2,*]

1 Institute of Chemical Biology, National Hellenic Research Foundation, 11635 Athens, Greece
2 Department of Materials Science Engineering, University of Ioannina, 45110 Ioannina, Greece
* Correspondence: npolit@eie.gr (N.P.); aavger@uoi.gr (A.A.)

Polymers are an essential class of materials; they are constantly evolving, and new monomers/polymers are being synthesized and/or produced every day. Polymeric materials can be used in both every-day and cutting-edge applications, including nanotechnology, energy, biotechnology, environment, aerospace, electronics, and agriculture industries. Currently, enormous challenges are being raised where polymers may offer solutions. Globally, there is a focus on circular economy and sustainability, high-performance and functional materials, and advanced manufacturing.

Block copolymers (BCPs) are a major part of polymeric materials overall. This class of materials is evolving through the development of new monomers and polymerization techniques, leading to biodegradable BCPs for novel applications. Self-assembly of BCPs can lead to improved applications in nanolithography and nanomanufacturing, as well as in various hierarchical architectures. In energy storage and conversion, particularly in photovoltaics, solid polymer electrolytes and BCPs are also essential. Another growing area of research into BCPs is innovative nanomedicine and targeted delivery of bioactive compounds, in which stimulus-responsive systems based on BCPs are being synthesized to deliver drugs, RNA, or proteins with high accuracy. The use of BCPs as vaccine platforms or in bioimaging is also relatively novel and currently under development. Finally, BCPs are also important for sustainability applications, especially as bio-based materials.

However, BCPs face new challenges every day, and scientists should focus their research on current and future trends in applications. Active areas of research include semiconductors, high-energy solid-state batteries, chemical recycling, personalized medical treatments, conversion of toxic materials, reutilization, and membranes.

This Special Issue is a collection of scientific contributions from research groups investigating diverse copolymers, highlighting recent advances in polymers, functionalities, and applications. The publication of thirteen (13) manuscripts in this Special Issue underscores the significance of block copolymers (BCPs) and demonstrates the sustained interest of the scientific community in the development and application of these materials.

A study on the compatibility, crystal structure, and mechanical properties of new-generation block copolymer PP-b-PE (polypropylene-b-polyethylene) and PPR (random polypropylene) blends has revealed that PPR crystallizes first, during the cooling process, inducing early crystallization of PP-b-PE, which increases the crystallization temperature of PPR/PP-b-PE blends [1].

Another study has explored the self-assembly of cylinder-forming BCPs under the guidance of pillar topographic patterns, utilizing mesoscale density functional theory simulations. This investigation also focuses on directed self-assembly structures of cylinder-forming BCPs with varying parameters, including pillar pitch and diameter. The insights from this study provide a theoretical framework for designing directed self-assembly systems capable of multiplying the density of contact holes, a crucial component of lithographic processes used in electronic devices [2].

Moreover, a series of novel renewable copolymers based on poly(ethylene succinate) (PESu) and poly(isosorbide succinate) (PISSu), with a varying isosorbide (Is)/PESu molar ratio, were synthesized in situ and studied. The designed copolymers were characterized in depth with respect to their structural and thermodynamic properties. Dielectric spectroscopy, along with proper analysis, enabled molecular dynamics mapping of both local and segmental types, which is presented for these materials for the first time. Based on the overall findings, the systems were homogeneous, with the two polymer segments thoroughly distributed. Upon analyzing the results, evaluating them by using widely adopted routes and models, and comparing those from different techniques, an in-depth structure–property relationship was found [3].

Other experimental work focused on interpenetrating hydrogel networks (IHNs) composed of p(HEMA) or p(MAA) and different pluronic block copolymers that were designed for potential use as medical device biomaterials or as platforms for controlled drug delivery. In this study, the physicochemical properties of IHNs provided a greater understanding of the physical state of the two polymer components as well as the interactions between them, and how these interactions affect retention within the network when immersed in an aqueous fluid. This approach provided invaluable knowledge of the practical uses of these specific IHNs as medical device biomaterials/coatings [4].

The high importance of BCPs in structural biology is evident in the production of functionally active membrane proteins (MPs) in an appropriate membrane environment. Polymer–lipid particles based on styrene and maleic acid (SMA) represent a promising type of membrane mimic, as they can extract properly folded MPs directly from their native lipid environment. The introduction of a novel SMA derivative with a negatively charged taurine moiety demonstrates that both polymers can form nanodiscs with a patch of lipid bilayer that can undergo phase transitions at temperatures close to those of the lipid bilayer membranes [5].

In the field of thermal insulation, a study of melamine–aniline–formaldehyde copolymer organoclay nanocomposite foams is presented, analyzing their structural and textural properties and investigating their thermal insulation and strength. Better thermal insulation behavior indicates the effect of a more uniformly packed microsphere arrangement and a more elastic network structure, including aniline–formaldehyde bridges, on the formation of void structures. The thermal insulation values decreased gradually with increasing organoclay content in resin–organoclay nanocomposites, while the mechanical stability indicated an increase in elasticity caused by the incorporation of aniline–formaldehyde bridges into the polymer network [6].

There is global demand for stimuli-responsive polymeric nanostructures for a wide range of applications. Self-assembling amphiphilic copolymers with stimuli-responsive characteristics—created by introducing a hydrophilic tuneable monomer, (2-dimethylamino)ethyl methacrylate (DMAEMA), together with a hydrophilic one, lauryl methacrylate (LMA)—within linear and branched copolymer topologies were investigated. The structural changes in the obtained self-organized supramolecular structures were thoroughly analyzed using aqueous media with varying pH and salinity values. These findings contribute to general knowledge of the synthesis of linear and hyperbranched amphiphilic polyelectrolyte-type copolymers and their self-assembly behavior, leading to the formation of nanostructures in solution. The pH- and salt-responsive properties of the PDMAEMA component, together with variations in copolymer composition, allow for fine-tuning of the formed nanoparticles' structures [7].

https://doi.org/10.3390/polym18030335

Additionally, responsive material and environmental applications could be possible for the CO_2-responsive polymer poly(N-[3-(dimethylamino)propyl]-acrylamide)-b-poly(methyl methacrylate) (PDMAPAm-*b*-PMMA), which is expected to be implemented in direct air capture (DAC) technology. In this work, a kinetic model, a crucial tool for optimizing polymer synthesis protocols and facilitating scaled-up production of the above-mentioned copolymer, is presented. In the end, the optimal parameters were determined for synthesizing a CO_2-responsive diblock copolymer of PDMAPAm-b-PMMA with the desired molar mass and PDMAPAm block-to-PMMA block molar mass ratio [8].

The mechanical properties of double hydrophilic block copolymers based on poly[oligo(ethylene glycol) methacrylate] (POEGMA) and poly(vinyl benzyl trimethylammonium chloride) (PVBTMAC) blocks were studied using small-amplitude oscillatory shear rheology. The authors showed that changing the block arrangement along the backbone from statistical to sequential results in a distinct change in the viscoelastic response, indicating the presence/absence of bulk-like regions. The tunable viscosity and shear-thinning behavior achieved by altering the copolymer composition and block arrangement along the backbone render these materials promising candidates for drug delivery applications. Overall, this study provides information about the viscoelastic properties of densely grafted macromolecular architectures and shows how the mechanical and dynamic properties can be tailored for different drug delivery applications by simply altering the ratio between the two homopolymers [9].

Further work on the combination of Brownian dynamics simulations and self-consistent field theory has demonstrated that stable assembled structures, such as vesicles, toroidal micelles, bowl-like micelles, sheet-like micelles, non-spherical vesicles, and cylindrical micelles, are dependent on the molecular parameters of amphiphilic comb-like copolymers. In this study, it was found that vesicle formation involves two intermediate states—sheet- and bowl-like micelles—and that the difference in their free energies is minimal, indicating coexistence. This study describes how to obtain different morphologies by adjusting the molecular parameters of amphiphilic comb-like copolymers, further confirming their potential for stable drug encapsulation and enhanced targeted drug delivery [10].

In the field of self-assembly behavior, a series of amphiphilic diblock copolymers, each consisting of a hydrophilic poly(N-vinyl pyrrolidone) (PNVP) block and a hydrophobic block derived from n-alkyl vinyl esters, namely, poly(vinyl butyrate) (PVBu), poly(vinyl decanoate) (PVDc), and poly(vinyl stearate) (PVSt), in aqueous solutions was investigated. Using dynamic and static light scattering techniques, the micellization behavior was studied. The effect of the hydrophobic block's nature in the copolymer composition, as well as the copolymer molecular weight on the self-assembly properties were thoroughly examined. The study also investigated the encapsulation of curcumin and indomethacin within the dry cores of the micellar structures. Combined with the observed stability of the micellar structures, these findings suggest that block copolymers have significant potential as drug carriers [11].

Another study on the self-assembly of a high-molecular-weight poly(styrene-*b*-isoprene) block copolymer was analyzed after chemical modification. The polymer was converted into poly(styrene-*b*-(ethylene-alt-propylene)) by hydrogenation and into poly(styrene-*b*-sulfonated isoprene) by mild sulfonation of the PI block. The obtained morphologies were examined by atomic force microscopy to analyze the effects of sample preparation parameters, including solvent, casting technique, and annealing temperature, which highlighted that significant morphological and topographical changes were observed across different parameters used. Each modification introduces new variables that can affect the final structure and properties of different types of copolymers [12].

https://doi.org/10.3390/polym18030335

Finally, a review of the different routes and methods for preparing hybrid materials based on nanostructured block copolymers (BCPs) and magnetic nanoparticles (MNPs) is included in this Special Issue. This work synthesizes studies on the synthesis and characterization of BCP/MNP hybrid systems using various strategies or methods, highlighting the main findings and potential applications of the resulting materials [13].

Throughout this Special Issue, it is clear that block copolymers play a key role in advancing polymer science. The progress of block copolymers, both in synthesis and in self-assembly, as well as diverse applications, is thoroughly demonstrated in this collection. As Guest Editors, we are confident that these studies will make a substantial contribution to the field of block copolymers.

Author Contributions: Writing—original draft preparation, N.P. and A.A.; writing—review and editing, N.P. and A.A. All authors have read and agreed to the published version of the manuscript.

Funding: N.P. would like to acknowledge the European Union for the funding, HORIZON-WIDERA-2022-TALENTS-02, grant number Project 101090325—PuDeRegels.

Conflicts of Interest: The authors declare no conflicts of interest.

References

1. Shao, W.; Liu, L.-Z.; Wang, Y.; Wang, Y.; Shi, Y.; Song, L. Investigation of Crystallization, Morphology, and Mechanical Properties of Polypropylene/Polypropylene-Polyethylene Block Copolymer Blends. *Polymers* **2023**, *15*, 4680. [CrossRef] [PubMed]
2. Huh, J. Directed Self-Assembly of Cylinder-Forming Block Copolymers Using Pillar Topographic Patterns. *Polymers* **2024**, *16*, 881. [CrossRef] [PubMed]
3. Bouyahya, C.; Klonos, P.A.; Zamboulis, A.; Xanthopoulou, E.; Ainali, N.M.; Majdoub, M.; Kyritsis, A.; Bikiaris, D.N. Preparation and Structural-Thermodynamical Investigation of Renewable Copolyesters Based on Poly (Ethylene Succinate) and Polyisosorbide. *Polymers* **2024**, *16*, 2173. [CrossRef] [PubMed]
4. Jones, D.S.; Westwood, M.; Li, S.; Andrews, G.P. Spectroscopic and Thermal Characterisation of Interpenetrating Hydrogel Networks (IHNs) Based on Polymethacrylates and Pluronics, and Their Physicochemical Stability under Aqueous Conditions. *Polymers* **2024**, *16*, 2796. [CrossRef] [PubMed]
5. Motov, V.V.; Kot, E.F.; Kislova, S.O.; Bocharov, E.V.; Arseniev, A.S.; Boldyrev, I.A.; Goncharuk, S.A.; Mineev, K.S. On the Properties of Styrene–Maleic Acid Copolymer–Lipid Nanoparticles: A Solution NMR Perspective. *Polymers* **2024**, *16*, 3009. [CrossRef] [PubMed]
6. Gürses, A.; Şahin, E. Preparation and Characterization of Melamine Aniline Formaldehyde-Organo Clay Nanocomposite Foams (MAFOCF) as a Novel Thermal Insulation Material. *Polymers* **2024**, *16*, 3578. [CrossRef] [PubMed]
7. Gerardos, A.M.; Forys, A.; Trzebicka, B.; Pispas, S. One-Pot Synthesis of Amphiphilic Linear and Hyperbranched Polyelectrolytes and Their Stimuli-Responsive Self-Assembly in Aqueous Solutions. *Polymers* **2025**, *17*, 701. [CrossRef] [PubMed]
8. Pashayev, E.; Georgopanos, P. Optimizing the Synthesis of CO_2-Responsive Polymers: A Kinetic Model Approach for Scaling Up. *Polymers* **2025**, *17*, 1115. [CrossRef] [PubMed]
9. Pipertzis, A.; Chroni, A.; Pispas, S.; Swenson, J. Viscoelastic Response of Double Hydrophilic Block Copolymers for Drug Delivery Applications. *Polymers* **2025**, *17*, 1857. [CrossRef] [PubMed]
10. Chen, Q.; Tian, K.; Zhu, R.; Ding, M.; Xu, Z. Self-Assembly of Amphiphilic Comb-like Copolymers into Micelles and Vesicles in Solution. *Polymers* **2025**, *17*, 1870. [CrossRef] [PubMed]
11. Plachouras, N.V.; Gkolemi, A.-M.; Argyropoulos, A.; Bouzoukas, A.; Papazoglou, T.-P.; Roka, N.; Pitsikalis, M. Amphiphilic Diblock Copolymers of Poly(N-vinyl pyrrolidone) and Poly(vinyl esters) Bearing N-Alkyl Side Chains for the Encapsulation of Curcumin and Indomethacin. *Polymers* **2025**, *17*, 2852. [CrossRef] [PubMed]
12. Politakos, N.; Kortaberria, G.; Avgeropoulos, A. Morphological Characterization of High Molecular Weight Poly(styrene-*b*-isoprene) or PS-*b*-PI and Its Hydrogenated, Sulfonated Derivatives: An AFM Study. *Polymers* **2025**, *17*, 3047. [CrossRef] [PubMed]
13. Kortaberria, G. Hybrid Materials Based on Self-Assembled Block Copolymers and Magnetic Nanoparticles—A Review. *Polymers* **2025**, *17*, 1292. [CrossRef] [PubMed]

https://doi.org/10.3390/polym18030335

polymers

MDPI

Article

Investigation of Crystallization, Morphology, and Mechanical Properties of Polypropylene/Polypropylene-Polyethylene Block Copolymer Blends

Wenjun Shao [1,2], Li-Zhi Liu [1], Ying Wang [1,2], Yuanxia Wang [1,*], Ying Shi [1,3,*] and Lixin Song [1]

1 Advanced Manufacturing Institute of Polymer Industry, Shenyang University of Chemical Technology, Shenyang 110142, China; shaowenjun0813@163.com (W.S.); violetlj@yahoo.com (L.-Z.L.); wy584436393@163.com (Y.W.); lxsong@syuct.edu.cn (L.S.)
2 School of Materials Science and Engineering, Shenyang University of Technology, Shenyang 110870, China
3 Dongguan HAILI Chemical Material Co., Ltd., Dongguan 523808, China
* Correspondence: wangyuanxia@aliyun.com (Y.W.); shiying86@aliyun.con (Y.S.)

Abstract: Polyethylene (PE)-based elastomers are the ideal choice for enhancing the compatibility of polypropylene/polyethylene (PP/PE) blends and improving the mechanical properties of PP-based materials. However, the issue of blend systems lies in the interplay between the crystallization processes. Therefore, we investigated the crystallization behavior during the cooling process of a new generation of PP/PE block copolymers (PP-b-PE) and random polypropylene (PPR, a copolymer of propylene and a small amount of ethylene or an alpha-olefin) blends using in-situ X-ray diffraction/scattering and differential scanning calorimetry (DSC) techniques. We also conducted mechanical performance tests on PPR/PP-b-PE blends at room temperature and low temperature (−5 °C). The results indicate that during the cooling process, the PP phase of PP-b-PE will follow the PPR to crystallize in advance and form a eutectic mixture, thereby enhancing the compatibility of PP/PE. Moreover, the PPR/PP-b-PE blend will form stable β-(300) crystals with excellent mechanical properties. Due to the improved compatibility of PP/PE with PP-b-PE, PE crystals are dispersed within PP crystals, providing bonding that improves the toughness of PPR under the low stiffness failure conditions of PPR/PP-b-PE blends, thereby enhancing their impact performance at low and room temperatures. This research has great significance for both recycling waste plastics and enhancing the low-temperature toughness of PPR.

Keywords: polypropylene; block copolymer; crystal structure; compatibility; X-ray

1. Introduction

It is widely known that despite the similarity of their chemical structure, polyethylene (PE) and polypropylene (PP) are incompatible [1–4]. Random polypropylene (PPR) is a PP that modified polypropylene obtained by adding a small amount of ethylene or α-olefin. The short-chain distribution of PP and other comonomers in the PPR molecular chain is irregular. Compared with traditional PP, PPR has excellent physical and chemical properties. Therefore, PPR is widely used in areas such as water pipes. However, PPR pipes are prone to brittle cracking in low temperatures, which limits their application in cold areas, and the brittle nature of PPR at low temperatures is also a PP/PE compatibility issue [5]. Adding a compatibilizer is the most common approach [6–10]. Compatibilizers typically include block or graft polymers, or compatibilizers with the same, or miscible, monomer units as the blend components [11–13]. Block copolymers can reduce the phase separation behavior, and may exhibit a homogeneous or microphase separation morphology, becoming commonly used compatibilizers [14–16]. Block copolymers with alternating hard and soft segments can effectively improve the rough phase structure and poor mechanical properties of materials, making them important materials for industrial products and academic research [17–19].

Previous studies have been conducted on the application of block copolymers [20,21]. For example, Fortelny et al. [22] researched PP/PE/PS tri-block copolymers and found the interface to be the main factor affecting mechanical properties. Chaffin K et al. [23] investigated the mechanical properties of PP/PE blends and found that the toughness of semi-crystalline blends depends on the existence of crystallite-anchored interfacial entanglement and proposed that a small amount of block copolymer is needed for melt stabilization of PE/iPP blends. James M et al. [24] used catalysts to synthesize iPP/PE di-block and tetra-block copolymers, showing that the addition of block copolymers increases the elongation at break of PP/PE blends. The fact that block copolymers can form different phase separations determines the interface adhesion of the material and the macroscopic mechanical properties of polyolefin blends [25,26]. Ethylene octene block copolymers (OBCs) possess several characteristics, such as low octene content, high melting point, and rapid crystallization, which make them an ideal choice for improving the compatibility of PP/PE block copolymers [27,28]. Hence, some researchers use OBCs as compatibilizers to enhance the mechanical properties of PPR [7,29–32]. For instance, Qilin Ren significantly improved the toughness of PPR by blending OBCs and PPR [33]. However, since OBCs are PE-based block copolymers while PPR mainly consists of PP, there remain phase interface issues in OBC/PPR blends, resulting in a reduction in tensile strength. To overcome this problem, it is better to choose PP-based block copolymers. In 2013, Dow Chemical Company introduced a new PP-based block copolymer called INTUNE4544, which combines polyethylene, polyolefin elastomer (POE), and polar materials such as ethylene vinyl alcohol copolymer (EVOH) and polyamide (PA) with polypropylene. INTUNE4544 not only has the high impact resistance of elastomers but also avoids the phase interface problems caused by PE-based block copolymers. The investigation of PP-based block copolymers and the PPR blend on the structural properties of PPR is of great significance for improving the performance and optimizing the application of PPR [34,35]. On the other hand, while the brittle fracture problem of PPR has been extensively studied, its low-temperature brittleness as a pipe material remains unsolved, particularly in colder regions where pipes are exposed for prolonged periods. Consequently, improving the low-temperature brittleness of PPR remains of significant importance. Moreover, the mechanical properties of semi-crystalline polymers are dependent on their crystal structure. Therefore, understanding how changes in crystal structure affect mechanical properties is critical to obtaining high-performance materials [34]. Currently, research on block copolymers is limited, particularly with regards to how their crystal structure during cooling affects the material's mechanical properties [35]. There is a lack of research on PP-based block copolymer modified PPR, and it is also important to have practical verification of the material that theoretically provides the most effective toughening of PPR. Real-time study of the crystalline structure of modified blends is crucial.

To address this issue, this article chose a new generation of PP-based block copolymers (PP-b-PE) to blend with PPR and investigate the effects of PP-b-PE on the mechanical properties of PPR at room temperature and low temperature (−5 °C). The changes in the crystal structure and interaction mechanism of PPR/PP-b-PE blends during cooling were studied in real-time using non-isothermal crystallization techniques (DSC) and synchrotron radiation X-ray (SAXS and WAXD). The relationship between the mechanical properties and crystal structure of PPR/PP-b-PE blends was explored through tensile and impact tests. The results of this research indicate that PP-b-PE can be utilized as a compatibilizer to enhance the compatibility of PP/PE and modify polyolefin materials. Furthermore, it has been demonstrated that PP-b-PE exhibits strong compatibility and interaction with PP-based polyolefin materials, even at low temperatures (−5 °C), which helps solve the low-temperature brittle cracking problem encountered in PPR.

2. Experimental

2.1. Preparation of Materials

The random polypropylene PPR503 from Sinopec Yangzi Petrochemical in Nanjing, China, and the block copolymer PP-b-PE, which is a combination of PP and PE produced by Dow in Delaware, USA, were used. The molecular weight (M_w) of PP-b-PE is 16,000, and the mass ratio of the PP-b-PE PE chain to the PP chain is 49:51 [36]. PPR and PP-b-PE were melted together using a micro twin screw extrusion equipment (WGT200, Shanghai, China) at a temperature of 200 °C and blended according to the proportions listed in Table 1. The mixture was then granulated.

Table 1. Samples with different proportions.

	Sample	PPR	PP-b-PE
1	PPR	100	0
2	PPR/PP-b-PE (90/10)	90	10
3	PPR/PP-b-PE (70/30)	70	30
4	PPR/PP-b-PE (50/50)	50	50
5	PPR/PP-b-PE (30/70)	30	70
6	PPR/PP-b-PE (10/90)	10	90
7	PP-b-PE	0	100

2.2. Thermal Analysis (DSC)

The thermodynamic behaviors of the blends were characterized using differential scanning calorimetry (TA, 200). The sample was initially heated from room temperature to 220 °C at a rate of 30 °C/min for 5 min to eliminate any thermal history of the crystal during processing. The sample was then cooled down to −30/20 °C at a rate of 5/10 °C/min and kept at this temperature for 3 min. Finally, the temperature was increased again to 200 °C at a rate of 10 °C/min.

2.3. Synchrotron Wide-Angle X-ray Diffraction (WAXD) Measurements

WAXD was conducted at the synchrotron radiation line station (BSRF 1W2A) at the Institute of High Energy Physics, Chinese Academy of Sciences, Beijing. For the WAXD experiments, the sample-to-detector distance was set at 89 mm and calibrated using aluminum oxide (Al_2O_3) as a standard. The detector used was a Mar 165 CCD (Mar USA Co., Palm Beach, FL, USA) with a light source wavelength of 1.54 Å and a recorded effective area of 165 mm. The resolution was 100 mm (FWHM), and the exposure time was 20 s. All 2D images were corrected for air scattering and beam fluctuations. Before any analysis, the collected data underwent air background correction. Data collection and processing were performed using the computer program "X-polar".

An in situ temperature rising and falling WAXD experimental device was also used in this study. The sample was fixed on a heating table and heated to 200 °C at a rate of 30 °C/min, maintained at this temperature for 10 min, and then cooled down to 40 °C at a rate of 4 °C/min. During the cooling process, a 2D image was collected every 11 s. The collector used was a Mar 165 CCD, and the distance from the sample to the detector was set at 100 mm. The obtained 2D scattering diagram was processed using x-polar software (version 1.6.3.0) and integrated within the range of 135°–225° to obtain the WAXD cooling process spectrum.

2.4. Small Angle X-ray Scattering (SAXS)

SAXS was also conducted at the synchrotron radiation line station (BSRF 1W2A) at the Institute of High Energy Physics, Chinese Academy of Sciences, Beijing. The detector used was a Mar 165 CCD (USA) with a light source wavelength of 1.54 Å, an acquisition time of 20 s, a resolution of 100 mm (FWHM), and a sample-to-detector distance of 2900 mm. The data were collected using a Genix3D Cu operated at 50 kV and 0.6 mA (30 W). The detector area was (83.9 × 106.5) mm^2, and the spot size was (0.8 × 0.8) mm^2. The peak position q_{max} can be related to the long period L by Bragg's law. Calculate the long period according to Formula (1). The collected data were corrected for air background prior to analysis using X-polar.

$$L = 2\Pi/q \quad (1)$$

where q is the scattering vector defined as $q = 4\pi (\sin \theta)/\lambda$, where λ is the X-ray wavelength, θ is half the scattering angle (2θ), and L is the long period.

2.5. Mechanical Scattering

2.5.1. Tensile Testing

A dumbbell-shaped sample rod with a length of 50 mm, neck length of 16 mm, and neck width of 4 mm was used to prepare samples with a thickness of 1 mm. The deformation test was conducted using an Instron 3365 tensile testing machine with a clamping distance of 20 mm. The specimens were stretched symmetrically at a constant strain rate of 100 mm/min at room temperature (25 °C), with the measuring point fixed in space. The average of five samples was collected for each material throughout the entire deformation process. All experimental data were averaged from five samples.

2.5.2. Impact Test

The impact strength was measured using a s1107 according to GB/T 1843-2008 [37]. The maximum pendulum energy was 5.5 J (−5 °C) and 50 J (25 °C), and the impact speed was 3.5 m/s. The sample had a 2mm notch with a cross-sectional area of 4 mm^2. The values obtained are presented as arithmetical means of the measurements on 10 specimens. All samples were kept at a constant temperature (−5 °C) for 24 h during the experiment.

2.6. Scanning Electron Microscopy (SEM)

Electron microscopy experiments were carried out with a high-resolution scanning electron microscope operated at 10 kV. All samples were gold sputtered prior to the SEM characterization. The Apreo-2C was used to observe the impact fracture surface and the morphological structure in the stress-whitened area of the PPR and PPR/PP-b-PE blend.

3. Results and Discussion

3.1. Non-Isothermal Crystallization Kinetics and Thermodynamic Behavior

In Figure 1, the DSC results of the PPR/PP-b-PE blends are presented. Figure 1a shows the cooling curve of the PPR/PP-b-PE blends. The crystallization temperatures of pure PPR and the PPR/PP-b-PE (90/10) blend were 102.3 °C and 102.1 °C, respectively, with almost no change observed. Figure 1a indicates that PPR has an initial crystallization temperature of around 110 °C, whereas the initial crystallization temperature of PP-b-PE is approximately 91 °C, suggesting that PPR undergoes crystallization earlier than PP-b-PE. As PPR is essentially PP, it exhibits good compatibility with PP in PP-b-PE. The ratio of PP to PE in PP-b-PE is approximately 1:1 [36], and the PPR/PP-b-PE (90/10) blend contains 95 wt% of PP. The early crystallization of PPR in PPR/PP-b-PE (90/10) blends promote the early crystallization of PP (PP-b-PE). These findings indicate that a small amount of PP-b-PE has almost no effect on the non-isothermal crystallization in PPR/PP-b-PE blends.

When the content of PP-b-PE was further increased to 30–50 wt%, a small heat dissipation peak appeared at a lower temperature (as shown in Figure 1a), indicating that the PPR/PP-b-PE blend had a small microphase separation.

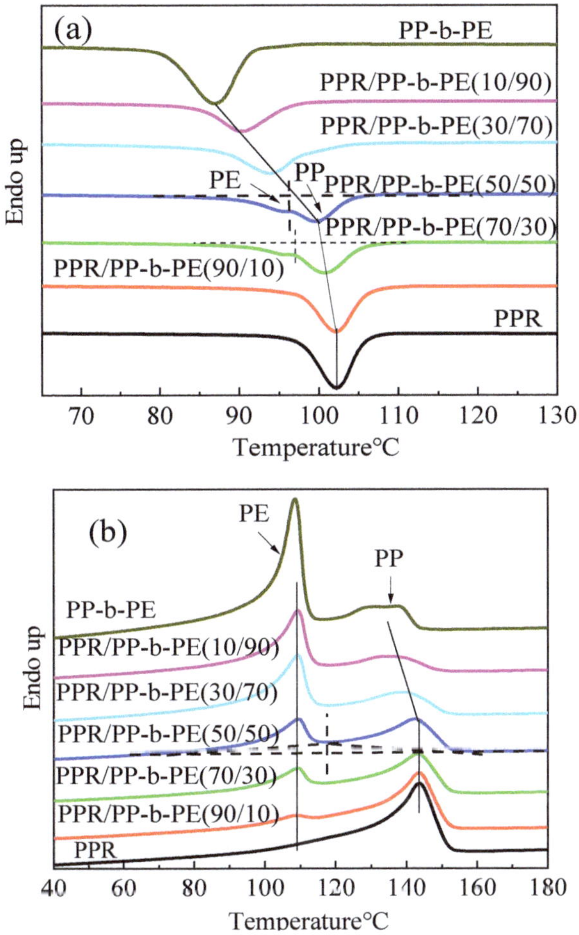

Figure 1. DSC curves of cooling (**a**) and heating (**b**) of PPR, PP-b-PE, and different proportions of PPR/PP-b-PE blends.

Further exploration of the crystallization process of phase separation was completed using PPR/PP-b-PE (50/50). According to the dashed line, the cooling curve and melting curve of the PPR/PP-b-PE (50/50) blend was divided into two parts, as shown in Figure 1. The crystallization and melting area ratios of PP peaks at high temperature and PE peaks at low temperature were calculated as 10:27 and 10:15, respectively. The results showed that the PP ratio of the PPR/PP-b-PE (50/50) blend in cooling crystallization was significantly higher than that in melting, indicating that both the PPR and PP phase from PP-b-PE contributed to the PP peak in cooling crystallization. This also showed that the PP segment in PP-b-PE had good compatibility with PPR and could co-crystallize. Since PPR has a higher crystallization temperature, PPR first crystallized in cooled crystallization and induced the PP phase (from PP-b-PE) to co-crystallize, thereby improving the compatibility of PPR and PP-b-PE. Additionally, the crystallization temperature of the PPR/PP-b-PE (10/90) blend was 102.1 °C, which was 15.2 °C higher than that of pure PP-b-PE. This suggests that a small amount of PPR (10 wt%) can also first crystallize and induce PP-b-PE to crystallize ahead of time. This happens because, as the content of PP-b-PE increases, the crystallization of PP-b-PE that is not induced by PPR to form nuclei becomes dominant, and a single peak is formed on the DSC cooling curve.

In Figure 1b, the second heating DSC traces of the PPR/PP-b-PE blends exhibit melting behavior. PP-b-PE has two melting peaks, of which the higher temperature is the wide melting peak of PP, which is platform-shaped, and the lower temperature is the sharp melting peak of PE. This is because PP-b-PE is a block copolymer of PP and PE. The molecular weight distribution of PP is wide and that of PE is narrow. PPR has only one melting peak at 143.8 °C. When the amount of PPR is 10 wt%, the crystallization temperature of the PE peak is 109.3 °C, which is higher than the Tm (PE) of PP-b-PE (108.5 °C). This is due to the early crystallization of PE induced by the first crystallization of PPR, forming a larger chain segment. However, when the PPR content increases, the melting point of PE almost does not change, because the PE of the generated larger sized chain segment is stable and does not change with the increase in PP-b-PE content. With an increase in PPR content, the melting peak of PP gradually narrows, and the strength gradually increases. The reason for the increased melting point is that PPR and PP-b-PE have good compatibility between the PPR and PP phases (from PP-b-PE), forming a relatively high molecular weight and uniform chain segment. Additionally, we conducted DSC testing while cooling down at a rate of 5 °C/min to −30 °C, and the results for Tg are summarized in Table 2. From Table 2, it can be observed that the Tg of PPR is −14.1 °C, while the Tg of PP-b-PE is −17.8 °C. As the content of PP-b-PE increases, the Tg gradually approaches that of PP-b-PE. Furthermore, the PPR/PP-b-PE blend exhibits only one Tg, which further indicates the good compatibility between PPR and PP-b-PE.

Table 2. Melting point and crystallization temperature of different samples.

Sample	Tc/°C	Tm (PE)/°C	Tm (PP)/°C	Tg/°C
PP-b-PE	86.9	108.5		−17.5
PPR/PP-b-PE (10/90)	90.6	109.3	136.5	−17.5
PPR/PP-b-PE (30/70)	93.9	109.1	140.0	−17.4
PPR/PP-b-PE (50/50)	99.6	109.1	142.8	−15.7
PPR/PP-b-PE (70/30)	100.7	108.9	143.3	−14.7
PPR/PP-b-PE (90/10)	102.1	107.8	143.4	−14.1
PPR	102.3	-	143.8	−13.8

In summary, the addition of 10% PP-b-PE has no effect on the crystallization temperature of PPR. However, when the content exceeds 30%, PPR co-crystallizes with PP in PP-b-PE. This indicates that PP-b-PE has good compatibility with PPR. Moreover, even a small amount of PPR can induce early crystallization of PE in PP-b-PE.

3.2. *In Situ Non-Isothermal Crystal Structure Change*

The selected circularly averaged one-dimensional (1D) wide angle X-ray diffraction (WAXD) intensity profiles at various temperatures after cooling are shown in Figure 2. WAXD was used to analyze the crystal structure of PPR/PP-b-PE blends during the cooling process. This study focused on the crystallization process of the PPR/PP-b-PE (50/50) blend during cooling. The initial pattern (T = 180 °C) showed diffuse scattering without any crystal reflections due to a completely amorphous polymer melt. When the temperature was reduced to 120 °C, a lower diffraction peak appeared near 14.2°, which corresponds to the PP (110) crystal plane and has a starting crystallization temperature much higher than PP-b-PE (as shown in the DSC curve in Figure 1a). This may be due to the strong crystallization ability of PPR. Within the experimentally determined crystallization temperature range of 120 °C to 40 °C, the diffraction intensity of the (110) peak increased with increasing crystallization temperature. When the temperature dropped to 111 °C, other PP α typical diffusion peaks of crystals (2θ angles of 14.1°, 16.8°, and 18.6°), corresponding to the reflections of the (110), (040), and (130) crystal planes, respectively, were observed. However, when the temperature cooled to 113 °C, a small new diffraction peak of the β-crystal of the PP phase (shown in the dotted line circle in Figure 2) appeared at 16.1°, corresponding to the (300) crystal plane. According to the cooling WAXD curves of PPR and PP-b-PE (Figure 3), it can be observed that during cooling crystallization, PP-b-PE and PPR only had a lower diffraction peak at 2θ angles of 15.3° corresponding to the reflections of (300) crystal planes. The β-(300) characteristic plane diffraction peak was not observed in Figure 3. This indicates that the PPR/PP-b-PE blends induced the formation of new β-crystals. With the further decrease in temperature, the relative intensity of the (300) plane of the β-crystal was gradually covered by the diffraction relative intensity of other crystal planes. Because the contribution of β (300) is relatively small, the scattering intensity was reduced. When the temperature dropped to 108 °C, another tiny characteristic peak of the PP β-crystal could be seen at about 20.3°, corresponding to the (301) crystal plane of the β-crystal. The characteristic peak of the β-crystal generated by the blend became increasingly sharper with decreasing temperature. As shown in Figure 3, the β-(301) crystal plane generated by the blend was provided jointly by PPR and PP-b-PE, this also proves the existence of simultaneous crystallization behavior of PPR and PP-b-PE.

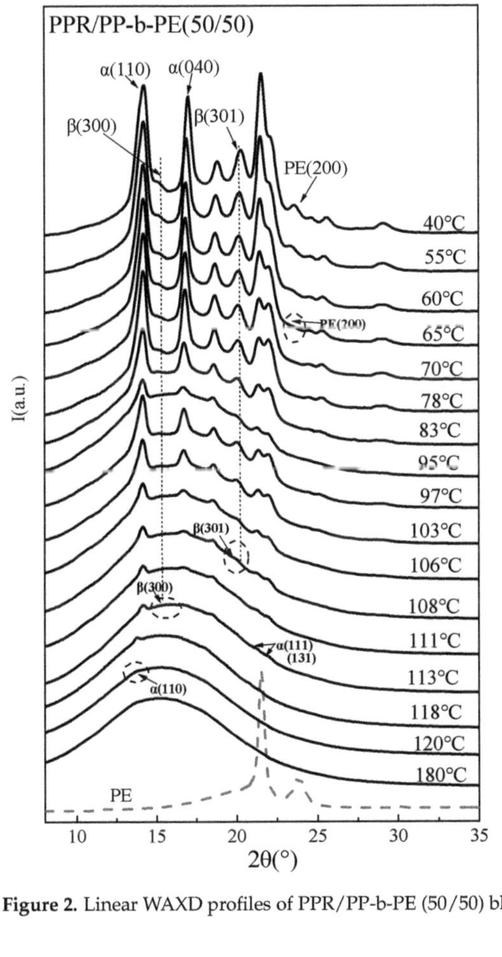

Figure 2. Linear WAXD profiles of PPR/PP-b-PE (50/50) blends cooling process at 4 °C/min.

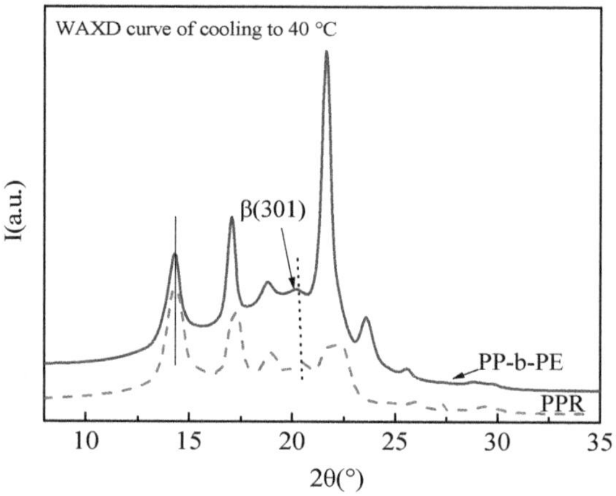

Figure 3. Linear WAXD profiles of PPR and PP-b-PE cooling to 40 °C at 4 °C/min.

The crystal structures of PPR/PP-b-PE blends were analyzed by WAXD, which revealed the presence of β-crystals with excellent mechanical properties. Furthermore, the good compatibility between PPR and PP phases in PP-b-PE led to the early crystallization of PP-b-PE this may help enhance the mechanical properties of the blend, which will be discussed below. The co-crystallization of PPR and PP phases in PP-b-PE during the cooling process, as demonstrated by WAXD, indicates good compatibility between PPR/PP-b-PE components. This finding is consistent with the conclusion drawn from DSC analysis.

3.3. Microcrystals and Arrangement Structure of Different Blending Systems

Figure 4 displays the SAXS static I-q curve of the PPR/PP-b-PE blend at room temperature, revealing that the scattering intensity of PP-b-PE is significantly higher than that of PPR. According to the formula, the intensity of SAXS is proportional to the square of the difference in scattering length density (SLD) between two phases. The SLD difference between PE amorphous and crystal is larger than that of PP, the scattering intensity of PE is significantly higher than that of PP. Since PP-b-PE is a PP and PE block copolymer, it is composed of two polymers, PP and PE, with different SLDs, so PP-b-PE has the highest scattering intensity. On the other hand, pure PPR only has a SLD contrast between amorphous and crystalline PP, so PPR has the lowest scattering intensity. This explains the higher scattering intensity of PP-b-PE compared to PPR. The scattering intensity of the PPR/PP-b-PE (90/10) blend was only slightly increased compared with PPR, and the peak position remained almost unchanged. This was because most of the blend was made up of PPR, with only a small amount of PE phase added, resulting in no significant change to the crystal structure. As the PP-b-PE content increased, the peak intensity gradually decreased, and a small scattering peak appeared at a higher angle in the PPR/PP-b-PE (50/50) blend (as indicated by the arrow in Figure 4). This is due to the increase in PP content and the decrease in SLD contrast between PP and PE. When the content of PP-b-PE increased to 30 wt%, the scattering intensity increased significantly and the scattering peak showed an obvious move to a lower angle ($q = 0.27$), which may be due to the penetration of PP-b-PE into the amorphous region of PPR, resulting in an increase in crystal spacing. When the content of PP-b-PE continued to increase, there was only an increase in scattering intensity while the position remained almost unchanged. Based on the DSC results, this may be because PP-b-PE and PPR co-crystallize, but excessive PP-b-PE will crystallize separately. The crystal spacing comes from the average of all crystal spacing, so the change in scattering peak position is not significant.

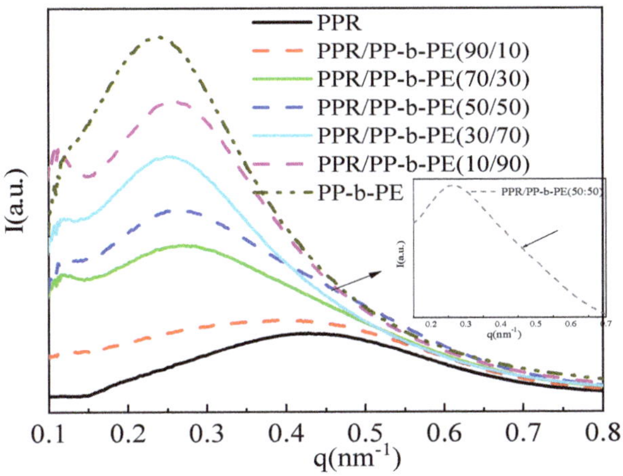

Figure 4. SAXS curves of PPR, PP-b-PE, and 10%, 30%, 50%, 70%, and 90% PPR/PP-b-PE blends.

In summary, the addition of 10 wt% PP-b-PE did not affect the crystal structure of PPR. However, when the content of PP-b-PE increased to 30–70 wt%, the excessive PP-b-PE penetrated the amorphous region of PPR, resulting in an increase in crystal spacing and a shift in the scattering peak position. Excessive PP-b-PE will crystallize separately, leading to a basically unchanged scattering peak position.

The in situ SAXS technique has been widely used to investigate the effects of cooling on flow-induced crystallization, including the early stages and final microstructure. The ordered melt exhibits both short-range (unit lattice) and long-range (alternating crystal and chip structure) order. To further understand the microscopic crystal structure during cooling, we conducted an in situ SAXS experiment to study the cooling crystallization process. Figure 5 displays the SAXS curve (I(q)-q) of the PPR/PP-b-PE (50/50) blend. Figure 5a is divided into two regions based on intensity changes.

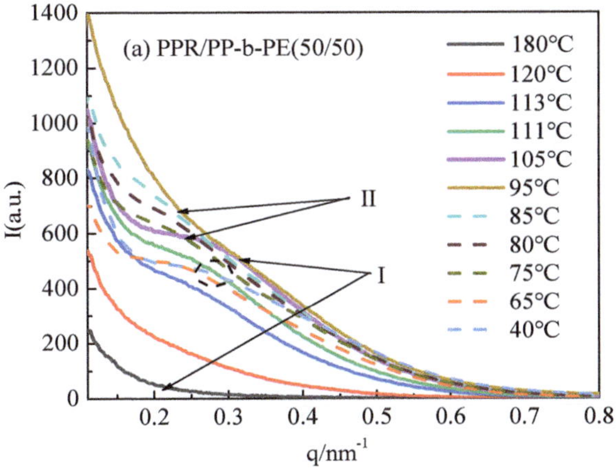

Figure 5. *Cont.*

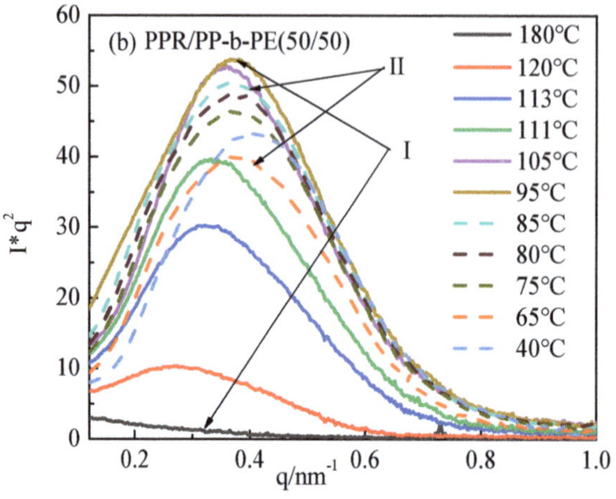

Figure 5. SAXS curves of PPR, PP-b-PE, and 10%, 30%, 50%, 70%, and 90% PPR/PP-b-PE blends (**a**) Linear-(I(q)-q) and (**b**) Lorentz-q^2*I-(q).

In region I, when the temperature is 180 °C, the materials are in a completely molten state and the lamellar structure is destroyed. The system hardly contributes to the intensity of SAXS, and the curve does not show any characteristic peaks. As the temperature decreases from 180 °C to 95 °C, the scattering peak intensity increases while shifting towards the high vector direction, and the characteristics of the scattering peak become more apparent. At 113 °C, the characteristic crystal peaks can be observed, indicating the formation of PP phase crystals and the folding of PP chain segments, forming a long-range ordered periodic structure. Small lamella structures are more likely to undergo structural changes as the temperature decreases, resulting in significant changes in scattering intensity. The most distinct characteristic peak is observed at 105 °C, indicating that the crystal structure of PP is relatively perfect, forming a regular lamellar structure. The peak position of the maximum value in the scattering curve corresponds to the long period of the semi-crystalline polymer. The size of the polymer's long period can be calculated using Formula (1). The Lorentz transformation method q^2I-(q) is typically used to plot the scattering intensity q to accurately determine the peak position (as shown in Figure 5b). The results are displayed in Figure 6. It can be observed from Figure 6 that the long period structure of 18.5 nm appears at 113 °C, and the long period decreases to 16.7 nm when the temperature drops to 105 °C. This decrease is due to the orderly arrangement and tight packing of PP crystal chains.

In region II, when the temperature is reduced to 95 °C, the scattering intensity reaches its maximum, but the characteristic peak disappears and only one peak shoulder appears. This phenomenon occurs because the PE phase crystal begins to form at 95 °C, resulting in a decrease in SLD contrast. On the other hand, the presence of PP crystals may affect the regularity of newly formed PE crystals and may also be one of the reasons this phenomenon occurs. As the temperature decreases further, the scattering intensity gradually decreases, and the peak position continues to shift towards the high vector direction. This decrease occurs because the scattering intensity of PE is significantly higher than that of PP. Due to the formation of PE crystals, the SLD becomes smaller, resulting in a reduction in scattering intensity. In addition, according to the DSC cooling curve (Figure 1), the PE crystal accounts for about 1/3 of the PP crystal, and although the proportion of crystal is small, the size of PE crystal is slightly larger than that of PP [38]. Therefore, after the formation of PE, it has little influence on the average size of the overall crystal, manifested as a small long-term change in the cooling region II.

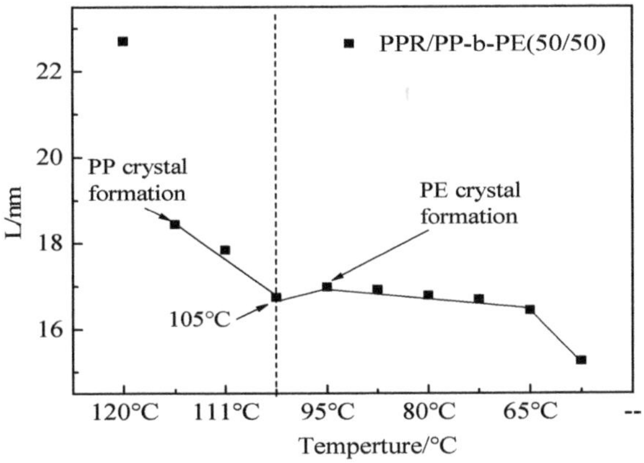

Figure 6. Long period curves of PPR/PP-b-PE (50/50) blends under difficult temperatures at 4 °C/min.

The study examined the crystallization behavior of PPR/PP-b-PE (50/50) blends using in situ cooling experiments with SAXS. The results showed that PP crystallized first and formed small lamellar structures before the temperature decreased to 105 °C. As the temperature decreased, crystal compact accumulation and long period reduction occurred quickly. When the temperature reached 105 °C, PE began to crystallize, but its crystal regularity decreased, and the scattering intensity reduced during the process. Since the crystal size of PE was larger than that of PP, there was little long-term change observed.

3.4. Mechanical Behavior

The effects of PP-b-PE components on the mechanical properties of the PPR/PP-b-PE blend systems were investigated. Table 3 summarizes the modulus, yield stress, impact strength, breaking strength, and elongation at break values for the PPR/PP-b-PE blend. The elongation at break of PPR/PP-b-PE (70/30) blends increased by 14.2% compared to pure PPR (457.5%), reaching a value of 522.6%. The yield stress of PPR/PP-b-PE (70/30) (19.9 MPa) only decreased by 3.1 MPa compared to pure PPR (23.0 MPa). As the PP-b-PE content increased further, the elongation at break of the PPR/PP-b-PE blends continued to increase significantly, while stiffness only slightly decreased. These results demonstrate that the addition of PP-b-PE improves interface adhesion and indicates that PPR has good compatibility with PP-b-PE.

Table 3. Mechanical data of samples.

Sample	Yield Stress	Elongation at Break	Impact Strength (KJ/m^2)		Breaking Strength	Modulus
	MPa	(%)	25 °C	−5 °C	MPa	GPa
PPR	23.0	457.5	15.8	4.1	32.7	7.2
PPR/PP-b-PE (90/10)	21.1	467.1	22.3	8.1	31.3	6.5
PPR/PP-b-PE (70/30)	19.9	522.6	58.0	9.8	33.6	6.2
PPR/PP-b-PE (50/50)	16.5	535.1	unbreak		28.4	4.8
PPR/PP-b-PE (30/70)	13.9	573.5	unbreak		26.0	3.9
PPR/PP-b-PE (10/90)	12.5	694.4	unbreak		22.1	3.3
PP-b-PE	13.1	739.5	unbreak		25.6	3.7

In addition to the tensile experiments, we also conducted impact tests on the blends, and the results are shown in Table 3. The increase in impact strength fell within the limits of experimental error. At room temperature (25 °C), when the amount of PP-b-PE added was 10–30 wt%, the impact strengths of PPR/PP-b-PE (90/10) (22.3 kJ/m^2) and PPR/PP-b-PE (70/30) (58.0 kJ/m^2) blends were 41.1% and 267.1% higher than that of pure PPR (15.8 kJ/m^2), respectively. This shows that PP-b-PE has a significant toughening effect on PPR at room temperature. We also conducted impact experiments at the glass transition temperature of PP (−5 °C), and the results are also summarized in Table 3. The impact strength of PPR was found to be low at 4.1 kJ/m^2. However, the impact strengths of PPR/PP-b-PE (90/10) and PPR/PP-b-PE (70/30) blends were significantly increased to 8.1 kJ/m^2 and 9.8 kJ/m^2, respectively, with an increase of 97.6% and 139%, respectively, compared to pure PPR. From Table 2, it can be seen that the Tg of the blended samples is all below −5 °C, and the Tg gradually decreases with the increase in PP-b-PE addition, which is also one of the reasons for the high impact strength. These results confirm that PP-b-PE also has a significant toughening effect on PPR at low temperature.

In conclusion, adding PP-b-PE can significantly increase the mechanical properties of PPR, owing to the good compatibility between PPR and the PP component of PP-b-PE, co-crystallization of PPR and part of PP-b-PE, and improved interface binding force. In blends, the phase interface is an important factor determining the material properties, including the mechanical properties, of the blends, thus improving the mechanical properties of PPR/PP-b-PE blends are essential. The better impact strength of PP-b-PE itself is also a contributing factor to this improvement. Additionally, the WAXD data analysis (Figure 4) indicates that PPR/PP-b-PE blends form β-crystals, which is another reason for the significant increase in the impact strength of PPR/PP-b-PE blends at low or room temperature.

In summary, studying the mechanical behavior of different samples has shown that the addition of PP-b-PE can significantly improve the impact strength and elongation at break of PPR at low and room temperatures without sacrificing stiffness. It is of great significance to toughen and improve the low-temperature brittleness of PPR.

3.5. SEM

To better observe the toughening effect of PP-b-PE on PPR, we further studied the SEM images of the impact fracture surface of PPR/PP-b-PE blends. Since samples with a content of 50 wt% or higher of PP-b-PE did not fracture, we only studied the impact fracture surface SEM images of samples with 10–30 wt% PP-b-PE content. From Figure 7, there is a significant relationship between the morphology and composition of different blends. The impact fracture surface of pure PPR is relatively flat and clear, while the addition of only 10 wt% PP-b-PE results in a very rough impact fracture surface with localized voids and domain distortions. However, in Figure 7b, distinct relatively flat and clear areas can still be observed. As the content of PP-b-PE continues to increase, the impact fracture surface becomes rougher, with larger void sizes, and more pronounced plastic flow can be observed, along with fiberization and drawing caused by tough fractures. This further indicates that increasing the content of PP-b-PE enhances the impact strength of the blend (from 22.3 KJ/m^2 to 58.0 KJ/m^2). Additionally, from Figure 7c, distinct intertwining between the matrices can be observed. Impact fractures can be seen as the fractures of polymer molecular chains, and when the entanglement density is high, the impact fracture's surface exhibits shear yielding as shown in Figure 7c, which requires a substantial amount of energy consumption. Therefore, adding 30 wt% PP-b-PE can increase the impact strength of the blend by 139% compared to PPR.

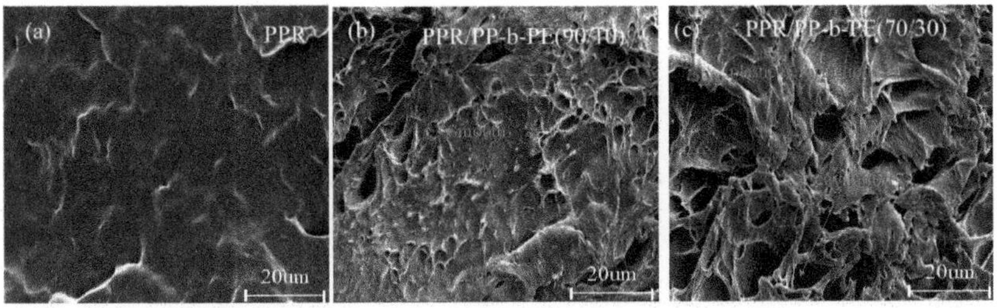

Figure 7. SEM images of the fracture surface of the PPR/PP-b-PE blend in specimens with different PP-b-PE content: (**a**) PPR (**b**) PPR/PP-b-PE (90/10); (**c**) PPR//PP-b-PE (90/10).

4. Conclusions and Discussion

The present study investigated the compatibility, crystal structure, and mechanical properties of new generation block copolymer PP-b-PE and PPR blends (hereafter referred to as PPR/PP-b-PE). Crystallization kinetics revealed that PPR crystallizes first during the cooling process, inducing early crystallization of PP-b-PE, which increases the crystallization temperature of PPR/PP-b-PE blends. When the PP-b-PE content exceeded 30 wt%, the PP crystallization peak of the PPR/PP-b-PE blend showed clear evidence of containing PP (from PP-b-PE) and PPR, indicating that PP-b-PaE and PPR have good compatibility.

The in situ non-isothermal WAXD/SAXS analysis identified a diffusion peak of the β-crystal in the PPR/PP-b-PE blends, with 2θ values of 16.1° and 20.8°, attributed to the crystal planes (300) and (301), respectively. The β-(301) crystal plane was supplied by PP-b-PE and PPR, while a new β-(300) crystal plane was generated, resulting in improved mechanical properties of the PPR/PP-b-PE blends. During cooling, PPR first crystallizes and induces early crystallization of PP-b-PE. As a result, the resulting PE is distributed in the PP crystal and plays a bonding role, enhancing the toughness of the PPR/PP-b-PE blends.

The mechanical properties test and SEM results demonstrated a significant increase in the impact strength of PPR/PP-b-PE blends, particularly when the PP-b-PE content was 30 wt%, resulting in increases of 267.1% (25 °C) and 139% (−5 °C), respectively. The impact strength also increased with the increase in PP-b-PE content. The addition of PP-b-PE can significantly increase the elongation at break of PPR/PP-b-PE blends at low and room temperature, with minimal damage to stiffness. These enhancements in mechanical properties indicate that PP-b-PE has good compatibility with PPR, which is crucial for addressing the compatibility problem between PP and PE and for the application of PPR tubes at low temperatures.

Author Contributions: Conceptualization, Y.S. and W.S.; Methodology, Y.S., Y.W. (Ying Wang), W.S. and L.-Z.L.; Formal analysis, Y.S., L.S. and W.S.; Investigation W.S.; Resources, Y.S. and Y.W. (Yuanxia Wang); Data curation, W.S.; Writing—original draft, W.S.; Writing—review & editing, W.S. All authors have read and agreed to the published version of the manuscript.

Funding: This research received no external funding.

Institutional Review Board Statement: Not applicable.

Data Availability Statement: All the data are available within this manuscript.

Acknowledgments: In this work, the Beijing Synchrotron Radiation Facility (BSRF) 1W2A line station was used for synchrotron radiation characterization. We sincerely thank the scientists of BSRF line station, especially Moguang and Zhihong Li, for their help.

Conflicts of Interest: Author Ying SHI was employed by the company Dongguan HAILI Chemical Material Co., Ltd. The remaining authors declare that the research was conducted in the absence of any commercial or financial relationships that could be construed as a potential conflict of interest.

References

1. Nedkov, T.; Lednický, F. Morphologies of polyethylene-ethylene/propylene/diene monomer particles in polypropylene-rich polyolefin blends: Flake structure. *J. Appl. Polym. Sci.* **2010**, *90*, 3087–3092. [CrossRef]
2. Nedkov, T.; Lednicky, F.; Mihailova, M. Compatibilization of PP/PE blends and scraps with Royalene: Mechanical properties, SAXS, and WAXS. *J. Appl. Polym. Sci.* **2010**, *109*, 226–233. [CrossRef]
3. Heino, M.; Kirjava, J.; Hietaoja, P. Compatibilization of polyethylene terephthalate/polypropylene blends with styrene-ethylene/butylene-styrene (SEBS) block copolymers. *J. Appl. Polym. Sci.* **2015**, *65*, 241–249. [CrossRef]
4. Nakamura, S.; Tokumitsu, K.; Kitamura, M.; Miyagawa, E.; Tanaka, A. A Study for Improvement of Mechanical Properties of Recycle PE/PP Blends. *Resour. Process.* **2007**, *54*, 167–174. [CrossRef]
5. Wu, Y.; Ge, Q.; Yang, F.; Wu, T.; Xiang, M. Improving the low-temperature toughness of PPR pipe by compounding with PERT. *J. Polym. Res.* **2021**, *28*, 143. [CrossRef]
6. Kim, B.C.; Hwang, S.S.; Lim, K.Y.; Yoon, K.J. Toughening of PP/EPDM blend by compatibilization. *J. Appl. Polym. Sci.* **2000**, *78*, 1267–1274. [CrossRef]
7. Zhang, Z.; Yu, F.; Zhou, N.; Zhang, H. Compatibilization by Olefin Block Copolymer (OBC) in Polypropylene/Ethylene-Propylene-Diene Terpolymer (PP/EPDM) Blends. *J. Macromol. Sci. B* **2015**, *54*, 159–176. [CrossRef]
8. Ubonnut, L.; Thongyai, S.; Praserthdam, P. Interfacial adhesion enhancement of polyethylene–polypropylene mixtures by adding synthesized diisocyanate compatibilizers. *J. Appl. Polym. Sci.* **2010**, *104*, 3766–3773. [CrossRef]
9. Silva-Vela, A.; Roudet, F.; Calderón, N.; Huanca-Zuñiga, P.; Tupayachy-Quispe, D.; Almirón, J. Study of the Mechanical Properties of Polymer Composites Based on Polyolefins with the Addition of Rice Husk and Compatibilizer. *Mater. Sci. Forum* **2022**, *1053*, 9–15. [CrossRef]
10. Penava, N.V.; Rek, V.; Houra, I.F. Effect of EPDM as a compatibilizer on mechanical properties and morphology of PP/LDPE blends. *J. Elastomers Plast.* **2012**, *45*, 391–403. [CrossRef]
11. Kotlar, H.K.; Gustafson, C.-G.; Børve, K.L. Polypropylene-phenol formaldehyde-based compatibilizers. II. Application in PP/PA6 75/25 (wt/wt) blends. *J. Appl. Polym. Sci.* **2000**, *75*, 355–360.
12. Zhang, Y.; He, J.; Liu, F. Synthesis of novel polycarbonate-based thermoplastic polyurethane elastomers compatibilizers with octadecyl side chains and their application in PC/PP blends. *Polym. Adv. Technol.* **2021**, *32*, 3070–3081. [CrossRef]
13. Lee, B.Y.; Lee, D.H.; Jang, K.S. Impact Modifiers and Compatibilizers for Versatile Epoxy-Based Adhesive Films with Curing and Deoxidizing Capabilities. *Polymers* **2021**, *13*, 1129. [CrossRef] [PubMed]
14. Marić, M.; Macosko, C.W. Block copolymer compatibilizers for polystyrene/poly(dimethylsiloxane) blends. *J. Polym. Sci. Polym. Phys.* **2002**, *40*, 346–357. [CrossRef]
15. Patel, R.; Park, J.T.; Hong, H.P.; Kim, J.H.; Min, B.R. Use of block copolymer as compatibilizer in polyimide/zeolite composite membranes. *Polym. Adv. Technol.* **2011**, *22*, 768–772. [CrossRef]
16. Auriemma, F.; De Rosa, C.; Scoti, M.; Di Girolamo, R.; Malafronte, A.; D'alterio, M.C.; Boggioni, L.; Losio, S.; Boccia, A.C.; Tritto, I. Structure and Mechanical Properties of Ethylene/1-Octene Multiblock Copolymers from Chain Shuttling Technology. *Macromolecules* **2019**, *52*, 2669–2680. [CrossRef]
17. Saito, M.; Ito, K.; Yokoyama, H. Mechanical Properties of Ultrathin Polystyrene-*b*-Polybutadiene-*b*-Polystyrene Block Copolymer Films: Film Thickness-Dependent Young's Modulus. *Macromolecules* **2021**, *54*, 8538–8547. [CrossRef]
18. Karaagac, E.; Koch, T.; Archodoulaki, V.-M. Choosing an Effective Compatibilizer for a Virgin HDPE Rich-HDPE/PP Model Blend. *Polymers* **2021**, *13*, 3567. [CrossRef]
19. Wang, P.; Gao, S.; Chen, X.; Yang, L.; Cao, T.; Fan, B.; Liu, J.; Hu, X. Effect of PCL-b-PEG Oligomer Containing Ionic Elements on Phase Interfacial Properties and Aggregated Structure of PLA/PCL Blends. *Macromol. Res.* **2022**, *30*, 438–445. [CrossRef]
20. Ma, Y.; He, A.; Liu, C. Crystallization kinetics, crystalline structures and properties of PB/PP blends regulated by poly(butene-block-propylene) copolymers. *Polymer* **2021**, *228*, 123901. [CrossRef]
21. Sun, X.; Kharbas, H.; Peng, J.; Turng, L.-S. A novel method of producing lightweight microcellular injection molded parts with improved ductility and toughness. *Polymer* **2015**, *56*, 102–110. [CrossRef]
22. Fortelný, I.; Michálková, D.; Kruliš, Z. An efficient method of material recycling of municipal plastic waste-ScienceDirect. *Polym. Degrad. Stab.* **2004**, *85*, 975–979. [CrossRef]
23. Yang, F.; Pan, L.; Du, H.Z.; Ma, Z.; Li, Y.S. Effect of Olefin-based Compatibilizers on the Formation of continuous Structure in Immiscible HDPE/iPP Blends. *Chin. J. Polym. Sci.* **2020**, *38*, 1248–1257. [CrossRef]
24. Eagan, J.M.; Xu, J.; Di Girolamo, R.; Thurber, C.M.; Macosko, C.W.; LaPointe, A.M.; Bates, F.S.; Coates, G.W. Combining polyethylene and polypropylene: Enhanced performance with PE/iPP multiblock polymers. *Science* **2017**, *355*, 814–816. [CrossRef] [PubMed]
25. Lin, Y.; Yakovleva, V.; Chen, H.; Hiltner, A.; Baer, E. Comparison of olefin copolymers as compatibilizers for polypropylene and high-density polyethylene. *J. Appl. Polym. Sci.* **2010**, *113*, 1945–1952. [CrossRef]
26. Souza, F.G., Jr.; Soares, B.G.; Manjunath, A.; Somashekar, R. Blends of styrene-butadiene-styrene tri-block copolymer/polyaniline-characterization by SAXS. *Polymer* **2006**, *476*, 240–247. [CrossRef]
27. Zhu, H.Y.; Tian, F.; Li, X.H.; Qiu, H.B.; Wang, J. Crystallization and Phase Behavior in Block Copolymer Solution: An in Situ Small Angle X-ray Scattering Study. *Chin. J. Polym. Sci.* **2019**, *37*, 1162–1168. [CrossRef]

28. Li, S.; Lv, Y.; Sheng, J.; Tian, H.; Ning, N.; Zhang, L.; Wu, H.; Tian, M. Morphology development of POE/PP thermoplastic vulcanizates (TPVs) during dynamic vulcanization. *Eur. Polym. J.* **2017**, *96*, 590–601. [CrossRef]
29. Jordan, A.M.; Kim, K.; Soetrisno, D.; Hannah, J.; Bates, F.S.; Jaffer, S.A.; Lhost, O.; Macosko, C.W. Role of Crystallization on Polyolefin Interfaces: An Improved Outlook for Polyolefin Blends. *Macromolecules* **2018**, *51*, 2506–2516. [CrossRef]
30. Xu, J.; Eagan, J.M.; Kim, S.-S.; Pan, S.; Lee, B.; Klimovica, K.; Jin, K.; Lin, T.-W.; Howard, M.J.; Ellison, C.J.; et al. Compatibilization of Isotactic Polypropylene (*i*PP) and High-Density Polyethylene (HDPE) with *i*PP–PE Multiblock Copolymers. *Macromolecules* **2018**, *51*, 8585–8596. [CrossRef]
31. Unger, T.; Klocke, L.; Herrington, K.; Miethlinger, J. Investigation of the rheological and mechanical behavior of Polypropylene/ultra-high molecular weight polyethylene compounds related to new online process control. *Polym. Test.* **2020**, *86*, 36–39. [CrossRef]
32. Ren, Q.; Zhang, Q.; Wang, L.; Yi, J.; Feng, J. Synergistic Toughening Effect of Olefin Block Copolymer and Highly Effective β-Nucleating Agent on the Low-Temperature Toughness of Polypropylene Random Copolymer. *Ind. Eng. Chem. Res.* **2017**, *56*, 5277–5283. [CrossRef]
33. Lin, Y.; Marchand, G.R.; Hiltner, A.; Baer, E. Adhesion of olefin block copolymers to polypropylene and high density polyethylene and their effectiveness as compatibilizers in blends. *Polymer* **2011**, *52*, 1635–1644. [CrossRef]
34. Ren, Q.; Fan, J.; Zhang, Q.; Yi, J.; Feng, J. Toughened polypropylene random copolymer with olefin block copolymer. *Mater. Des.* **2016**, *107*, 295–301. [CrossRef]
35. Karaagac, E.; Koch, T.; Archodoulaki, V.M. The effect of PP contamination in recycled high-density polyethylene (rPE-HD) from post-consumer bottle waste and their compatibilization with olefin block copolymer (OBC). *Waste Manag.* **2020**, *119*, 285–294. [CrossRef] [PubMed]
36. Hua, X.; Cheng, J.; Wang, Y.; Liu, L.-Z.; Wang, Y.; Shi, Y. Crystallization, structure and properties of PP-b-PE block copolymer. *Polymer* **2023**, *284*, 126310. [CrossRef]
37. Liu, J.; Liu, J. Characterization of maleic anhydride/styrene melt-grafted random copolypropylene and its impact on crystallization and mechanical properties of isotactic polypropylene. *Polym. Bull.* **2019**, *76*, 4369–4387. [CrossRef]
38. Wang, Y.; Shi, Y.; Shao, W.; Ren, Y.; Dong, W.; Zhang, F.; Liu, L.-Z. Crystallization, Structures, and Properties of Different Polyolefins with Similar Grafting Degree of Maleic Anhydride. *Polymers* **2020**, *12*, 675. [CrossRef]

polymers

MDPI

Article

Directed Self-Assembly of Cylinder-Forming Block Copolymers Using Pillar Topographic Patterns

June Huh

Department of Chemical and Biological Engineering, Korea University, Seoul 02841, Republic of Korea; junehuh@korea.ac.kr

Abstract: We conducted a computational study on the self-assembly behavior of cylinder-forming block copolymers, directed by a guide pattern of hexagonally or tetragonally arrayed pillars, using mesoscale density functional theory simulations. By adjusting the spacing (L_p) and diameter (D) of the pillars in relation to the intrinsic cylinder-to-cylinder distance (L_2) of the cylinder-forming block copolymer, we investigated the efficiency of multiple-replicating cylinders, generated by the block copolymer, through the pillar-directed self-assembly process. The simulations demonstrated that at specific values of normalized parameters $\tilde{L}_2 = L_2/L_p$ and $\tilde{D} = D/L_p$ coupled with suitable surface fields, triple and quadruple replications are achievable with a hexagonally arrayed pillar pattern, while only double replication is attainable with a tetragonally arrayed pillar pattern. This work, offering an extensive structure map encompassing a wide range of possible parameter spaces, including $\tilde{L}_2$ and $\tilde{D}$, serves as a valuable guide for designing the contact hole patterning essential in nanoelectronics applications.

Keywords: block copolymer; directed self-assembly; hole density multiplication; contact hole

1. Introduction

Directed self-assembly (DSA), a process involving the structural organization of self-assembling molecules such as block copolymers (BCPs) under the guidance of external processing factors, is a challenging yet crucial technique in various applications, including nanoelectronics. A prime example is the use of chemo-epitaxy or grapho-epitaxy, which involves chemical or topographic patterns to effectively direct BCP self-assembly into desired patterned structures with enhanced pattern precision that is unattainable without the guidance of these prepatterns [1–3].

In DSA for nanoelectronics, a meaningful approach, along with line–space patterns [4,5], involves fabricating nanosized contact holes that serve as connection points for electrical signals across different layers or components in semiconductor devices, such as integrated circuits (ICs) [6–9]. Indeed, numerous recent studies demonstrate that DSA effectively complements extreme ultraviolet lithography (EUV), a premier method in the semiconductor industry [10–12]. This synergy, where DSA aids in multiplying, healing, or rectifying EUV patterns, is particularly beneficial given the challenges EUV faces, including stochastic defects, high operational costs, and other limitations [13–16]. As a specific implementation of a DSA system for contact hole patterning, cylinder-forming BCPs or their blends can be utilized as self-assembling molecules under the guidance of either chemical or topographic patterns. In this case, the pattern transfer from the cylindrical domains of BCPs creates the contact hole patterns. Particularly, when it comes to multiplying contact holes, an effective strategy is to design guide patterns that are congruent with the BCP pattern (i.e., cylinders on a hexagonal array) but have a larger pitch. This larger pitch should ideally match the intrinsic pitch of the BCP pattern, allowing for precise alignment and replication. The pattern multiplication based on this "hexagons within hexagons" strategy also necessitates the appropriate tuning of system or material parameters including the interaction energies between the prepattern and BCP components.

This work conducts a computational investigation into the self-assembled structures formed by cylinder-forming BCPs under the guidance of topographic pillar patterns. We explore two types of pillar arrays: those arranged in hexagonal and tetragonal configurations. In this DSA system, it is postulated that arrays of contact holes, formed through prepatterning, are capped with hollow cylinders or pillars that serve as a guiding pattern (Figure 1). In semiconductor manufacturing, this type of nanofabrication is indeed feasible; hollow cylinders, typically made from materials such as silicon dioxide (SiO_2) or similar compounds [17–20], are strategically placed over contact holes to serve various functions. In particular, one such function is to guide the assembly of material in subsequent layers, which is the specific focus of this study. Upon removal of these pillars after the DSA process, a density multiplication of contact holes can be achieved.

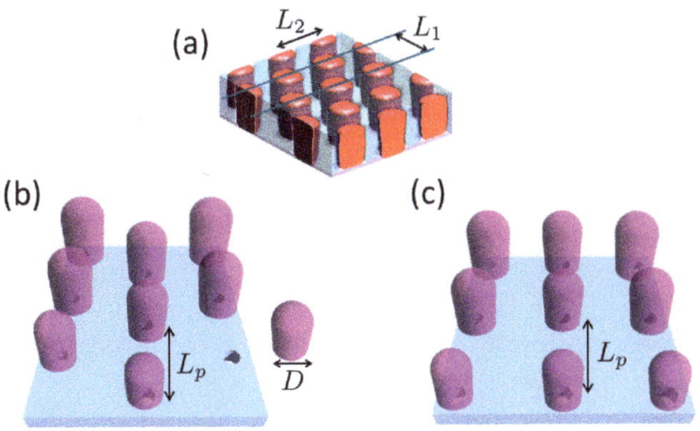

Figure 1. Illustration of (**a**) a cylinder-forming BCP with a natural period $L_1 = \sqrt{3}L_2/2$, (**b**) pillars arranged in a hexagonal array, and (**c**) pillars in a tetragonal array, where L_p represents the center-to-center distance between pillars and D denotes the diameter of a pillar. The holes beneath the pillars represent the contact holes that were prepatterned.

For the model DSA system, we employ a simulation of mesoscale density functional theory simulation based on a modified diffusion equation using Landau–Ginzburg–Cahn–Hilliard (LGCH) theory [21,22]. This simulation was used to investigate the effects of various DSA parameters, including those related to the BCP material (e.g., intrinsic pitch of the BCP) and the prepattern (e.g., pillar-to-pillar distance), on the resulting DSA pattern. This approach enables the construction of DSA structure maps within the associated parameter space.

2. Simulation Methods

We model the self-assembly process of a BCP, that is, the phase separation process between monomer types A and B that constitute the BCP, using simulations of mesoscale density functional theory. The key parameter describing phase separation in this mesoscale density functional is the order parameter $\Psi(\mathbf{r})$, which represents the local deviation of the A-monomer fraction $\phi(\mathbf{r})$ from its average value, i.e., the A-monomer fraction in the BCP f at a given position $\mathbf{r}$ (thus, $\Psi(\mathbf{r}) = \phi(\mathbf{r}) - f$). The diffusion equation governing the dynamics of $\Psi(\mathbf{r})$ with time t is given by

$$\frac{\partial \Psi(\mathbf{r})}{\partial t} = M\nabla^2 \frac{\partial (F + F_{surf})}{\partial \Psi(\mathbf{r})} + \xi(\mathbf{r}) \quad (1)$$

Here, M is a mobility constant which we set to unity for simplicity, F represents the free energy functional of the BCP, F_{surf} accounts for the free energy associated with the interaction between the BCP component (A and B) and the surface (pillar, substrate, and film top) in contact with the BCP, and ξ represents the thermal noise. The free energy functional, F, is approximated using the Landau-type expression

$$F(\Psi) = \int d\mathbf{r}\left[-\frac{\tau}{2}\Psi^2(\mathbf{r}) + \frac{\mu}{3!}\Psi^3(\mathbf{r}) + \frac{\lambda}{4!}\Psi^4(\mathbf{r}) + \frac{D}{2}\left\{\nabla\Psi(\mathbf{r})\right\}^2\right] + \frac{b}{2}\int d\mathbf{r}_1 \int d\mathbf{r}_2 G(\mathbf{r}_1 - \mathbf{r}_2)\Psi(\mathbf{r}_1)\Psi(\mathbf{r}_2) \tag{2}$$

Here, τ represents a parameter related to the Flory interaction parameter between the A- and B-monomers (χ), functioning similarly to a temperature-like parameter. The parameters μ, λ, and D are integral to the chain architecture of the BCP, shaping the conformational contributions. The last term in the expression, characterized by the Green function G with a period control parameter b, accounts for long-range repulsion, thus penalizing long-wavelength inhomogeneity. The Green function for the Laplacian is given by $\nabla^2 G(\mathbf{r}_1 - \mathbf{r}_2) = -\delta(\mathbf{r}_1 - \mathbf{r}_2)$. For a diblock copolymer architecture, the molecular parameters τ, μ, λ, D, and b are approximated as

$$\tau = 2(\chi - \chi_s) + \frac{3^{1/2}}{Nf^{3/2}(1-f)^{3/2}},\ \mu = \Gamma_3/N,\ \lambda = \Gamma_4(0,0)/N, \quad D = \frac{1}{12f(1-f)},\ b = \frac{9}{N^2 f^2 (1-f)^2} \tag{3}$$

where χ_s is the χ at the spinodal, N is the number of statistical monomers in a BCP, and Γ_3 and $\Gamma_4(0,0)$ are Leibler vertex functions computed from the monomer correlation functions [23–25]. The surface free energy, F_{surf}, is expressed as

$$F_{surf} = \sum_{\{\alpha\}} \int d\mathbf{r} s_\alpha(\mathbf{r})\Psi(\mathbf{r}) \tag{4}$$

Here, $s_\alpha(\mathbf{r})$ is a surface field at the position $\mathbf{r}$ in contact with the surface α (where $\alpha = p$ for pillar, s for substrate, t for film top). This field is proportional to the difference in interfacial tension, $\sigma_{\alpha A} - \sigma_{\alpha B}$, where $\sigma_{\alpha A}$ and $\sigma_{\alpha B}$ represent the interfacial tension between surface α and component A and between surface α and component B, respectively.

The geometries of the BCP mesophase and the guide patterns used in the simulation, featuring cylindrical A-domains hexagonally embedded in a B-matrix along with pillars arranged in both hexagonal and tetragonal arrays, are illustrated in Figure 1. The natural period of a cylinder-forming BCP, $L_1 = \sqrt{3}L_2/2$, and the diameter of the pillar, D, are the main variables investigated in this study, while the center-to-center distance between adjacent pillars, L_p, remains constant throughout the investigation. The geometry of a single pillar is composed of a cylindrical region with a height of $0.8L_p$ and a capping region formed by a hemisphere with a radius of $D/2$, where the block copolymer (BCP) fills only the cylindrical region. The fraction, f, of the minor component A in the cylinder-forming BCP and the degree of incompatibility, χN, are set to $f = 0.3$ and $\chi N = 18$, respectively. In our investigation, the variables L_2 and D are explored within the normalized ranges of $\tilde{L}_2 = L_2/L_p = 1/3.0$ to $1/1.7$ and $\tilde{D} = D/L_p = 1/1.9$ to $1/1.3$, respectively. These ranges were selected based on their potential to yield interesting DSA structures. For pillars, the meaningful physical sizes are encompassed in the range of $0 < \tilde{D} < 1$, where $\tilde{D} = 1$ signifies the case in which pillars are in contact with their adjacent pillars. Within this range, we focused on $1/1.9 < \tilde{D} < 1/1.3$, as the DSA systems within this specific range of pillar size exhibited notable results concerning hole density multiplication. Preliminary simulations were also conducted for other pillar sizes ($0 < \tilde{D} < 1/1.9$ or $1/1.3 < \tilde{D} < 1$), but these did not yield structures featuring hole density multiplication. Since the natural period L_2 of a BCP is dependent on N, altering N serves as a means to adjust $\tilde{L}_2$, given that L_p remains constant. The range of$\tilde{L}_2$, spanning from $1/3.0$ to $1/1.7$, is equivalent to N ranging from 100 to 320, which roughly translates to the molecular weights from

10 kg/mol to 32 kg/mol, assuming that the monomer has a molecular weight of 100 g/mol. The surface field for the top of the BCP film is set to be neutral between blocks A and B, ($s_t = 0$), while the surface field for the pillar and substrate can be designed to be attractive to majority B ($s_p > 0, s_s > 0$) or neutral.

The diffusion equation Equation (1) is numerically integrated in the discrete space using an explicit method through a numerical algorithm based on the cell dynamics simulation (CDS) approach initially proposed by Oono and Puri [26,27]. This method has been widely utilized for simulating the phase separation phenomena in various multicomponent polymeric systems at mesoscale, including polymer blends and block copolymers, both in bulk and confined spaces, as well as mixtures with particles [28–31]. All the DSA systems are modeled in an initially disordered state by assigning uniformly distrubuted random numbers between $-\eta$ and η to the order parameter at each grid point: $\Psi(\mathbf{r}) \sim u(-\eta, \eta)$ where $u(-\eta, \eta)$ denotes a uniform distribution ranging from $-\eta$ to η. The parameter η is set to be 0.1. The surface boundary is treated by a Neumann boundary with vanishing gradient of the order parameter, that is, $\frac{\partial \Psi}{\partial n}|_s = 0$, where n represents the direction that is normal for the surface boundary (s). This boundary condition implies that there is no flux of the order parameter across the boundary, which corresponds to a reflecting boundary. To numerically update the order parameter at each cell or grid, we utilized $L_x \times L_y \times L_z = 96 \times 84 \times 60$ grids for the DSA systems with hexagonally arrayed pillars and $96 \times 96 \times 60$ grids for those with tetragonally arrayed pillars. Here, L_α represents the lattice dimension in the α-direction, measured in units of grids. Periodic boundary conditions were applied in the x and y directions, while reflective boundary conditions were implemented in the z direction and at the surface of pillars. Our CDS model, based on the CHC diffusion equation, adheres to a stability condition, formulated as follows: [32]

$$\Delta t < \frac{\Delta x^4}{18M - 3(A-1)\Delta x^2} \tag{5}$$

where Δt is the time step, Δx is the grid spacing, M is the diffusivity in Equation (1), and A is related to the parameter τ (approximately $1 + \tau$) in Equation (3). Given that Δx, $M = 1$, and A ranges from 1.1 to 1.25, this condition indicates that the time increment for the stability should be $\Delta t < 0.056$. All simulations were performed using self-developed code for the CDS algorithm with a time step $\Delta t = 0.05$.

3. Results

In the following, we present the DSA structure maps within the variable space of $\tilde{L}_2 = L_2/L_p$ and $\tilde{D} = D/L_p$ where the geometrical parameters for the cylinder-forming BCP and pillars, L_2 and D, are varied, while the parameter for prepatterned holes, L_p, remains constant. In practice, this means adjusting the molecular weight of the BCP by controlling b through the adjustment of N in Equations (2) and (3) as well as the size of pillars (D) that are placed over the prepatterned contact holes, thereby keeping L_p fixed.

We begin with the case of hexagonally arrayed pillars. Figure 2a presents the structure map of the BCP self-assembled under the guidance of the hexagonally arrayed pillars that are attractive to the majority B block ($s_p = 0.5$) on a neutral substrate ($s_s = 0$). The film thickness h of the BCP is set to $\tilde{h} = h/L_p = 0.8$. This case of surface fields ($s_p = 0.5, s_s = 0.0$) roughly corresponds to the condition where the pillar is coated with homopolymer B, the same component as the major block B in the cylinder-forming BCP, while the substrate is coated with a random copolymer composed of A and B components. Well-organized cylinder arrays are predominantly found when $\tilde{L}_2 = 1/3^{1/2}$ and $\tilde{L}_2 = 1/2$, showcasing two distinct hexagonal orientations. Ideally, as shown in Figure 2b,c, a 30° orientation difference between the hexagons formed by the pillars and those by the BCP cylinders (referred to as H30) is expected when $\tilde{L}_2 = 1/\sqrt{3}$, whereas a 0° orientation difference (referred to as H0) should be observed when $\tilde{L}_2 = 1/2$. In the former case, the number of replicated holes increases from the original 4 prepatterned holes (beneath the pillars) to 12, which includes 8 BCP cylinders in addition to the 4 prepatterned holes, as observed at ($\tilde{D} = 1/1.5, \tilde{L}_2 = 1/\sqrt{3}$) (Figure 2b). In the latter case, the number increases to 16, consisting of 12 BCP cylinders plus the original 4 prepatterned holes, as identified at ($\tilde{D} = 1/1.9, \tilde{L}_2 = 1/2$) (Figure 2c). It is noted that both hexagon orientations, H30 and H0, are found not only within their ideal regions (associated with $\tilde{L}_2 = 1/\sqrt{3}$ for H30 and $\tilde{L}_2 = 1/2$ for H0) but also in regions where $\tilde{L}_2$ is slightly smaller or larger than these ideal values by compressing or expanding the intrinsic cylinder-to-cylinder distance of the BCP. Furthermore, of interest is the dodecagon structure surrounding a pillar, which results in a sevenfold replication of holes (increasing from 4 to 28 holes), observed at ($\tilde{D} = 1/1.3, \tilde{L}_2 = 1/3$) (Figure 2d). Unlike the homogeneous tessellation patterns seen with H30 and H0, this case shows a heterogeneous tessellation characterized by the nonuniform spacings between the holes (both pillar-to-cylinder and cylinder-to-cylinder), leading to a pattern that is not a single, uniform tessellation.

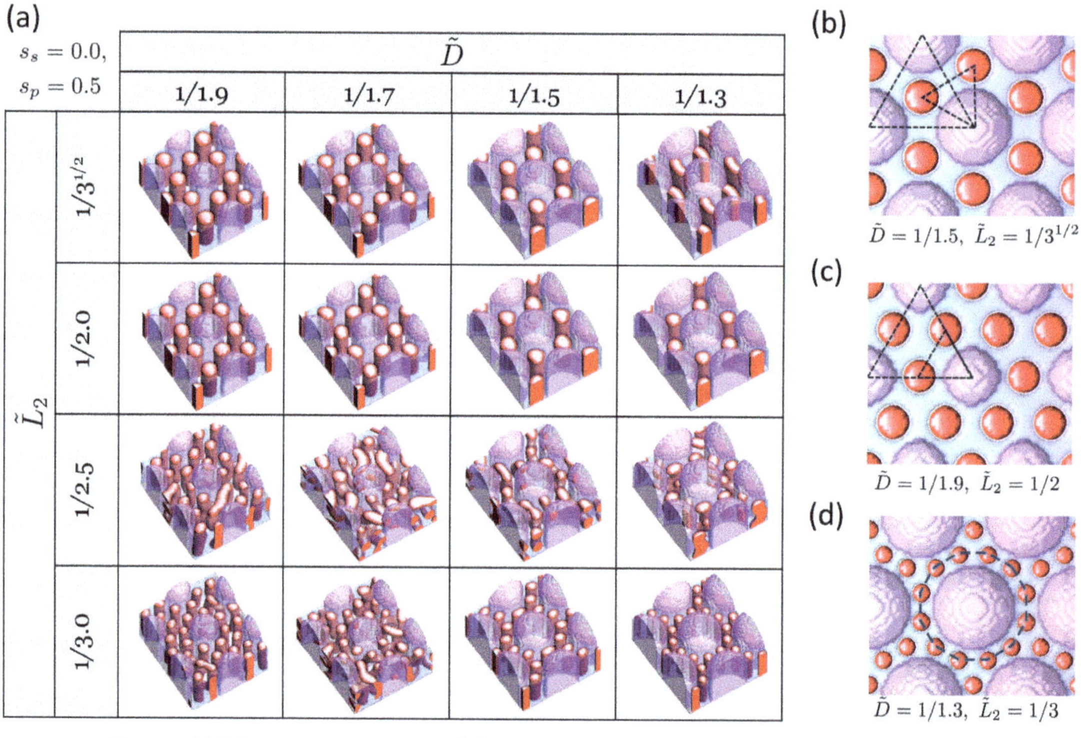

Figure 2. (**a**) DSA structure map in the ($\tilde{D}, \tilde{L}_2$) variable space for cylinder-forming BCP, guided by hexagonally arrayed pillars selective for the majority B block ($s_p = 0.5$) on a neutral substrate ($s_s = 0$). The minority A domains of the block copolymer (BCP) are represented by red color, and the minority B domains are depicted by translucent blue, while the pillars are represented by translucent purple color. Notable structures (top views) observed include the following: (**b**) threefold replication of holes, expanding from 4 to 12 holes (H30), (**c**) fourfold replication of holes, expanding from 4 to 16 holes (H0), (**d**) sevenfold replication of holes, expanding from 4 to 28 holes. The dashed triangles in (**b**,**c**) depict the difference in hexagon orientation between the hexagons formed by the pillars and those by the BCP cylinders. The dashed dodecagon in (**d**) represents the BCP cylinders surrounding a pillar.

Another interesting aspect is the dynamics of structure evolution, specifically, how the surface fields from pillars arranged in a hexagonal pattern direct the self-assembly of the BCP. Figure 3 illustrates the temporal evolution of the BCP structure shown in Figure 2c, evolving into the H0 structure starting from the early stages of self-assembly. In the initial stages, phase separation starts with AB layering encircling the pillar (Figure 3a). As the layering thickens and the layers start to converge (Figure 3b), cylinders emerge, notably with the B component organized into a hexagonal lattice exhibiting an H30 orientation (Figure 3c). As the self-assembly process progresses, the B domains become interconnected, breaking the continuous A domains into discrete sections, resulting in the inversion of cylinders from B to A (Figure 3d,e), eventually leading to the formation of cylinders composed of the A component with an H0 orientation (Figure 3f).

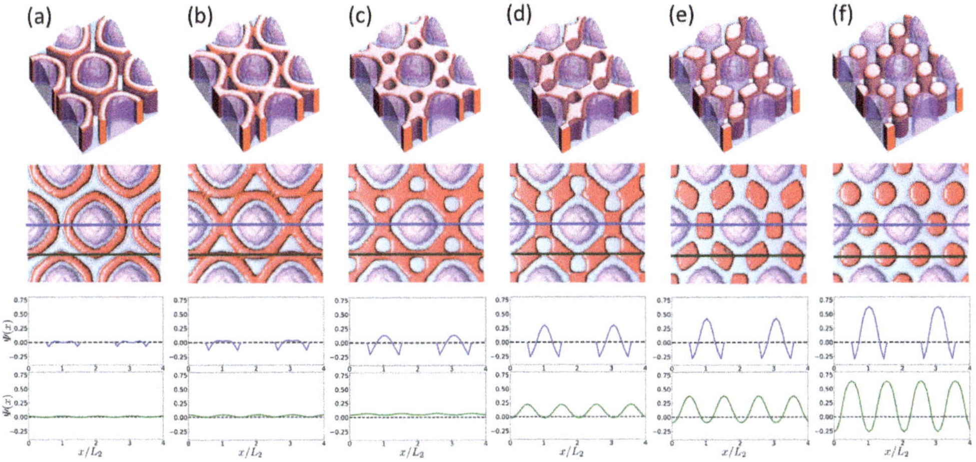

Figure 3. Time evolution of BCP self-assembly guided by hexagonally arrayed pillars with $s_p = 0.5$ on a neutral substrate ($s_s = 0$) under the geometric conditions of $\tilde{D} = 1/1.9$ and $\tilde{L}_2 = 1/2$, depicted in arbitrary time units, t. Phases at different times are shown: (**a**) t = 100, (**b**) t = 300, (**c**) t = 600, (**d**) t = 1000, (**e**) t = 1200, and (**f**) t = 2000. For each of (**a**–**f**), the first two images at the top display the angular and top views of the DSA structures, respectively. The minority A domains of the block copolymer (BCP) are represented by red color, and the minority B domains are depicted by translucent blue, while the pillars are represented by translucent purple color. The bottom two plots for each time frame illustrate the order parameter profiles ($\Psi(x) = \psi(x) - f$) along the blue and green lines indicated in the top-view images. The dashed line represents the case of a homogeneous state ($\Psi(x) = 0$).

Figure 4a presents the DSA structure map for the same case as Figure 2 but with $s_p = 0.5$ and $s_s = 0.5$, i.e., both pillar and substrate are attractive to the major component B. This setup corresponds to both the pillar and the substrate being coated with homopolymer B, which, in terms of the coating process, offers advantages over the previous case ($s_p = 0.5$, $s_s = 0.0$), which requires multiple steps for surface modification. Certainly, coating with single-molecular species simplifies the process by enabling completion in a single step. Compared to the neutral substrate case shown in Figure 2, the effect of the surface field from the substrate is obvious, as seen in Figure 4. For instance, at ($\tilde{D} = 1/1.5, \tilde{L}_2 = 1/\sqrt{3}$), an H30 structure is also formed, but slight modulations from the perfect cylindrical shape are observed (Figure 4b). The substrate effect becomes more pronounced as the area of substrate increases, which corresponds to a decrease in D. At ($\tilde{D} = 1/1.9, \tilde{L}_2 = 1/2$), where H0 is formed in the case of the neutral substrate, the B-attractive substrate induces the formation of connected cylinders that do not come into contact with the substrate due to the formation of the B layer on the substrate (Figure 4c). These imperfections can be mitigated by either reducing the surface fields at both the pillar and the substrate to be weakly attractive to the major B block or by increasing the film thickness, thereby enhancing the pillar wall effect relative to the substrate effect. This is demonstrated in Figure 5, where the impacts of surface fields and film thickness are examined. In practice, a weaker B attraction can be achieved by coating them with a random copolymer composed of components A and B, with a higher fraction of B [33,34].

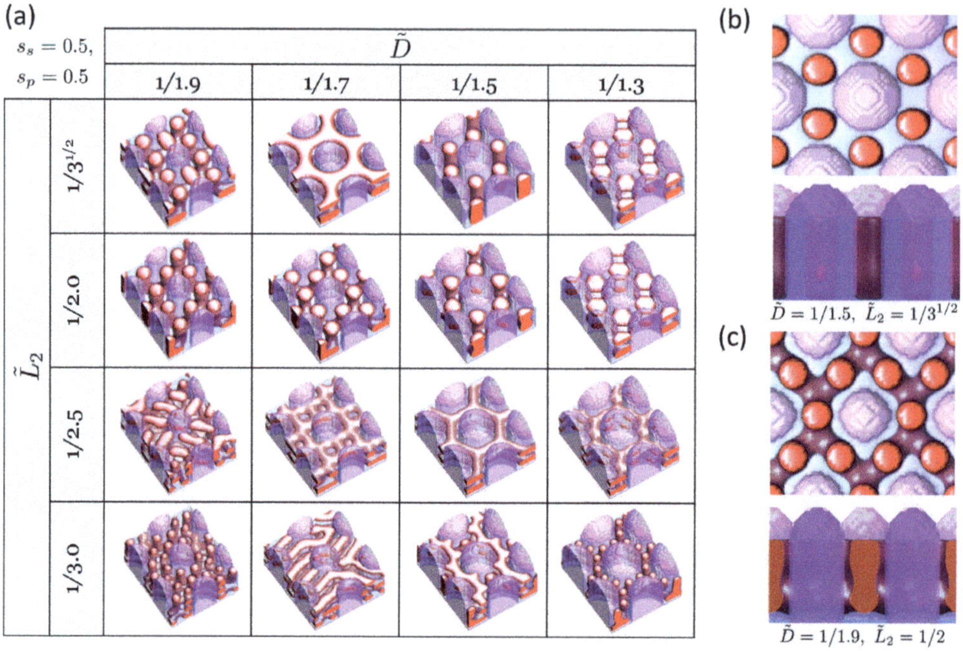

Figure 4. (**a**) DSA structure map in the ($\tilde{D}, \tilde{L}_2$) variable space for cylinder-forming BCP, guided by hexagonally arrayed pillars where both pillars and substrate are selective for the majority B block ($s_p = 0.5$, $s_s = 0.5$). The minority A domains of the block copolymer (BCP) are represented by red color, and the minority B domains are depicted by translucent blue, while the pillars are represented by translucent purple color. Notable structures for comparison with Figure 2 include the following: (**b**) top view and side view of DSA structure at ($\tilde{D} = 1/1.5, \tilde{L}_2 = 1/\sqrt{3}$), (**c**) top view and side view of DSA structure at ($\tilde{D} = 1/1.9, \tilde{L}_2 = 1/2$).

Next, we explore the case involving an alternative guide-pattern geometry: pillars arranged on a tetragonal lattice. This configuration has garnered both academic and practical interest due to the inherent challenges in achieving tetragonal symmetry through self-assembly as compared to the more commonly observed hexagonal symmetry [35,36]. Figure 6a presents the structure map for structures formed by DSA under the guidance of tetragonally arrayed pillars that are selective for the majority B block ($s_p = 0.5$) on a neutral substrate ($s_s = 0$). Notably, when $\tilde{L}_2 = 1$, the DSA structure forms a single and uniform tessellation with twofold-replicated holes (increasing from four to eight holes), exhibiting tetragonal symmetry (Figure 6b). One crucial aspect of BCPs, especially when they need to fill a confined space, is their tendency to stretch uniformly as they form mesophases. This underpins the preference for a hexagonal array over a tetragonal array, where BCPs must stretch further to fill the corners in the tetragon compared to the hexagon, even though tetragonally arrayed cylinders have a smaller interfacial area than hexagonally arrayed cylinders. This excessive and undesirable chain stretching of BCPs at the corners of the tetragon can be significantly mitigated by introducing pillars, as illustrated in the schematic in Figure 6b. Another interesting observation includes the formation of heterogeneous tessellations. We identified two instances of such patterns, observed at ($\tilde{D} = 1/1.5, \tilde{L}_2 = 1/2.0$) and at ($\tilde{D} = 1/1.5, \tilde{L}_2 = 1/2.5$), where fivefold and sevenfold replications of holes were found, respectively. These notable DSA structures, identified for the tetragonally arrayed pillar case, can similarly be achieved on substrates with pillars coated in a single material (uniform surface fields) by precisely adjusting the film thickness and surface fields, akin to the hexagonally arrayed pillar case.

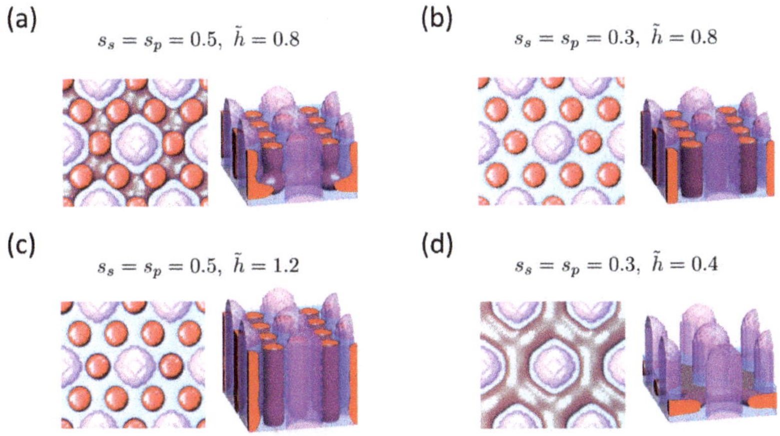

Figure 5. Effect of surface fields and film thickness on the DSA structure guided by hexagonally arrayed pillars with $\tilde{D} = 1/1.9, \tilde{L}_2 = 1/2$. Displayed structures correspond to the following: (**a**) $s_s = s_p = 0.5$, $\tilde{h} = 0.8$, (**b**) $s_s = s_p = 0.3$, $\tilde{h} = 0.8$, (**c**) $s_s = s_p = 0.5$, $\tilde{h} = 1.2$, and (**d**) $s_s = s_p = 0.3$, $\tilde{h} = 0.4$. The minority A domains of the block copolymer (BCP) are represented by red color, and the minority B domains are depicted by translucent blue, while the pillars are represented by translucent purple color.

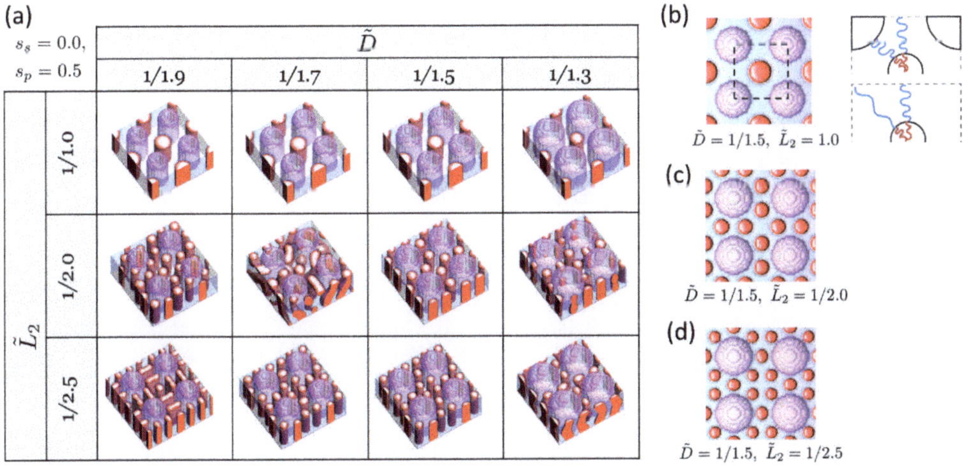

Figure 6. (**a**) DSA structure map in the $(\tilde{D}, \tilde{L}_2)$ variable space for cylinder-forming BCP, guided by tetragonally arrayed pillars selective for the majority B block ($s_p = 0.5$) on a neutral substrate ($s_s = 0$). The minority A domains of the block copolymer (BCP) are represented by red color, and the minority B domains are depicted by translucent blue, while the pillars are represented by translucent purple color. Notable structures (top views) observed include the following: (**b**) twofold replication of holes, expanding from 4 to 8 holes, (**c**) fivefold replication of holes, expanding from 4 to 20 holes, (**d**) sevenfold replication of holes, expanding from 4 to 28 holes. The schematic in (**b**) illustrates howBCPs, represented by red and blue lines, stretch to fill the space, both in the presence and absence of pillars.

4. Concluding Remarks

In conclusion, this study explores the self-assembly of cylinder-forming BCPs under the guidance of pillar topographic patterns, utilizing mesoscale density functional theory simulations. The investigation encompasses DSA structures of cylinder-forming BCPs across various parameters, including the pitch and diameter of the pillars. Under hexagonally arrayed pillars, we observed that homogeneous tessellations featuring threefold and fourfold hole replications can be realized through the careful adjustment of the pitch and diameter of the pillars, as well as the surface interactions. A previous study employing hexagonally arrayed circles as chemo-patterns for the DSA of cylinder-forming BCPs also reported threefold and fourfold hole multiplication. Similar to the grapho-epitaxial assembly in our work, this multiplication occurs when the distance between circles is $\sqrt{3}$ times and twice larger, respectively, than the natural period L_2 of the BCP [37]. On the other hand, for tetragonally arrayed pillars, a single instance of homogeneous tessellation was identified, achieving twofold hole replication.

While not all aspects of our simulations could be directly validated by experiments due to the scope and available data, certain key findings show promising alignment with experimental observations conducted by our collaborators at Samsung Electronics [38]. This alignment reinforces the credibility of our simulation approach, particularly in modeling the complex dynamics of DSA systems. The insights from this study provide a theoretical framework for designing DSA systems capable of density multiplication of contact holes, a crucial component in the lithographic processes used for electronic devices.

Author Contributions: Conceptualization, J.H.; methodology, J.H.; validation, J.H.; formal analysis, J.H.; investigation, J.H.; resources, J.H.; data curation, J.H.; writing—original draft preparation, J.H.; writing—review and editing, J.H.; visualization, J.H.; supervision, J.H.; project administration, J.H.; funding acquisition, J.H. All authors have read and agreed to the published version of the manuscript.

Funding: This research was supported by Samsung Electronics.

Institutional Review Board Statement: Not applicable.

Informed Consent Statement: Not applicable.

Data Availability Statement: Data are contained within the article.

Conflicts of Interest: The author declares no conflicts of interest.

References

1. Ji, S.; Wan, L.; Liu, C.-C.; Nealey, P.F. Directed self-assembly of block copolymers on chemical patterns: A platform for nanofabrication. *Prog. Polym. Sci.* **2016**, *54–55*, 76–127. [CrossRef]
2. Cummins, C.; Lundy, R.; Walsh, J.J.; Ponsinet, V.; Fleury, G.; Morris, M.A. Enabling future nanomanufacturing through block copolymer self-assembly: A review. *Nano Today* **2020**, *35*, 100936. [CrossRef]
3. Nicaise, S.M.; Tavakkoli, S.M.; Berggren, K.K. Self-assembly of block copolymers by graphoepitaxy. In *Directed Self-Assembly of Block Co-Polymers for Nano-Manufacturing*; Gronheid, R., Nealey, P., Eds.; Elsevier: Amsterdam, The Netherlands, 2015; pp. 199–232.
4. Seino, Y.; Kasahara, Y.; Sato, H.; Kobayashi, K.; Miyagi, K.; Minegishi, S.; Kodera, K.; Kanai, H.; Tobana, T.; Kihara, N.; et al. A novel simple sub-15 nm line-and-space patterning process flow using directed self-assembly technology. *Microelectron. Eng.* **2015**, *134*, 27–32. [CrossRef]
5. Giammaria, T.J.; Gharbi, A.; Paquet, A.; Nealey, P.; Tiron, R. Resist-Free Directed Self-Assembly Chemo-Epitaxy Approach for Line/Space Patterning. *Polymers* **2020**, *10*, 2443. [CrossRef]
6. Yi, H.; Bao, X.-Y.; Tiberio, R.; Wong, H.-S.P. A General Design Strategy for Block Copolymer Directed Self-Assembly Patterning of Integrated Circuits Contact Holes using anAlphabet Approach. *Nano Lett.* **2015**, *15*, 805–812. [CrossRef] [PubMed]
7. Yi, H.; Bao, X.-Y.; Zhang, J.; Bencher, C.; Chang, L.-W.; Chen, X.; Tiberio, R.; Conway, J.; Dai, H.; Chen, Y.; et al. Flexible Control of Block Copolymer Directed SelfAssembly using Small, Topographical Templates: Potential Lithography Solution for Integrated Circuit Contact Hole Patterning. *Adv. Mater.* **2012**, *24*, 3107–3114. [CrossRef]
8. Hu, X.-H.; Zhang, R.; Zhang, X.; Wu, Z.; Zhou, J.; Li, W.; Xiong, S. Focused solar annealing for block copolymer fast self-assembly. *Heliyon* **2024**, *10*, e24016. [CrossRef] [PubMed]

9. Wu, H.; Lou, J.J.-C.; Xiao, H.; Cai, Y.; Yang, Y. Advanced IC Manufacturing Processes. In *Handbook of Integrated Circuit Industry*; Wang, Y., Chi, M.-H., Lou, J.J.-C., Chen, C.-Z., Eds.; Springer: Singapore, 2023; pp. 1865–1872.
10. Griffiths, R.A.; Williams, A.; Oakland, C.; Roberts, J.; Vijayaraghavan, A.; Thomson, T. Directed self-assembly of block copolymers for use in bit patterned media fabrication. *J. Phys. D Appl. Phys.* **2013**, *46*, 503001. [CrossRef]
11. Gronheid, R.; Singh, A.; Younkin, T.R.; Delgadillo, P.R.; Nealey, P.; Chan, B.T.; Nafus, K.; Negreira, A.R.; Somervell, M. Rectification of EUV-patterned contact holes using directed self-assembly. In Proceedings of the Advances in Resist Materials and Processing Technology XXX, San Jose, CA, USA, 24–28 February 2013; Volume 8682.
12. Guo, J.; Janes, D.W.; Mignot, Y.; Johnson, R.C.; Chi, C.; Liu, C.-C.; Meli, L.; Kuroda, T.; DiPaola, D.A.; Tanaka, Y.; et al. LCDU improvement of EUV-patterned vias with DSA. In Proceedings of the Novel Patterning Technologies for Semiconductors, MEMS/NEMS, and MOEMS 2019, San Jose, CA, USA, 24–28 February 2019; Volume 10958.
13. Mulkens, J.; Slachter, B.; Kubis, M.; Tel, W.; Hinnen, P.; Maslow, M.; Dillen, H.; Ma, E.; Chou, K.; Liu, X.; et al. Holistic approach for overlay and edge placement error to meet the 5 nm technology node requirements. In Proceedings of theMetrology, Inspection, and Process Control for Microlithography XXXII, San Jose, CA, USA, 25 February–1 March 2018; Volume 10585.
14. De Bisschop, P. Stochastic effects in EUV lithography: Random, local CD variability, and printing failures. *J. Micro/Nanolithogr. MEMS MOEMS* **2017**, *16*, 041013.
15. De Bisschop, P. Stochastic printing failures in extreme ultraviolet lithography. *J. Micro/Nanolithogr. MEMS MOEMS* **2018**, *17*, 041011. [CrossRef]
16. Verstraete, L.; Suh, H.S.; van Bel, J.; Timi, P.H.; Vallat, R.; Bezard, P.; Vandereyken, J.; Beggiato, M.; Tamaddon, A.-H.; Beral, C.; et al. Mitigating stochastics in EUV lithography by directed self-assembly. In Proceedings of theNovel Patterning Technologies 2023, San Jose, CA, USA, 26 February–2 March 2023; Volume 12497.
17. Tinck, S.; Altamirano-Sanchez, E.; De Schepper, P.; Bogaerts, A. Formation of a Nanoscale SiO_2 Capping Layer on Photoresist Lines with an $Ar/SiCl_4/O_2$ Inductively Coupled Plasma: A Modeling Investigation. *J. Plasma Process. Polym.* **2014**, *11*, 52–62. [CrossRef]
18. He, Y.; Che, Y.; Que, L. A Top–Down Fabrication Process for Vertical Hollow Silicon Nanopillars. *J. Microelectromech. Syst.* **2016**, *25*, 662–667. [CrossRef]
19. Wi, J.-S.; Lee, H.-S.; Lim, K.; Nam, S.-W.; Kim, H.-M.; Park, S.-Y.; Lee, J.J.; Hong, C.D.; Jin, S.; Kim, K.-B. Fabrication of Silicon Nanopillar Teradot Arrays by Electron-Beam Patterning for Nanoimprint Molds. *Small* **2008**, *4*, 2118–2122. [CrossRef] [PubMed]
20. Barrios, C.A.; Canalejas-Tejero, V. A top-down approach for fabricating three-dimensional closed hollow nanostructures with permeable thin metal walls. *Beilstein J. Nanotechnol.* **2017**, *8*, 1231–1237. [CrossRef] [PubMed]
21. Pinna, M.; Guo, X.; Zvelindovsky, A.V. Block Copolymer Nanocontainers. *ACS Nano* **2010**, *4*, 2845–2855. [CrossRef] [PubMed]
22. Huh, J.; Jung, J.Y.; Lee, J.U.; Cho, H.; Park, S.; Park, C.; Jo, W.H. Supramolecular Assembly of End-Functionalized Polymer Mixtures Confined in Nanospheres. *ACS Nano* **2011**, *5*, 115–122. [CrossRef] [PubMed]
23. Leibler, L. Theory of Microphase Separation in Block Copolymers. *Macromolecules* **1980**, *13*, 1602–1617. [CrossRef]
24. Fredrickson, G.H. Surface Ordering Phenomena in Block Copolymer Melts. *Macromolecules* **1987**, *20*, 2535–2542. [CrossRef]
25. Mayes, A.M.; Olvera de la Cruz, M.; McMullen, W.E. Asymptotic Properties of Higher-Order Vertex Functions for Block Copolymer Melts. *Macromolecules* **1993**, *26*, 4050–4051. [CrossRef]
26. Oono, Y.; Puri, S. A Study of phase-separation dynamics by use of cell dynamical systems. I. Modeling. *Phys. Rev. A* **1988**, *38*, 434–453. [CrossRef]
27. Oono, Y.; Puri, S. A Study of phase-separation dynamics by use of cell dynamical systems. II. Two-dimensional demonstrations. *Phys. Rev. A* **1988**, *38*, 1542–1565. [CrossRef] [PubMed]
28. Roths, T.; Friedrich, C.; Marth, M.; Honerkamp, J. Dynamics and rheology of the morphology of immiscible polymer blends-on modeling and simulation. *Rheol. Acta* **2002**, *38*, 211–222. [CrossRef]
29. Ginzburg, V.V. Nanoparticle/Polymer Blends: Theory and Modeling. In *Encyclopedia of Polymer Blends*; Isayev, A.I., Ed.; Wiley-VCH: Weinheim, Germany, 2010; pp. 233–268.
30. Pinna, M.; Zvelindovsky, A.V. Large scale simulation of block copolymers with cell dynamics. *Eur. Phys. J. B* **2012**, *85*, 210. [CrossRef]
31. Diazm J.; Pinna, M.; Zvelindovsky, A.V.; Asta, A.; Pagonabarraga, I. Cell Dynamic Simulations of Diblock Copolymer/Colloid Systems. *Macromol. Theory Simul.* **2017**, *26*, 1600050. [CrossRef]
32. Teixeira, P.I.C.; Mulder, B.M. Comment on "Study of phase-separation dynamics by use of cell dynamical systems. I. Modeling". *Phys. Rev. E* **1997**, *55*, 3789–3791. [CrossRef]
33. Han, E.; Stuen, K.O.; Leolukman, M.; Liu, C.-C.; Nealey, P.F.; Gopalan, P. Perpendicular Orientation of Domains in Cylinder-Forming Block Copolymer Thick Films by Controlled Interfacial Interactions for Block Copolymer Melts. *Macromolecules* **2009**, *42*, 4896–4901. [CrossRef]
34. Kim, S.-W.; Kim, E.; Lee, H.; Berry, B.C.; Kim, H.-C.; Ryu, D.Y. Thickness-dependent ordering of perpendicularly oriented lamellae in PS-b-PMMA thin films. *Polymer* **2015**, *74*, 63–69. [CrossRef]
35. Tang, C.; Lennon, E.M.; Fredrickson, G.H.; Kramer, E.J.; Hawker, C.J. Evolution of Block CopolymerLithography to Highly OrderedSquare Arrays. *Science* **2008**, *332*, 429–432. [CrossRef]
36. Vayer, M.; Guliyeva, A.; Warmont, F.; Takano, A.; Matsushita, Y.; Sinturel, C. Transition between tetragonal and hexagonal pattern in binary blends of ABC block copolymers with different chain lengths. *Eur. Polym. J.* **2020**, *138*, 109986. [CrossRef]

37. Singh, A.; Chan, B.T.; Cao, Y.; Lin, G.; Gronheid, R. Using chemo-epitaxial directed self-assembly for repair and frequency multiplication of EUVL contact hole patterns. In Proceedings of theAlternative Lithographic Technologies VI, San Jose, CA, USA, 23–27 February 2014; Volume 9049.
38. Kim, E. *Personal Communication*; Samsung Electronics: Suwon, Republic of Korea, 2017.

polymers MDPI

Article

Preparation and Structural-Thermodynamical Investigation of Renewable Copolyesters Based on Poly (Ethylene Succinate) and Polyisosorbide

Chaima Bouyahya [1,2], Panagiotis A. Klonos [1,3,*], Alexandra Zamboulis [1], Eleftheria Xanthopoulou [1], Nina Maria Ainali [1], Mustapha Majdoub [2], Apostolos Kyritsis [3] and Dimitrios N. Bikiaris [1,*]

1 Laboratory of Polymer Chemistry and Technology, Department of Chemistry, Aristotle University of Thessaloniki, 541 24 Thessaloniki, Greece; bouyahyachaima@fsm.u-monastir.tn (C.B.); azampouli@chem.auth.gr (A.Z.); nsainali@chem.auth.gr (N.M.A.)
2 Laboratoire des Interfaces et Matériaux Avancés, Université de Monastir, Monastir 5000, Tunisia; mustapha.majdoub@fsm.rnu.tn
3 Department of Physics, National Technical University of Athens, Zografou Campus, 157 80 Athens, Greece; akyrits@central.ntua.gr
* Correspondence: pklonos@central.ntua.gr (P.A.K.); dbic@chem.auth.gr (D.N.B.)

Abstract: A series of novel renewable copolymers based on poly(ethylene succinate) (PESu) and poly(isosorbide succinate) (PISSu), with the Isosorbide (Is)/PESu molar ratio varying from 5/95 to 75/25, were synthesized in-situ and studied in this work. A sum of characterization techniques was employed here for the structural and thermo-dynamical characterization. The sophisticated technique of dielectric spectroscopy, along with proper analysis, enabled the molecular dynamics mapping of both the local and segmental types, which is presented for such materials for the first time. With increasing the Is fraction, shorter copolymeric entities were gradually formed. Based on the overall findings, the systems were found to be homogeneous, e.g., exhibiting single glass transitions, with the two polymer segments being found to be excellently distributed. The latter is indirect, although strong, evidence for the successful copolymerization. The thermal degradation mechanism for the copolymers was exhaustingly explored employing analytical pyrolysis. The systems exhibited, in general, good thermal stability, according to the thermogravimetric analysis. Confirming one of the initial scopes for the present systems, isosorbide plays here the role of hardener (PISSu) over the soft polymer (PESu), and this is reflected in the monotonic increase of the glass transition temperature, T_g, from −16 to ~56 °C. The introduction of Is results in an increase in constraints (hardening of the matrix), while there seems to be an overall densification of the polymer (decrease of the free volume).

Keywords: poly(ethylene succinate); poly(isosorbide succinate); renewable copolyesters; structure-properties relationships; molecular dynamics

1. Introduction

For more than a century, polymers, commonly referred to as plastics, have been applied to a vast number of human needs, namely, from industry to biomedicine, academia, and everyday life [1,2]. This mainly owes to the fact that polymeric materials combine, in general, a sum of wanted properties regarding the requirements of applications with the low cost of synthesis and processing [3,4]. As expected, due to the growing aspects/needs of humanity for new and more targeted applications (e.g., aeronautics, drug delivery, 3–4D printing, etc.), more and more specialized materials are necessary. In parallel, along the fast development of society, there have grown significant concerns for the large amount of 'plastic waste' on earth [5,6], which has been, up to now, due to the extensive use of polymers synthesized from fossil-based sources and employing non-ecofriendly processes [7]. A main problem for this accumulation of traditional polymer wastes is their lack or weak degradability in nature [8–11]. Obviously, this is combined with the lack of people's general environmental consciousness [12]. During the last couple of decades, such consciousness has grown, and, regarding polymers, a serious attempt to turn toward the so-called 'green materials' and 'circular economy' is taking place [13–15].

A partial solution to these issues is attempted via the use of polymers based on renewable resources [16–20], for example, by exploiting agricultural products and byproducts. Such can be, for example, acids formed during metabolic processes (lactic acid, succinic acid, furan-dicarboxylic acid, adipic acid, caprolactone, vanillic acid, itaconic acid, etc.) [16,21–25]. Among others, the polymers prepared by such resources, being mainly polyesters [26,27], exhibit both desired properties, such as thermochemical stability, good mechanical and permeation performance, and good and moderate physical/enzymatic degradation rates in soil conditions (compostability) [28]. As a next step, there has been progress by many groups, including ours, to combine such biobased polymers in the form of polymeric blends [29–31] and copolymers [32–37], as well as composites/nanocomposites [38–41]. This combination has opened room for new innovative materials envisaged for even more specific applications due to the achievement of complex and, even, new properties of polymers. For example, this work has resulted in the almost 'at will' manipulation of thermal degradation, mechanical performance, crystallinity, permeation of small molecules [42], thermal and electrical conductivity [43], as well as physical degradation [44]. In the same context, the study of these new materials has been fruitful from the point of view of basic research.

In our recent published works, we have shown that by carefully chosen synthetic routes, the compatibilization of different polyesters can be achieved by in situ polymerization of a polymer block in the presence of either another existing block or simultaneously with the polymerization of the second block [34,36,45,46]. The compatibilization and successful copolyester synthesis result in many effects, for example, the tuning of the glass transition, crystallization and melting, the manipulation of free volume [47], and, subsequently, the micro- and macro-scopic performance of the polymers.

Following such routes, in the present study, we synthesized and investigated a series of copolymers based on two aliphatic renewable polyesters. On the one hand, poly(n-alkylene succinate) [48–50] with $n = 2$, i.e., poly(ethylene succinate) (PESu), is used as the first block. On the other hand, poly(isosorbide succinate) (PISSu) [51,52] was synthesized as the second block.

We should report that isosorbide is a diol exhibiting a 'V'-like shape consisting of two sterically different secondary hydroxyl groups with the so-called 'exo-' and 'endo-' configurations, respectively, as well as two 'cis-' connected tetrahydrofuran rings [53–55]. Isosorbide is actually a commercial biobased monomer, produced either from D-glucose or, even, from starch [56]. Isosorbide has been involved in part in the step-type synthesis of many polymers, including polyesters, polyurethanes, polyamides, polycarbonates, etc. [56–59].

Coming back to the present systems, for PISSu, prior to polymerization, the isosorbide (Is) monomers were practically mixed with the succinic acid at various Is/ethylene glycol molar ratios between 5/95 and 75/25. For comparison, we also synthesized the two homopolymers PESu and PISSu, i.e., of molar ratios 0/100 and 100/0, respectively. For the structural investigation of these newly synthesized co-polyesters, we employed infrared spectroscopy, nuclear magnetic resonance spectroscopy, gel permeation chromatography, thermogravimetric analysis, and pyrolysis gas chromatography-mass spectrometry analysis, while for the thermal transitions, we employed conventional calorimetry. Finally, to investigate molecular mobility (local and segmental), we employed the sophisticated technique of broadband dielectric spectroscopy. Upon analysis of the results and evaluation by widely adopted routes and models, we combined the results by the different techniques to conclude the in-depth structure-properties relationship. Especially, the molecular dynamics for these novel systems are shown here for the first time.

2. Materials and Methods

2.1. Materials

Succinic acid (SA, purum, ≥99.5%), ethylene glycol (EG, 99.8%), and isosorbide (Is, 98%) were purchased from Sigma Aldrich (St. Louis, MO, USA). Titanium isopropoxide (TIS, ≥97%), used as a catalyst, was provided by Alfa Aesar (Kandel, Germany). Any other employed reagents were of analytical grade.

2.2. Synthesis of Copolyesters

The PESu-PISSu (Figure 1) copolyesters were prepared by a two-step in-situ melt polycondensation method (esterification and polycondensation). During the esterification process, appropriate amounts of SA, EG, and Is with 400 ppm of titanium isopropoxide (TIS) catalyst were added to a 250 mL round-bottom flask equipped with a mechanical stirrer and a vacuum apparatus. The reaction mixtures were degassed and purged three times with dry nitrogen. The reaction mediums were then heated up to 180 °C for a period of time until all the reagents were completely melted. Then, the reaction mixtures were further heated at a slow rate to 220 °C and maintained there for 4 h. During the polycondensation process, a high vacuum was applied, and the temperature was increased slowly to 250 °C for 2 h.

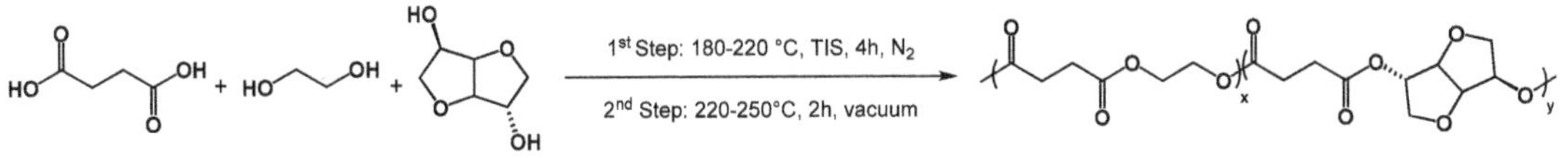

Figure 1. Schematic of the synthesis of poly(ethylene succinate)-co-(isosorbide succinate).

In total, eight (8) polyesters were prepared, namely, six (6) copolyesters with Is/EG mol ratios of 4.5/95.5, 9.6/90.4, 18.6/81.4, 35/65, 50/50, and 75/25, and two (2) homopolymers, namely, neat PESu (0/100) and neat PISSu (100/0). For the sake of ease, a specific code name is used henceforth. The list of samples, along with the code names and main properties, can be found in Tables 1 and 2 (sections in the following).

Table 1. Code naming of the materials under investigation and properties of main interest: intrinsic viscosity, [η], molecular weight, M_n, polydispersity index, *PDI*, average sequence length, L_i, and degree of randomness, *R*, respectively (details in the main text).

					Microstructure		
Sample Code Name	Is/PESu Ratio	[η] (dL/g)	M_n (g/mol)	*PDI*	L_{EGSu}	L_{IsSu}	*R*
PESu	0/100	0.51	83 k	1.71	-	-	-
co045	4.5/95.5	0.56	62 k	1.84	-	-	-
co096	9.6/90.4	0.43	39 k	2.15	-	-	-
co186	18.6/81.4	0.50	33 k	2.32	-	-	-
co350	35/65	0.18	12 k	1.83	2.7	1.7	0.97
co500	50/50	0.28	8 k	1.97	1.9	2.1	1.02
co750	75/25	0.17	7 k	1.78	1.4	3.9	0.99
PISSu	100/0	0.16	5 k	1.81	-	-	-

Table 2. Main thermal events as observed by DSC: glass transition temperature, T_g; corresponding changes in heat capacity, Δc_p; melt crystallization temperature and enthalpy change, T_c and ΔH_c, respectively; cold crystallization temperature and enthalpy change, T_{cc} and ΔH_{cc}, respectively; and melting temperature and enthalpy change, T_m and ΔH_m, respectively.

	DSC Cooling			DSC Heating					
Sample Code Name	T_c (°C)	ΔH_c (J/g)	T_g (°C)	T_g (°C)	Δc_p (J/g·K)	T_{cc} (°C)	ΔH_{cc} (J/g)	T_m (°C)	ΔH_m (J/g)
PESu	34	1	−16	−13	0.65	41/81	51	102	53
co045	-	-	−12	−9	0.63	61	12	94	12
co096	-	-	−9	−6	0.60	-	-	-	-
co186	-	-	−2	1	0.59	-	-	-	-
co350	-	-	3	5	0.56	-	-	-	-
co500	-	-	19	20	0.52	-	-	-	-
co750	-	-	40	43	0.49	-	-	-	-
PISSu	-	-	51	53	0.46	-	-	-	-

2.3. *Fourier-Transform Infrared Spectroscopy (FTIR)*

For the FTIR study we employed a Perkin-Elmer spectrometer (Thermo Fisher Scientific, Waltham, MA, USA). Each sample was scanned 32 times with a resolution of 4 cm^{-1} from 4000 to 650 cm^{-1}. All spectra were subsequently baseline-corrected and normalized.

2.4. *Nuclear Magnetic Resonance Spectroscopy (NMR)*

Nuclear magnetic resonance spectra were recorded at room temperature, in deuterated chloroform, on an Agilent 500 spectrometer (Agilent Technologies, Santa Clara, CA, USA). Spectra were calibrated using the residual solvent peaks. The average sequence length and degree of randomness were calculated according to Equations (1) and (2).

$$L_{EGSu} = 1 + \frac{2 \times I_{EGSuEG}}{I_{EGSuIs-EG} + I_{EGSuIs-Is}} \text{ and } L_{IsSu} = 1 + \frac{2 \times I_{IsSuIs}}{I_{EGSuIs-EG} + I_{EGSuIs-Is}} \quad (1)$$

$$R = \frac{1}{L_{EGSu}} + \frac{1}{L_{IsSU}} \quad (2)$$

where L_{EGSu} and L_{IsSu} are the average sequence lengths of the copolymer segments containing EG and Is respectively, I_{EGSuEG} is the integration of the C(O) resonance signals of succinic acid with EG units on both sides, I_{IsSuIs} is the integration of the C(O) resonance signals of succinic acid with Is units on both sides, $I_{EGSuIs\text{-}EG}$ is the integration of the C(O) resonance signals of the carbon atom adjacent to EG in units of succinic acid with EG and Is on each side, $I_{EGSuIs\text{-}Is}$ is the integration of the C(O) resonance signals of the carbon atom adjacent to Is in units of succinic acid with EG and Is on each side, and R is the degree of randomness.

2.5. *Gel Permeation Chromatography (GPC)/Size Exclusion Chromatography (SEC)*

The number and weight (average molecular weight, respectively, Mn and Mw) and dispersity (PDI) of synthesized materials were obtained by GPC/SEC. To that end, we employed an Agilent 1260 Infinity II LC system (Agilent Technologies, Santa Clara, CA, USA) equipped with an isocratic G7110B pump, an automatic vialsampler G7129A, a Refractive Index Detector (RID) G7162A, and a PLgel 5 µm (50 × 7.5 mm) guard column combined with 2 PLgel 5 µm (300 × 7.5 mm) MIXED-C columns. Polystyrene standards with Mns varying between ~0.5 and ~1600 kg/mol were used for the determination of the calibration curve. The corresponding solutions were prepared at a concentration of 3 mg/mL and filtered through PTFE filters of ~0.45 µm in pore size. The injection volume was 20 μL, whereas the total elution time of each sample was 30 min. The temperature of the columns and the RID were both fixed at 40 °C.

2.6. Differential Scanning Calorimetry (DSC)

The thermal transitions, with a focus on the glass transition, were studied by DSC in the temperature range of −100 to 160 °C and in a high-purity nitrogen atmosphere. The measurements were conducted by means of a TA Q200 calorimeter (TA Instruments, New Castle, DE, USA) properly calibrated for temperature and heat capacity. Pieces of the prepared samples of ~8 mg in mass were closed in TA-standard aluminum pans. All samples were subjected to a first heating at 160 °C to erase any thermal history and maximize thermal contact with the aluminum pan. Then, a cooling and subsequent heating scan were performed at the fixed rate of 10 K/min.

2.7. Broadband Dielectric Spectroscopy (BDS)

The technique of dielectric spectroscopy was employed to assess the local and segmental molecular dynamics. To that end, we employed a Novocontrol (GmbH, Germany) BDS setup, i.e., an Alpha frequency response analyzer combined with a Quatro liquid-nitrogen cryosystem. The measurements were performed on samples in the form of a sandwich-like capacitor, with the polymer sample (insulator) being melted between two well-polished, brash disk-like plates (electrodes, one with a diameter of 14 mm and the other of 30 mm). The specimen diameter was determined by the upper electrode diameter (14 mm). To prevent the contact of the electrodes and keep a constant electrode's distance, thin silica spacers from Novocontrol (~100 μm) were employed. The melting of the polymers was performed on a hot plate by heating the samples up to T equal to the melting temperature (T_m), the latter being previously estimated by DSC. The samples were kept at T_m for ~30 s, removed from the hot plate, and immediately placed on a large, cold metallic surface in order to be cooled down as soon as possible. After this process, the capacitor was placed in a Novocontrol BDS 1200 measurement cell and inserted into the cryostat. Prior to the recordings, the sample was cooled in the cryostat by a cold nitrogen gas flow down to −140 °C. The complex dielectric permittivity, ε^*, was isothermally recorded as a function of frequency, from 0.1 Hz to 1 MHz, for a large range of temperatures, from −140 to 120 °C, during heating at steps of 5 or 10 K, depending on the followed process.

2.8. Thermogravimetric Analysis (TGA)

The thermal degradation profile of the synthesized copolymers was evaluated by thermogravimetric analysis (TGA), employing the thermal analyzer NETZSCH STA 449F5 (Germany). At a heating rate of 20 °C/min, the samples were heated in the temperature range of 30 to 600 °C under N_2 atmosphere (99.9%) at 30 cm 3 min^{-1}.

2.9. Pyrolysis–Gas Chromatography-Mass Spectrometry Analysis (Py–GC/MS)

For Py-GC/MS analysis [60] of the PESu-PISSu copolymers, a small amount of each composition/sample was initially placed into the "Double-Shot" EGA/PY 3030D Pyrolyzer (Frontier Laboratories Ltd., Fukushima, Japan) employing a CGS-1050Ex carrier gas selector. For the pyrolysis analysis (flash pyrolysis), each sample was placed into the sample cup and, subsequently, allowed to fall into the Pyrolyzer furnace. The temperature of the preselected pyrolysis was set at 450 °C, while the GC oven temperature was programmed at 50 °C for 2 min. This was followed by a step-like elevation of the temperature to 200 °C at 5 °C/min, at which the temperature was kept stable for a period of 8 min. Then, the temperature was increased to 300 °C at 20 °C/min, where it was preserved for 5 min. Sample vapors generated in the furnace were split (at a ratio of 1/50), a portion moved to the column at a flow rate of 1 mL/min and a pressure of 53.6 kPa, and the remaining portion exited from the system via the vent. The pyrolyzates were separated using a temperature-programmed capillary column of a Shimadzu QP-2010 Ultra Plus (Kyoto, Japan) gas chromatogram and analyzed by the mass spectrometer MS-

QP2010SE of Shimadzu (Kyoto, Japan). An Ultra-ALLOY® metal capillary column from Frontier Laboratories LTD (Fukushima, Japan) was used, containing 5% diphenyl and 95% dimethylpolysiloxane stationary phases, a column length of 30 m, and a column ID of 0.25 mm. For the mass spectrometer, the following conditions were used: ion source heater 200 °C, interface temperature 300 °C, vacuum 10^{-4}–10^{0} Pa, m/z range 40–500 amu (atomic mass unit), and scan speed 10,000. The ion gas chromatograms and spectra retrieved by each experiment were subjected to further interpretation through Shimadzu and Frontier post-run software. The chromatograms and spectra retrieved by each experiment were subjected to further interpretation through Shimadzu (NIST11.0) and Frontier (F-Search software 4.3) post-run software. The identification/recognition were done depending on the similarity percentage (minimum value of 80%) between the average mass spectra on the entire chromatogram.

3. Results and Discussion

3.1. Materials—Scope—Structure

The initial scope of these materials (described in Table 1) is the tuning of PESu properties (soft polymer, low glass transition temperature polymer) by the addition of another harder monomer: isosorbide. The copolymerization involving the in-situ polymerization of the different monomers simultaneously aims at achieving the as-uniform as possible distribution (compatibilization) of the two polymers.

The structural characterization begins with FTIR, the results of which are shown in Figure 2. The main absorbance peaks are described within the plot. These are the CH_2 symmetric and asymmetric stretching, the various O-C-O and C-O-C vibrations, and the vibration of the ester bond (-C=O), the latter being the more polar site of PESu and is expected to be involved within interpolymer physical associations/interactions. Regarding PESu, the FTIR data are in accordance with previous findings in the literature [61,62].

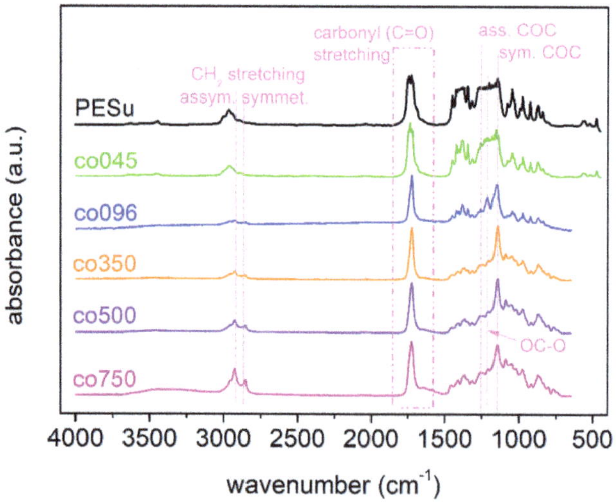

Figure 2. Comparative FTIR spectra for PESu and selected PESu/PISSu copolymers. The most significant absorbance bands along the corresponding molecular-bond origins are marked within the plot.

The NMR spectra of the homo- and copolymers are presented in Figure 3. In the 1H NMR spectrum of PESu, two resonance signals corresponding to OCH_2 C (4.29 ppm) and $C(O)CH_2$ B (2.66 ppm) groups are observed. In the spectra of PISSu, we can note the following signals: 5.20 and 5.16 ppm (OCH 4&5), 4.83 and 4.47 ppm (CH 8&9), 3.9 and 3.8 ppm (CH_2 6&7), 2.7 ppm ($C(O)CH_2$ 3). Signals of both PESu and PISSu segments are observed in the copolymers, with proportional intensities to the EG/IS ratio, thus confirming the successful polymerization. Similarly, in the 13C spectra of the copolymers, signals of both PESu and PISSu segments are observed: 171.8 ppm (C(O) A), 171.4 and 171.2 ppm (C(O) 1&2), 85.7 and 80.6 ppm (CH 8&9), 78.1 and 74.1 ppm (OCH 4&5), 73.1 and 70.3 ppm (CH_2 6&7), 62.2 ppm (OCH_2 C), and 28.6–28.8 ppm ($C(O)CH_2$ 3&B).

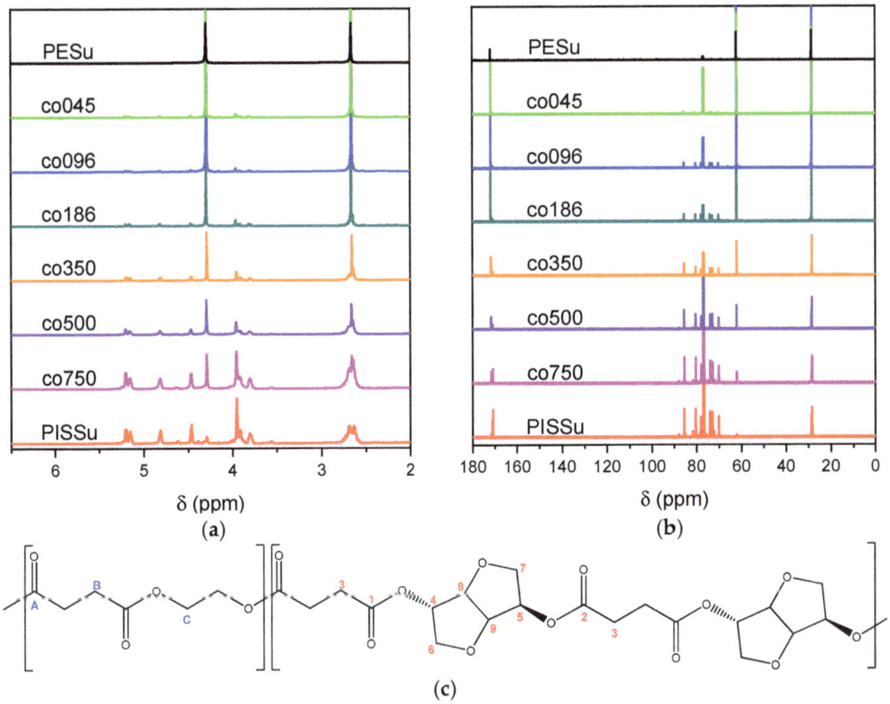

Figure 3. Nuclear magnetic resonance spectra of the synthesized polymers, (**a**) ^{1}H and (**b**) ^{13}C, and (**c**) the structure of the copolymers.

Regarding the composition of the copolymers (calculated by comparing the integration of the -OCH- signals of the PISSu segments and the -OCH_2- signals of the PESu segments), we can note that the Is content is slightly lower compared to the feed ratio, and this trend is more pronounced in the copolymers with a higher Is content. Finally, the microstructure of the copolymers was studied by NMR (Table 1), and co350, co500, and co750 were found to be random copolymers (*R* = 1). For the other copolymers synthesized with a lower Is content, i.e., co045, co096, and co0186, the studied NMR signals could not be safely integrated; the copolymers are probably constituted by large PESu blocks interrupted by random PISSu units. More specifically, the ester carbon atoms exhibit slightly different shifts depending on whether the adjacent units are an EG or an Is unit. Therefore, besides the EG-SA-EG and Is-SA-Is units, where succinic acid is esterified by the same alcohol, EG or Is, discrete signals corresponding to EG-SA-Is or Is-SA-EG units were detected. Assuming similar relaxation times for all C=O ester groups, the average sequence length, *L*, and degree of randomness of the copolymers were calculated [63].

In Figure 4, we present the results by GPC-SEC for M_n/M_w and PDI. Neat PESu exhibits an M_n of ~83 kg/mol and a moderate *PDI*~1.7. During the copolymerization, the M_n drops systematically, with the Is fraction increasing. This can be explained as follows: Is hydroxyl groups have a lower reactivity compared to EG. Since the time periods and temperatures of the polymerization were kept constant for all systems, we expected the formation of gradually shorter copolyesters when increasing Is due to its lower reactivity, as indicated by the systematic M_n drop. We have reported similar correlations of M_n alterations with successful/homogeneous in situ copolymerization within various other copolyesters (e.g., in [34–36,46]). At the same time, within all copolyesters, *PDI* is increased, interestingly, in a non-monotonic way regarding the Is ratio. Obviously, the situation of the chemo- and thermodynamical processes taking place is quite complex to easily interpret the complex *PDI*(Is) trend.

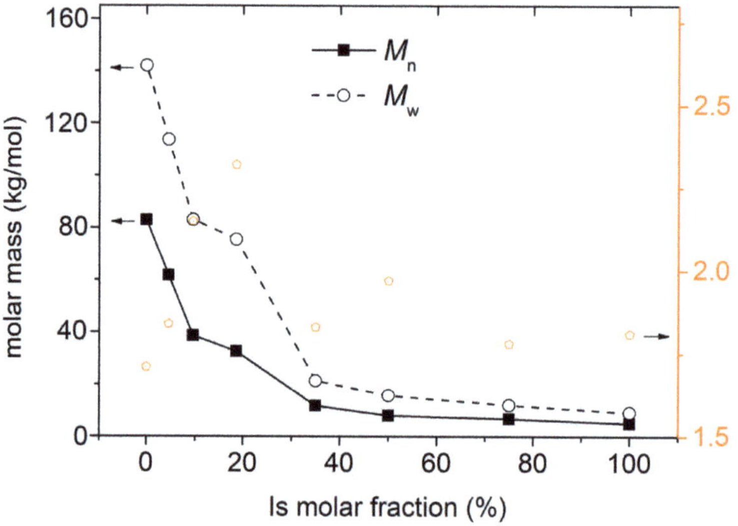

Figure 4. The isosorbide molar fraction dependence of (left axis) the molar mass values M_n and M_w and (right axis) of the polydispersity index.

3.2. Glass Transition—Thermal Events

Coming to the thermal response, we present in Figure 5 the comparative DSC traces for all compositions. We recall that prior to the cooling shown in Figure 5a, all samples had been well melted at ~160 °C; thus, any thermal history was erased and the thermal contact between the polymer and the aluminum pan was optimized. Neat PESu exhibits a weak crystallization peak at ~34 °C (Table 2), whereas PISSu is amorphous. The presence of Is in the copolymers, at all ratios, seems to preclude the weak crystallization of PESu, as manifested by the complete lack of exothermal peaks in the copolymers.

During cooling as well as the subsequent heating (Figure 5b), all samples demonstrate single glass transition steps. This provides, by itself, solid evidence for the homogeneity of the systems. From the data on heating, we evaluated the characteristic temperature of the glass transition, T_g, as the T of the half-heat capacity change, $\Delta c_p/2$. The T_g values are listed in Table 2 and are shown in Figure 5a as a function of the Is molar fraction. While for neat PESu, T_g equals −16 °C, in the copolymers, T_g increases monotonically up to 40 °C, tending to approach the value of neat PISSu (51 °C). To check this systematic behavior, we fitted mathematical models describing the degree of miscibility judging from the T_g. Our results seem to coincide quite well with the prediction by the well-known Fox equation [64] for completely miscible polymers in the form of blends. A similar systematic dependence is observed for the Δc_p of the glass transition (the so-called 'glass transition strength'), dropping from 0.65 J/g·K down to 0.46 J/g·K.

From the point of view of segmental polymer mobility, both results (T_g and Δc_p, Figure 6) indicate that these PISSu/PESu can be considered homogeneous systems of excellently mixed polymers (miscible), wherein the 'hard' polymer (isosorbide-based) systematically affects the softer one (PESu). In practice, we gain the first significant proofs of the successfully synthesized copolymers and the achievement of the initial scopes. The simultaneous increase of T_g and decrease of Δc_p is a result of constrained polymer chain mobility [65,66] (even the degrees of freedom for each chain), which actually suggests the hardening of the polymeric matrix. Obviously, this will have a severe impact on the materials' performance, such as the mechanical permeation of small molecules (e.g., air, humidity, gases), heat transport, etc., as well as future processing.

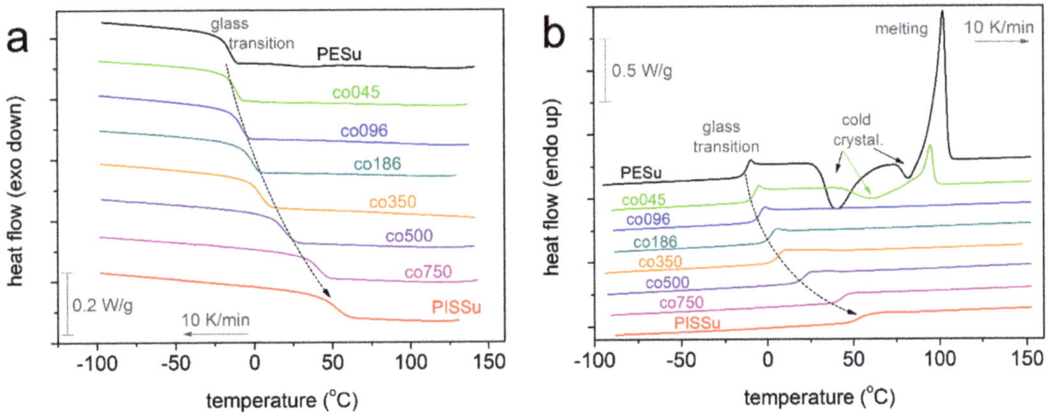

Figure 5. Raw DSC results for all samples are indicated during (**a**) cooling and (**b**) subsequent heating. The recorded heat flow is shown upon normalization to each sample mass. The added curved arrows mark the effect on the glass transition step by increasing the isosorbide molar fraction.

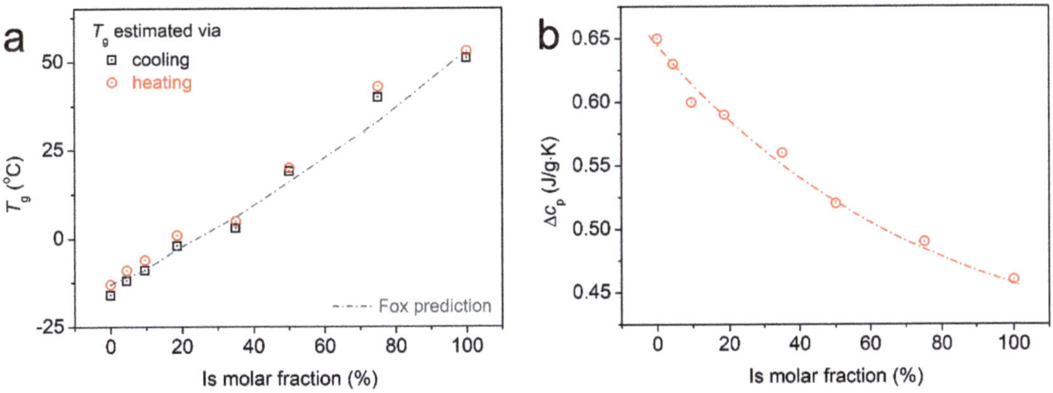

Figure 6. The isosorbide molar fraction dependence of (**a**) T_g and (**b**) Δc_p. The added line in (**a**) corresponds to the prediction by the Flory equation, whereas the added line in (**b**) is just an eye-guiding line.

Due to the hardening of the systems in combination with the increased rigidity of the copolymer chains, both due to the existence of the hard PISSu segments and the simultaneous shortening of the chain lengths, we would expect a suppression of crystallizability. This is found to be true, as observed by the almost immediate vanishing of melt/cold crystallization and melting events in Figure 5a,b. The result is another supportive clue on the copolymers' homogeneity, as, at least on the micrometric scale, there seems to be no PESu separation from PISSu.

3.3. Molecular Dynamics

The effects on molecular dynamics of local and segmental types are assessed by BDS. Upon keeping fixed thermal treatment and initial erasing of the thermal history for all samples, we basically exploit the information supplied by the imaginary part of dielectric permittivity $\varepsilon''(f,T)$. ε'' is considered to be related to the dielectric losses, whereas the dipolar relaxation mechanisms related to the molecular mobility are recorded as peaks of the isothermally recorded $\varepsilon''(f)$ spectra.

In Figure 7, we present a selection of representative BDS data for the overall temperature range of investigation (−140 to 120 °C). Therein, mainly three distinct types of relaxations are recorded. At $T < T_g$, the dipolar response is mainly weak and is dominated by two local-like relaxations, i.e., the ones named as β and β_W. These relaxation mechanisms originate from the more localized molecular motions [35,67–69] (and references therein). At $T > T_g$, the dielectric signal increases significantly by ~1 order of magnitude due to the implementation of additional dipole moments due to the mobilization of the overall polymer chains and their cooperative motions (transition from the glassy to rubbery state of the polymers). The corresponding relaxation mechanism (peak) is widely addressed as the 'main α relaxation' and is considered the dielectric and dynamic analogue of the calorimetric glass transition [70].

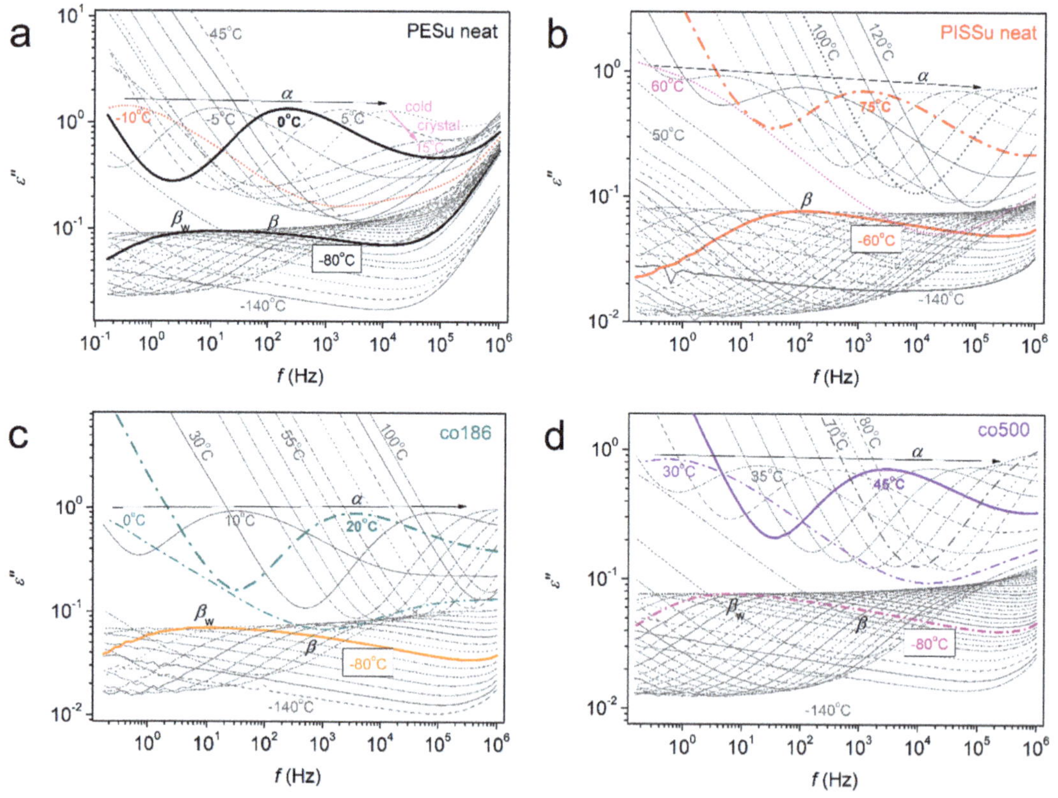

Figure 7. The raw BDS data for the indicated selected samples, namely, in terms of the isothermal $\varepsilon''(f)$ plots at all temperatures of measurement. The results are shown for (**a**) neat PESu, (**b**) neat PISSu, (**c**) co186 and (**d**) co500. The local (β, β_W) and segmental (α) relaxations, recorded as peaks of ε'', are indicated at selected temperatures.

To facilitate a better comparison between the different samples and, simultaneously, with the calorimetric findings (Figure 5b), we replotted the isothermal results [$\varepsilon''(f)$] in the form of isochronal plots [$\varepsilon''(T)$]. In particular, this is done for the selected relatively high frequency of ~3 kHz. These spectra can be seen in Figure 8. From these data, it can be easily seen that while the local dynamics (β-type relaxations) are barely affected by the copolymers' composition, the segmental dynamics (α relaxation) migrate systematically toward higher temperatures with an increasing Is/EG ratio. This is in qualitative agreement with the effects recorded on the calorimetric glass transition, as shown in Figures 5b and 6a. This migration suggests the retardation, or else deceleration, of segmental dynamics.

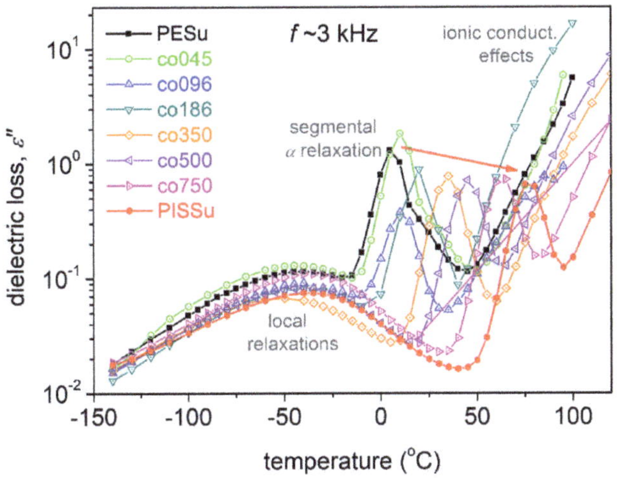

Figure 8. Comparative isochronal $\varepsilon''(T)$ results for all compositions shown at 3.162 kHz. The added red arrow marks the impact of the isosorbide molar fraction on the main α relaxation.

In order to enrich our conclusions with more information regarding molecular dynamics, such as the dynamical T_g and fragility, the width of the relaxation time range, and dielectric strength, we performed a specific analysis of the complex spectra in terms of fitting mathematical models [70,71]. To this end, we employed the Havriliak–Negami (HN) [72] function, mathematically described in Equation (3).

$$\varepsilon^*(f) = \frac{\Delta\varepsilon}{\left[1 + \left(\frac{if}{f_0}\right)^{\alpha_{HN}}\right]^{\beta_{HN}}} + \varepsilon_\infty \quad (3)$$

In the HN function, the following parameters are involved: the dielectric strength, $\Delta\varepsilon$, a characteristic frequency related to the frequency of maximum dielectric loss, f_0, a value of the real part of the dielectric permittivity (ε') at $f >> f_0$, ε_∞, the shape parameter that denotes the width of relaxation times range, αHN, and, finally, another shape parameter that evaluates the symmetry or asymmetry of the $\varepsilon''(f)$ peak, β_{HN}. For $\beta_{HN} < 1$, the HN term is asymmetric, which is the common case for an amorphous, unaffected segmental α relaxation. We note that in the case of a symmetric relaxation, β_{HN} equals 1, and the HN equation becomes practically the so-called Cole-Cole function [73]. During the fitting, there is also the need to simulate the effects arising from conductivity-related effects, i.e., at the higher frequency side of the measurement window. For that, we may employ the additional linear term $-i{\cdot}\sigma_0/(2\pi f)^C\varepsilon_0$ where σ_0 is related to the specific DC conductivity and the C exponent parameter describes the Ohmic (for C = 1) and non-Ohmic contributions in conductivity (for C < 1). In case of additional polarization effects, such as the 'electrodes polarization' additional terms may be needed [71,74].

In Figure 9, we show representative examples of the performed fittings for both homopolymers and copolymers, as well as both local and segmental relaxations. As expected [70,71], the local-β-type relaxation could be sufficiently fitted by symmetric terms, whereas the segmental α relaxations were fitted by asymmetric HN terms.

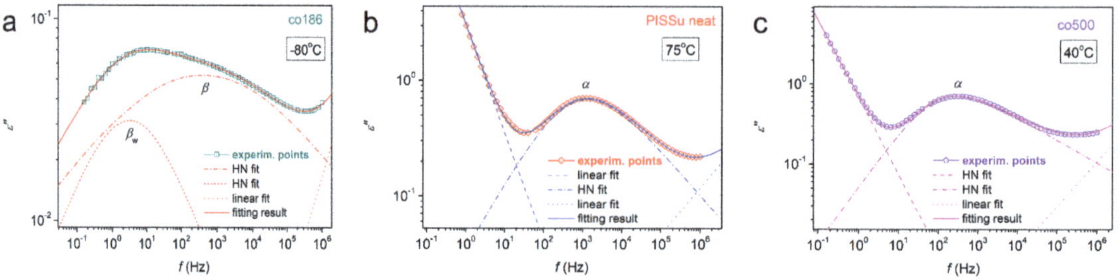

Figure 9. Examples of analysis of the $\varepsilon''(f)$ spectra for three cases of samples [(**a**) co186, (**b**) PISSu neat and (**c**) co500] and temperatures in terms of model functions (details in the main text).

From the values for the peak maxima, fmax, coming from this fitting, we were able to construct the dielectric relaxation map (time scale) by plotting log$f_{\max}$ as a function of the reciprocal temperature, 1000/T, for all samples under investigation. This map, also called Arrhenius plots, is presented in Figure 10. Therein, we have included data for neat poly(isosorbide) (PIS), as adapted from a recent work by our groups [36].

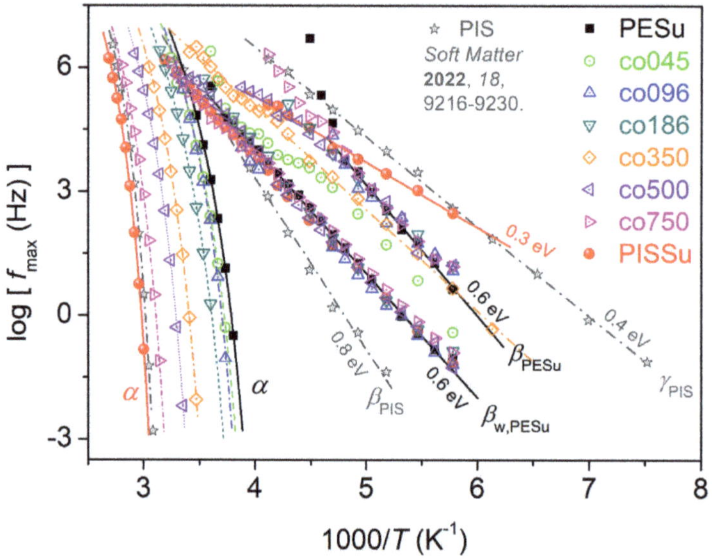

Figure 10. The overall dielectric relaxation map for all samples in terms of time scale, i.e., the reciprocal temperature dependence of the frequency maxima for all the resolved dipolar relaxations (local and segmental). For comparison, we have added results for neat poly (isosorbide) and PIS (grey stars), adapted from a recent work [36]. The straight and curved lines connecting the experimental points are fittings of the Arrhenius and the Vogel–Tammann–Fulcher–Hesse [75] equations, respectively.

Comparing with previous findings on poly(n-alkylene succinates) [35,64,68], we could identify the local relaxations recorded here. Both processes were found to originate from the localized motions of PESu, namely the β_{PESu} and the $\beta_{w,PESu}$. β_{PESu} has been proposed to monitor the dipole moments arising from the crankshaft rotations of the ester groups at the polymer backbone. Such relaxations in many polyesters have been found to depend strongly on segmental dynamics, in particular in poly(n-alkylene succinates) to even scale with the T_g [68], despite their local character. Regarding the slightly slower $\beta_{w,PESu}$ relaxation, there are only two previous recordings in the literature on poly(n-alkylene succinates) to compare with [35,64,68]. The relaxation has been found to be stronger for small n- and weaker for larger n-alkylene sequencies. Based on similarities recorded between this process and other local processes arising from hydrophilic sites with attached water molecules located at solid surfaces and/or hydrophilic polymer groups [76–78], we have proposed such origins for our $\beta_{w,PESu}$ [68].

For PESu and the copolymers, both processes exhibit independence from the Is fraction and show similar and almost constant activation energies, E_{act}~0.6 eV. Interestingly, in neat PISSu, the local relaxation (β_{PESu}-like process) is mainly accelerated, exhibiting a suppressed E_{act}~0.3 eV. Taking into account the data for γ_{PIS} relaxation, it seems that β in PISSu could be the superposition of β_{PESu} and γ_{PIS}. Contrariwise, $\beta_{w,PESu}$ demonstrates a similar time scale with PESu and PESu/PISSu.

None of the local processes was found to coincide with the time scale of local relaxations of neat PIS (grey stars in Figure 10).

The focus is now turned to segmental dynamics. To facilitate the discussion, we created Figure 11a, which shows the dynamics map solely of the α relaxations. It is quite clear from a simple glance that by increasing the amount of the 'hard phase' (Is), the α relaxations are systematically decelerated, i.e., migrate toward higher temperatures/lower frequencies. The migration, at the higher Is ratio (75%), sets the segmental mobility similar to that of a neat PIS. Even more decelerated (and constrained) is the segmental dynamics of neat PISSu. Employing known routes, from the data in Figure 11a, we could evaluate the dielectric glass transition temperature, $T_{g,diel}$, and the fragility index of α relaxation, mα [68,79]. $T_{g,diel}$ is actually the T extrapolation of the time scale of α to the equivalent frequency of conventional DSC (i.e., ~0.0016 Hz), arising from the corresponding relaxation time at $T = T_g$ (τ_{rel}~100 s).

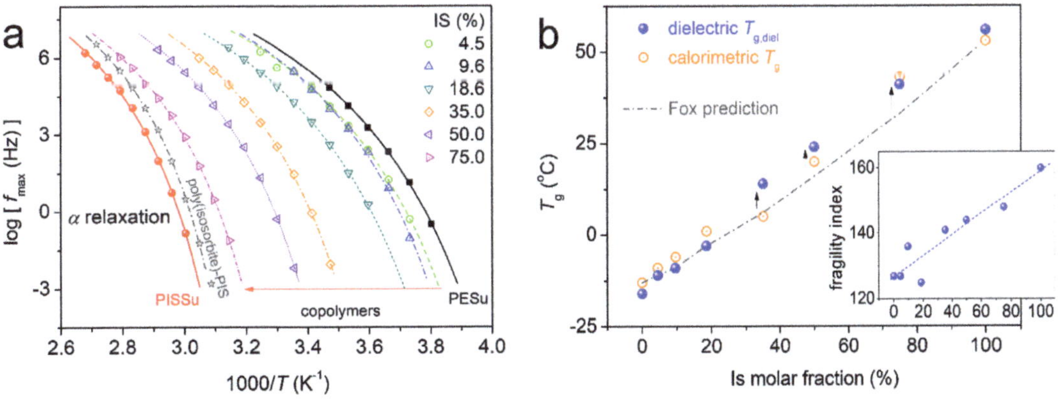

Figure 11. The segmental dynamics characteristics, in terms of (**a**) the time scale for α relaxation and (**b**) the isosorbide fraction dependence of the dielectric and calorimetric T_g, as well as the fragility index of α relaxation, m_α (inset to **b**).

$T_{g,diel}$ and m_α were plotted in Figure 11b as a function of the Is molar fraction. Similarly to the calorimetric T_g, the $T_{g,diel}$ increases almost linearly with the Is fraction, from −16 to 56 °C. The calorimetric and dielectric data are in quite good agreement with each other. Taking a careful look at the main Figure 11b, there seems to be an interesting variation of the T_g(Is) above the Fox prediction for the higher EG/IS ratios. This may suggest the involvement of more interactions between the two polymers that lead to constraints on the chains' mobility or/and the chain-chain associations. The fragility index in the inset to Figure 11b shows an in general increasing m_α(Is) trend (from 127 to 160), nevertheless not strictly monotonic. The increasing of T_g(Is) in terms of the increasing of the dynamical constraints seems initially controversial to the ma(Is). The dynamical fragility of 'simple' homopolymers could be used as a measure of the chains' cooperativity [79]. In our case, this seems to increase, despite the increasing constraints imposed by the addition of the 'hard polymer (PISSu)' to the softer one (PESu). During the copolymers' synthesis, the average M_n decreases (Table 1), which would also lead to a tendency for m_α to drop. This would be due to an increase in the average cooperativity length [80,81], or, in other words, the inter-chain distance and the free volume would increase [82]. Obviously, this is not the main case here. So, taking into account the overall findings, we may conclude that the data revealed in Figure 11 suggest the implementation of more than one parameter for the molecular mobility effects. In particular, with the addition of Is, both constraints increase (hardening of the matrix); however, there seems to be an overall densification of the polymer (decrease of the free volume).

3.4. Thermal Stability and Degradation Mechanism of Copolymers

Aiming to investigate the thermal stability of the synthesized copolymers, TGA analysis was employed. The TGA thermograms of PESu, PISSu, and their copolymers are presented in Figure 12. The thermal degradation evolved in one step for all samples, in particular, starting after 350 °C and being almost completed at around 480 °C. As noted from the recorded thermograms, the studied copolymers exhibited relatively good thermal stability, with no significant mass loss up to 300 °C, implying by these means their potential to undergo shape-forming processes without degrading. Seeking further systematic behavior of thermal degradation, from the data in Figure 12 (and inset), it can be stated that there was no direct effect of the PESu/PISSu composition. The weaker thermal stability was observed for co750, whereas the higher one was observed for co096, presenting even higher thermal stability than the starting polyester building blocks. Obviously, this suggests the multiple parameters involving, beside the structure, the interchain associations (bond strengths) as well as, in connection to the previous section, the potential alterations in the free volume.

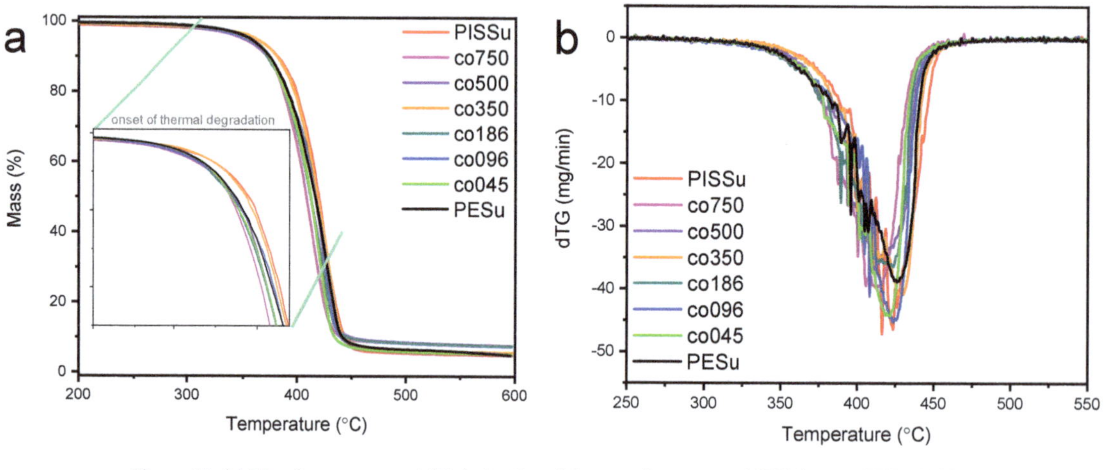

Figure 12. (**a**) Mass loss curves and (**b**) derivative of the mass loss curves (dTG) for neat PESu, PISSu, and the studied copolymers.

While TGA was used to study the effect of different contents in homo-polyesters on the thermal stability of the synthesized polymers, the technique of pyrolysis combined with gas chromatography/mass spectrometry (Py–GC/MS) was employed to evaluate in detail the mechanism of the thermal degradation of copolymers and, moreover, to determine whether and how the latter was altered by changing the EG/Is ratio. Analytical pyrolysis has been extensively investigated to explore the polymers' thermal degradation mechanisms, since it delivers significant data on the macromolecules after the analysis and identification of their evolved thermal degradation products [83]. In the context of the present study, flash pyrolysis was performed at 450 °C, corresponding roughly to the end of the degradation for all studied polymers and being a temperature at which most of the degradation products were possibly formed. Understanding the decomposition process of newly synthesized polymers is crucial because it aids in selecting the most appropriate methodologies or additives to ameliorate their thermal stability.

The total ion chromatograms (TICs) of PISSu and PESu polyesters, along with their copolyesters, after pyrolysis at 450 °C are illustrated in Figure 13 (mass spectra not shown for briefness reasons). As depicted, the chromatograms of the studied samples exhibit relatively simple patterns, with some differences being present in the intensity of the peaks corresponding to each different block of polyester. Each peak corresponds to a mass spectrum that resembles a thermal decomposition product. As for the assigned compounds identified by their mass spectra and the available library, the most important pyrolysates are summed and listed in Table 3. Our research group has already thoroughly explored the thermal degradation pattern of several succinate copolymers [83–85], as well as the mechanism of thermal degradation of poly(isosorbide furanoate) (PIsF) [58].

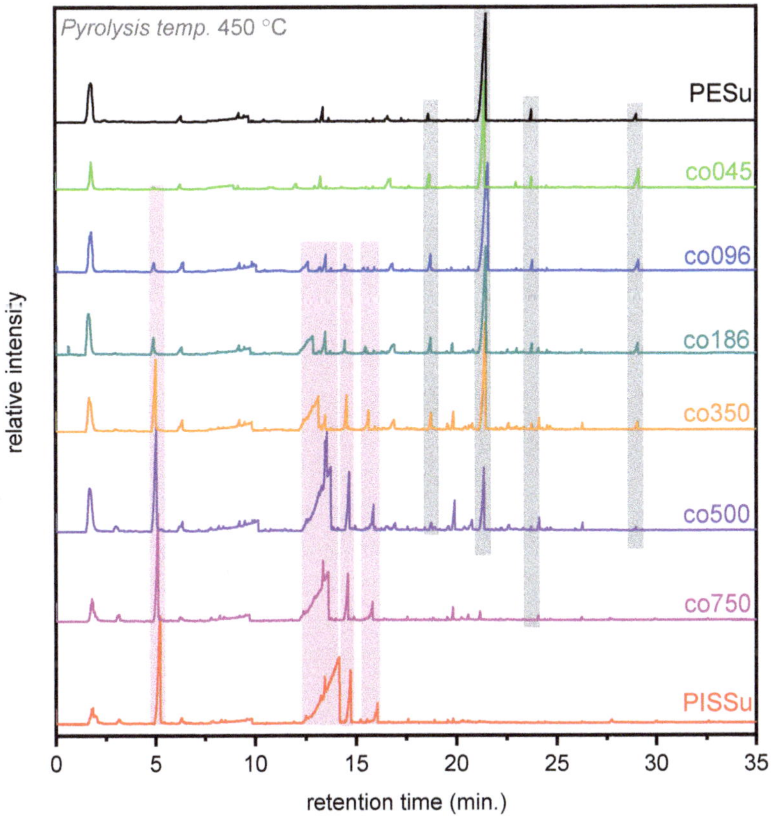

Figure 13. Overall ion chromatograms (TICs) of the PESu-PISSu copolymers in different ratios after pyrolysis at 450 °C.

Table 3. Analysis of evolved gaseous compounds during pyrolysis at 450 °C of neat PESu, PISSu, and their copolymers at different content ratios.

Rt (min.)								MW (amu)	Assigned Structure
PISSu	**co750**	**co500**	**co350**	**co186**	**co096**	**co045**	**PESu**		
1.72	1.72	1.68	1.67	1.64	1.77	1.78	1.78	44	CO_2 or acetaldehyde
1.92	1.91	n.d.	n.d.	n.d.	n.d.	n.d.	n.d.	86	(S)-tetrahydrofuran-3-ol 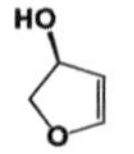 or (R)-tetrahydrofuran-3-ol 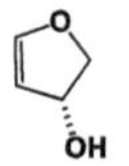
5.20	5.11	5.04	5.00	4.91	4.64	n.d.	n.d.	110	3a,6a-dihydrofuro[3,2-b]furan 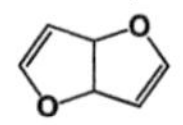
n.d.	n.d.	6.32	6.35	6.34	6.42	6.28	6.33	118	propanoic acid, 2-hydroxyethyl ester 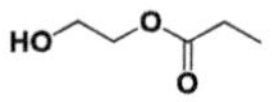
n.d.	n.d.	9.20	9.20	9.22	9.24	9.25	9.21	100	succinic anhydride 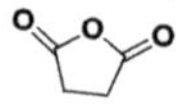
9.67	9.69	9.69	9.50	9.49	n.d.	n.d.	n.d.	126	(3R)-hexahydrofuro[3,2-b]furan-3-ol 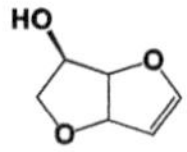 or (3S)-hexahydrofuro[3,2-b]furan-3-ol 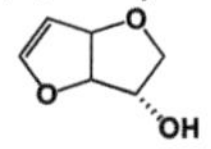
n.d.	n.d.	n.d.	13.47	13.49	13.51	13.25	13.40	172	butanedioic acid, diethyl ester 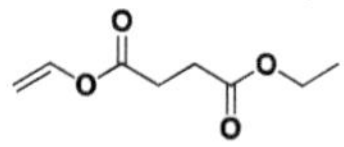
13.42	13.36	13.32	13.49	13.51	13.50	13.55	n.d.	146	hexahydrofuro[3,2-b]furan-3,6-diol (isosorbide) 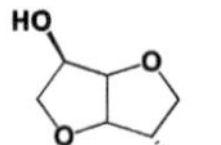
14.70	14.60	14.64	14.53	14.45	14.49	n.d.	n.d.	142	(3S)-3-methoxyhexahydrofuro[3,2-b]furan 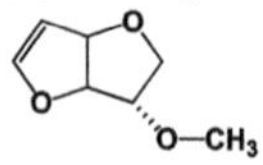
16.05	15.80	15.87	15.61	15.49	n.d.	n.d.	n.d.	160	(3R,6S)-6-methoxyhexahydrofuro[3,2-b]furan-3-ol 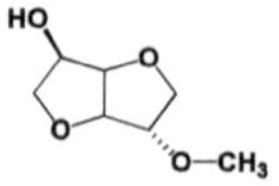

Table 3. *Cont.*

Rt (min.)								MW (amu)	Assigned Structure
PISSu	co750	co500	co350	co186	co096	co045	PESu		
n.d.	n.d.	16.94	16.89	16.91	16.83	16.72	16.61	202	allyl (2-hydroxyethyl) succinate 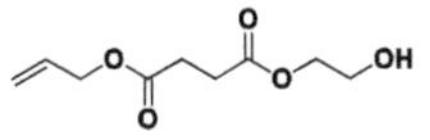
n.d.	n.d.	18.71	18.71	18.71	18.72	18.63	18.62	188	2-hydroxyethyl vinyl succinate 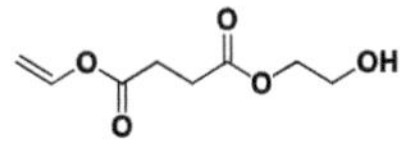
n.d.	19.81	19.88	19.83	19.80	19.75	n.d.	n.d.	229	(((3S,6R)-6-hydroxyhexahydrofuro[3,2-b]furan-3-yl)oxy)methyl acrylate 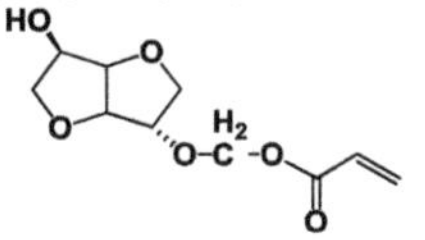
n.d.	21.15	21.36	21.42	21.47	21.56	21.40	21.49	260	2-(acryloyloxy)ethyl (2-hydroxyethyl) succinate 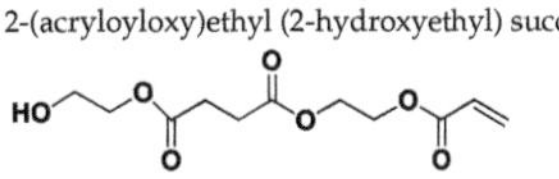
n.d.	n.d.	23.71	23.72	23.74	23.76	23.70	23.73	290	2-hydroxyethyl (2-((4 oxobutanoyl)oxy)ethyl) succinate 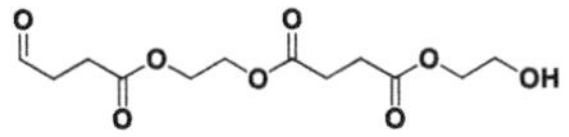
n.d.	24.03	24.10	24.10	24.08	n.d.	n.d.	n.d.	306	4-((((3S,6R)-6-(hydroxymethoxy)hexahydrofuro[3,2-b]furan-3-yl)oxy)methoxy)-4-oxobutanoic acid 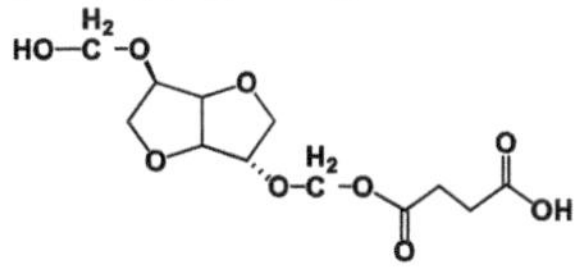
n.d.	26.21	26.25	26.24	n.d.	n.d.	n.d.	n.d.	328	2,3,3a,6a-tetrahydrofuro[3,2-b]furan-3-yl (2-((vinyloxy)methoxy)ethyl) succinate 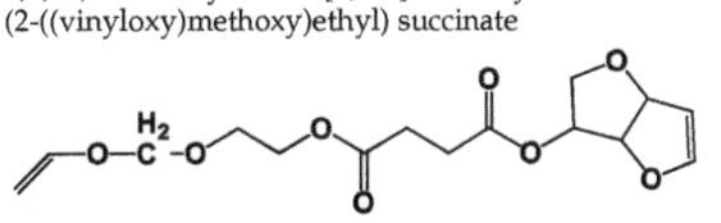
n.d.	n.d.	28.95	29.02	29.05	29.10	29.10	29.01	332	2-((4-(2-hydroxyethoxy)-4-oxobutanoyl)oxy)ethyl vinyl succinate 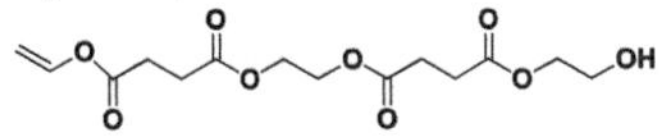

At lower retention times (Rt), volatile molecules with lower molecular weights, such as carbon dioxide or acetaldehyde, were observed. The presence of carbon dioxide indicated the occurrence of homolytic processes and the decarboxylation of carboxyl-end groups in the synthesized polymers under study [86]. As the Rt proceeded, larger and more complex structures were observed. The chromatogram of PESu, as already reported in previous works, illustrated a fairly simplistic pattern, with some major pyrolytic peaks (annotated with gray color) that were identified as propanoic acid, 2-hydroxyethyl ester (Rt = 6.33 min), succinic anhydride (Rt = 9.21 min), butanedioic acid, diethyl ester (Rt = 13.40 min), allyl (2-hydroxyethyl) succinate (Rt = 16.61 min), 2-hydroxyethyl vinyl succinate (Rt = 18.62 min), 2-(acryloyloxy)ethyl (2-hydroxyethyl) succinate (Rt = 21.49 min), 2-hydroxyethyl (2-((4 oxobutanoyl)oxy)ethyl) succinate (Rt = 23.73 min), 2-((4-(2-hydroxyethoxy)-4-oxobutanoyl) oxy)ethyl vinyl succinate (Rt = 29.01 min). These compounds could be categorized as hydroxyl-, vinyl-, methyl-, and aldehyde-terminated, which can be primarily products of β-hydrogen and, to a lesser extent, α-hydrogen scission reactions. After the incorporation of Is units in the macromolecular chain of PESu, new peaks emerge in the recorded TICs. It is important to mention that the new Is-derived peaks increased in number as well as in intensity as the PISSu molar ratio in the copolymers increased, whereas the intensity of the PESu-derived peaks decreased significantly. The identified compounds assigned to PISSu counterpart are the following (annotated with pink color): tetrahydrofuran-3-ol (Rt = 1.92 min), 3a,6a-dihydrofuro[3,2-b]furan (Rt = 5.20 min), hexahydrofuro[3,2-b]furan-3-ol (Rt = 9.67 min), isosorbide (Rt = 13.42 min), 3-methoxyhexahydrofuro[3,2-b]furan (Rt = 14.70 min), and (3R,6S)-6-methoxyhexahydrofuro[3,2-b]furan-3-ol (Rt = 16.05 min). For higher retention times, pyrolysis compounds were not recorded. Noteworthy was also the formation of new peaks for almost all the studied copolymers, which were comprised of both copolymers that were absent in PESu and PISSu polyesters. These newly formed peaks were detected for Rt > 19 min and were primarily vinyl-, carboxyl-, and hydroxyl-terminated compounds, namely (((3S,6R)-6-hydroxyhexahydrofuro[3,2-b]furan-3-yl)oxy)methyl acrylate (Rt = 19.8 min), 4-((((3S,6R)-6-(hydroxymethoxy)hexahydrofuro[3,2-b]furan-3-yl)oxy)methoxy)-4-oxobutanoic acid (Rt = 24.01 min) and 2,3,3a,6a-tetrahydrofuro furan-3-yl(2-((vinyloxy)methoxy)ethyl) succinate (Rt = 26.25 min). The detection of the aforementioned compounds could also be considered a validation of the successful copolymerization of the studied polymers, nicely supplementing the indirect evidence reported in the previous sections.

A large sum of interesting data regarding these novel PESu/PISSu copolyesters was recorded, as expected, with a few points still open. Further work on these materials (mechanical tests, compostability, permeation to water and gases) would be illuminating, both from the point of view of basic research and that of future applications, processing, and disposal aspects.

4. Conclusions

We demonstrate a sum of structural and thermodynamic data on a newly synthesized series of copolymers based on two renewable polymers, PESu and PISSu. The synthesized copolymers exhibited, in general, good thermal stability, as indicated by TGA thermograms, which is quite an exploitable parameter regarding processing and specific applications. The thermal degradation mechanism for the studied copolymers was also explored in high detail employing analytical pyrolysis. With the increasing of the Isosorbide(Is)/EG molar ratio, i.e., gradually from 5/95 to 75/25 during the in-situ copolymerization, the molar mass (copolymer chain length, M_n) decreases from ~80 down to ~5 kg/mol. Despite that, the 'single' glass transition temperature increases monotonically from −16 to 40 °C, while, at the same time, the crystallizability of PESu vanishes. Both direct and indirect results are dependent on the homogeneity of these systems. During the molecular dynamics mapping, the local mobility was found to be dominated by the ester dipolar motions exhibiting mild effects, while the dynamical view resembles mainly that of PESu rather than that of PISSu. On the other hand, in agreement with the calorimetric findings, the dynamical glass transition was found to monotonically increase when increasing Is. Interestingly, this increase follows quite well the predictions of models on miscible polymers in the form of blends. Simultaneously, the dynamical fragility increases, supplying indirect evidence for a decrease in free volume, or, in other words, a densification of the copolymers. Combining all the results, we gained firm evidence for the 'hardener' role played by Is on the softer PESu, which was actually the main scope of these PESu/PISSu systems, envisaging specific applications for the future.

Author Contributions: Conceptualization, methodology, P.A.K.; validation, A.Z., M.M., A.K. and D.N.B.; formal analysis, investigation, C.B., P.A.K., A.Z., E.X. and N.M.A.; resources, M.M., A.K. and D.N.B.; writing—original draft preparation, P.A.K.; writing—review and editing, C.B., A.Z.,

E.X., N.M.A., M.M., A.K. and D.N.B.; visualization, C.B., P.A.K. and N.M.A.; supervision, D.N.B. All authors have read and agreed to the published version of the manuscript.

Funding: This work was developed within the scope of the ERASMUS+ project between the University of Monastir and the Aristotle University of Thessaloniki.

Institutional Review Board Statement: Not applicable.

Data Availability Statement: The original contributions presented in the study are included in the article, further inquiries can be directed to the corresponding authors.

Acknowledgments: C.B., M.M. and D.N.B. would like to acknowledge the ERASMUS+ project between the University of Monastir and the Aristotle University of Thessaloniki. P.A.K. thanks Dr. Daniel Fragiadakis from the Naval Research Laboratory of the Polymer Physics Section (Washington, DC, USA) for the free provision of the sophisticated data analysis software *'Grafity'* (http://grafitylabs.com/ accessed on 1 January 2024).

Conflicts of Interest: The authors declare no conflicts of interest.

References

1. Pearce, A.K.; O'Reilly, R.K. Polymers for biomedical applications: The importance of hydrophobicity in directing biological interactions and application efficacy. *Biomacromolecules* **2021**, *22*, 4459–4469. [CrossRef] [PubMed]
2. Rowan, S.J. 100th Anniversary of Macromolecular Science Viewpoints. *ACS Macro Lett.* **2021**, *10*, 466–468. [CrossRef] [PubMed]
3. Chanda, M.; Roy, S.K. *Industrial Polymers, Specialty Polymers, and Their Applications*; CRC Press: Boca Raton, FL, USA, 2008. [CrossRef]
4. Andrady, A.L.; Neal, M.A. Applications and societal benefits of plastics. *Philos. Trans. R. Soc. B* **2009**, *27*, 1977–1984. [CrossRef] [PubMed]
5. Kalali, E.N.; Lotfian, S.; Shabestari, M.E.; Khayatzadeh, S.; Zhao, C.; Nezhad, H.Y. A critical review of the current progress of plastic waste recycling technology in structural materials. *Curr. Opin. Green Sustain. Chem.* **2023**, *40*, 100763. [CrossRef]
6. Nayanathara Thathsarani Pilapitiya, P.G.C.; Ratnayake, A.S. The world of plastic waste: A review. *Clean. Mater.* **2024**, *11*, 100220. [CrossRef]
7. Mori, R. Replacing all petroleum-based chemical products with natural biomass-based chemical products: A tutorial review. *RSC Sustain.* **2023**, *1*, 179–212. [CrossRef]
8. Sisti, L.; Totaro, G.; Marchese, P. *PBS Makes Its Entrance into the Family of Biobased Plastics*; John Wiley & Sons: Hobocan, NJ, USA, 2016. [CrossRef]
9. Lendlein, A.; Sisson, A. (Eds.) *Handbook of Biodegradable Polymers: Isolation, Synthesis, Characterization and Applications*; Wiley-VCH: Weinheim, Germany, 2011. [CrossRef]
10. Tanaka, M.; Sato, K.; Kitakami, E.; Kobayashi, S.; Hoshiba, T.; Fukushima, K. Design of biocompatible and biodegradable polymers based on intermediate water concept. *Polym. J.* **2015**, *47*, 114–121. [CrossRef]
11. Pan, P.; Inoue, Y. Polymorphism and isomorphism in biodegradable polyesters. *Prog. Polym. Sci.* **2009**, *34*, 605–640. [CrossRef]
12. Kim, N.; Lee, K. Environmental consciousness, purchase intention, and actual purchase behavior of eco-friendly products: The moderating impact of situational context. *Int. J. Environ. Res. Public Health* **2023**, *20*, 5312. [CrossRef]
13. Sheldon, R.A.; Norton, M. Green chemistry and the plastic pollution challenge: Towards a circular economy. *Green Chem.* **2020**, *22*, 6310–6322. [CrossRef]
14. Hong, M.; Chen, E.Y.X. Chemically recyclable polymers: A circular economy approach to sustainability. *Green Chem.* **2017**, *19*, 3692–3706. [CrossRef]
15. Tazmeen, T.; Mir, F.Q. Sustainability through materials: A review of green options in construction. *Results Surf. Interfaces* **2024**, *14*, 100206. [CrossRef]
16. Balla, E.; Daniilidis, V.; Karlioti, G.; Kalamas, T.; Stefanidou, M.; Bikiaris, N.D.; Vlachopoulos, A.; Koumentakou, I.; Bikiaris, D.N. Poly(lactic acid) a versatile biobased polymer of next decades with multifunctional properties. From monomer synthesis, polymerization techniques and molecular weight increase to PLA applications. *Polymers* **2021**, *13*, 1822. [CrossRef]
17. Ainali, N.M.; Kalaronis, D.; Evgenidou, E.; Kyzas, G.Z.; Bobori, D.C.; Kaloyianni, M.; Yang, X.; Bikiaris, D.N.; Lambropoulou, D.A. Do poly(lactic acid) microplastics instigate a threat? A perception for their dynamic towards environmental pollution and toxicity. *Sci. Total Environ.* **2022**, *832*, 155014. [CrossRef]
18. Papageorgiou, G.Z. Thinking green: Sustainable polymers from renewable resources. *Polymers* **2018**, *10*, 952. [CrossRef]
19. Lahcini, M.; Qayouh, H.; Yashiro, T.; Simon, P.; Kricheldorf, H.R. Syntheses of poly(butylene succinate) by means of non-toxic catalysts. *J. Macrom. Sci. A Pure Appl. Chem.* **2010**, *47*, 503–509. [CrossRef]
20. Papageorgiou, G.Z.; Papageorgiou, D.G.; Terzopoulou, Z.; Bikiaris, D.N. Production of bio-based 2,5-furan dicarboxylate polyesters: Recent progress and critical aspects in their synthesis and thermal properties. *Eur. Polym. J.* **2016**, *83*, 202–229. [CrossRef]

21. Zhu, Y.; Romain, C.; Williams, C.K. Sustainable polymers from renewable resources. *Nature* **2016**, *540*, 354–362. [CrossRef] [PubMed]
22. Gandini, A.; Lacerda, T.M. Polymers from renewable resources: Macromolecular materials for the twenty-first century? *Macromol. Eng.* 2022, *in press*. [CrossRef]
23. Psochia, E.; Papadopoulos, L.; Gkiliopoulos, D.J.; Francone, A.; Grigora, M.E.; Tzetzis, D.; Vieira de Castro, J.; Neves, N.M.; Triantafyllidis, K.S.; Sotomayor Torres, C.M.; et al. Bottom-up development of nanoimprinted PLLA composite films with enhanced antibacterial properties for smart packaging applications. *Macromol* **2021**, *1*, 49–63. [CrossRef]
24. Zhang, T.; Howell, B.A.; Dumitrascu, A.; Martin, S.J.; Smith, P.B. Synthesis and characterization of glycerol-adipic acid hyperbranched polyesters. *Polymer* **2014**, *55*, 5065–5072. [CrossRef]
25. Robert, T.; Friebel, S. Itaconic acid—A versatile building block for renewable polyesters with enhanced functionality. *Green Chem.* **2016**, *18*, 2922–2934. [CrossRef]
26. Kunduru, K.R.; Hogerat, R.; Chosal, K.; Shaheem-Mualim, M.; Farah, S. Renewable polyol-based biodegradable polyesters as greener plastics for industrial applications. *Chem. Eng. J.* **2023**, *459*, 141211. [CrossRef]
27. Zhang, Q.; Song, M.; Xu, Y.; Wang, W.; Wang, Z.; Zhang, L. Bio-based polyesters: Recent progress and future prospects. *Prog. Polym. Sci.* **2021**, *120*, 101430. [CrossRef]
28. Kijchavengkul, T.; Auras, R. Compostability of polymers. *Polym. Int.* **2008**, *57*, 793–804. [CrossRef]
29. Yu, L.; Dean, K.; Li, L. Polymer blends and composites from renewable resources. *Prog. Polym. Sci.* **2006**, *31*, 576–602. [CrossRef]
30. Lazaridou, M.; Klonos, P.A.; Barmpa, E.D.; Kyritsis, A.; Bikiaris, D.N. Thermal transitions and molecular mobility in polymeric blends based on polylactide (PLA) and poly(3,3-ethylene dithiodipropionate) (PEDPA). *Polymer* **2023**, *277*, 125970. [CrossRef]
31. Engler, L.G.; Farias, N.C.; Crespo, J.S.; Gately, N.M.; Major, I.; Pezzoli, R.; Devine, D.M. Designing sustainable polymer blends: Tailoring mechanical properties and degradation gehaviour in PHB/PLA/PCL blends in a seawater environment. *Polymers* **2023**, *15*, 2874. [CrossRef] [PubMed]
32. Chrysafi, I.; Pavlidou, E.; Christodoulou, E.; Vourlias, G.; Klonos, P.A.; Kyritsis, A.; Bikiaris, D.N. Effects of poly(hexylene succinate) amount on the crystallization and molecular mobility of poly(lactic acid) copolymers. *Thermochim. Acta* **2021**, *698*, 178883. [CrossRef]
33. Atkinson, R.L.; Monaghan, O.R.; Elsmore, M.T.; Topham, P.D.; Toolan, D.T.W.; Derry, M.J.; Taresco, V.; Stockman, R.A.; De Focatiis, D.S.A.; Irvine, D.J.; et al. RAFT polymerisation of renewable terpene (meth)acrylates and the convergent synthesis of methacrylate–acrylate–methacrylate triblock copolymers. *Polym. Chem.* **2021**, *12*, 3177–3189. [CrossRef]
34. Klonos, P.A.; Terzopoulou, Z.; Zamboulis, A.; Valera, M.A.; Mangas, A.; Kyritsis, A.; Pissis, P.; Bikiaris, D.N. Direct and indirect effects on molecular mobility in renewable polylactide–poly(propylene adipate) block copolymers as studied via dielectric spectroscopy and calorimetry. *Soft Matter* **2022**, *18*, 3725–3737. [CrossRef] [PubMed]
35. Klonos, P.A.; Bikiaris, N.D.; Zamboulis, A.; Valera, M.A.; Mangas, A.; Kyritsis, A.; Terzopoulou, Z. Segmental mobility in sustainable copolymers based on poly(lactic acid) blocks built onto poly(butylene succinate) in situ. *Soft Matter* **2023**, *19*, 7846–7858. [CrossRef] [PubMed]
36. Bouyahya, C.; Bikiaris, N.D.; Zamboulis, A.; Kyritsis, A.; Majdoub, M.; Klonos, P.A. Crystallization and molecular mobility in renewable semicrystalline copolymers based on polycaprolactone and polyisosorbide. *Soft Matter* **2022**, *18*, 9216–9230. [CrossRef]
37. Tang, S.; Lin, B.L.; Tonks, I.; Eagan, J.M.; Ni, X.; Nozaki, K. Sustainable copolymer synthesis from carbon dioxide and butadiene. *Chem. Rev.* **2024**, *124*, 3590–3607. [CrossRef] [PubMed]
38. Xie, F. Sustainable polymer composites: Functionality and applications. *Funct. Compos. Mater.* **2021**, *2*, 15. [CrossRef]
39. Hejna, A. Renewable, degradable, and recyclable polymer composites. *Polymers* **2023**, *15*, 1769. [CrossRef] [PubMed]
40. Paraye, P.; Sarviya, R.M. Advances in polymer composites, manufacturing, recycling, and sustainable practices. *Polym. Plast. Technol. Mater.* **2024**, *63*, 1474–1497. [CrossRef]
41. Bikiaris, N.D.; Koumentakou, I.; Samiotaki, C.; Meimaroglou, D.; Varytimidou, D.; Karatza, A.; Kalantzis, Z.; Roussou, M.; Bikiaris, R.D.; Papageorgiou, G.Z. Recent advances in the investigation of poly(lactic acid) (PLA) nanocomposites: Incorporation of various nanofillers and their properties and applications. *Polymers* **2023**, *15*, 1196. [CrossRef] [PubMed]
42. Wang, S.; Li, X.; Wu, H.; Tian, Z.; Xin, Q.; He, G.; Peng, D.; Chen, S.; Yin, Y.; Jiang, Z.; et al. Advances in high permeability polymer-based membrane materials for CO_2 separations. *Energy Environ. Sci.* **2016**, *9*, 1863–1890. [CrossRef]
43. Zhai, S.; Zhang, P.; Xian, Y.; Zeng, J.; Shi, B. Effective thermal conductivity of polymer composites: Theoretical models and simulation models. *Int. J. Heat Mass Transf.* **2018**, *117*, 358–374. [CrossRef]
44. Silva, R.R.A.; Marques, C.S.; Arruda, T.R.; Teixeira, S.C.; De Oliveira, T.V. Biodegradation of polymers: Stages, measurement, standards and prospects. *Macromol* **2023**, *3*, 371–399. [CrossRef]
45. Terzopoulou, Z.; Zamboulis, A.; Bikiaris, D.N.; Valera, M.A.; Mangas, A. Synthesis, properties, and enzymatic hydrolysis of poly(lactic acid)-co-poly(propylene adipate) block copolymers prepared by reactive extrusion. *Polymers* **2021**, *13*, 4121. [CrossRef]
46. Bikiaris, N.D.; Klonos, P.A.; Kyritsis, A.; Barmpalexis, P. Structural and thermodynamical investigation in triblock copolymers of polylactide and poly(ethylene glycol), PLA-*b*-PEG-*b*-PLA, envisaged for medical applications. *Mater. Today Commun.* **2024**, *38*, 107799. [CrossRef]
47. White, R.P.; Lipson, J.E.G. Polymer free volume and its connection to the glass transition. *Macromolecules* **2016**, *49*, 3987–4007. [CrossRef]

48. Ranucci, E.; Liu, Y.; Söderqvist Lindblad, M.; Albertson, A.C. New biodegradable polymers from renewable resources. High molecular weight poly(ester carbonate)s from succinic acid and 1,3-propanediol. *Macromol. Rapid Commun.* **2000**, *21*, 680–684. [CrossRef]
49. Hartlep, H.; Hussmann, W.; Prayitno, N.; Meynial-Salles, I.; Zeng, A.P. Study of two-stage processes for the microbial production of 1,3-propanediol from glucose. *Appl. Microbiol. Biotechnol.* **2002**, *60*, 60–66. [CrossRef] [PubMed]
50. Bikiaris, D.N.; Papageorgiou, G.Z.; Papadimitriou, S.A.; Karavas, E.; Avgoustakis, K. Novel biodegradable polyester poly(propylene succinate): Synthesis and application in the preparation of solid dispersions and nanoparticles of a water-soluble drug. *AAPS PharmSciTech* **2009**, *10*, 138–146. [CrossRef]
51. Qu, D.; Sun, S.; Gao, H.; Bai, Y.; Tang, Y. Biodegradable copolyester poly(butylene-co-isosorbide succinate) as hot-melt adhesives. *RSC Adv.* **2019**, *9*, 11476–11483. [CrossRef]
52. Smiga-Matuszowicz, M.; Korytkowska-Walach, A.; Nowak, B.; Pilawka, R.; Lesiak, M.; Sieron, A.L. Poly(isosorbide succinate)-based in situ forming implants as potential systems for local drug delivery: Preliminary studies. *Mater. Sci. Eng. C* **2018**, *91*, 311–317. [CrossRef]
53. Fleche, G.; Huchette, M. Isosorbide. Preparation, properties and chemistry. *Starch Stärke* **1986**, *38*, 26–30. [CrossRef]
54. Fang, W.; Xu, F.; Zhang, Y.; Zhang, Z.; Yang, Z.; Wang, W.; He, H.; Luo, Y. Acylamido-based anion-functionalized ionic liquids for efficient synthesis of poly(isosorbide carbonate). *Catal. Sci. Technol.* **2022**, *12*, 1756–1765. [CrossRef]
55. Jiang, Y.Y.; Wu, G.H.; Guo, W.; Guan, X.F.; Zhang, M.Y.; Zhang, H.X. An environmentally sustainable isosorbide-based plasticizer for biodegradable poly(butylene succinate). *J. Polym. Eng.* **2022**, *42*, 331–342. [CrossRef]
56. Laanesoo, S.; Bonjour, O.; Parve, J.; Parve, O.; Matt, L.; Vares, L.; Jannasch, P. Poly(alkanoyl isosorbide methacrylate)s: From amorphous to semicrystalline and liquid crystalline biobased materials. *Biomacromolecules* **2021**, *22*, 640–648. [CrossRef] [PubMed]
57. Feng, L.; Zhu, W.; Zhou, W.; Li, C.; Zhang, D.; Xiao, Y.; Zheng, L. A designed synthetic strategy toward poly(isosorbide terephthalate) copolymers: A combination of temporary modification, transesterification, cyclization and polycondensation. *Polym. Chem.* **2015**, *6*, 7470–7479. [CrossRef]
58. Terzopoulou, Z.; Kasmi, N.; Tsanaktsis, V.; Doulakas, N.; Bikiaris, D.N.; Achilias, D.A.; Papageorgiou, G.Z. Synthesis and Characterization of Bio-Based Polyesters: Poly(2-methyl-1,3-propylene-2,5-furanoate), Poly(isosorbide-2,5-furanoate), Poly(1,4-cyclohexanedimethylene-2,5-furanoate). *Materials* **2017**, *10*, 801. [CrossRef] [PubMed]
59. Wang, Y.; Davey, C.J.E.; Van der Maas, K.; Van Putten, R.J.; Tietema, A.; Parsons, J.R.; Gruter, G.J.M. Biodegradability of novel high T_g poly(isosorbide-co-1,6-hexanediol) oxalate polyester in soil and marine environments. *Sci. Total Environ.* **2022**, *815*, 152781. [CrossRef]
60. Pico, Y.; Barcelo, D. Pyrolysis gas chromatography-mass spectrometry in environmental analysis: Focus on organic matter and microplastics. *Trends Anal. Chem.* **2020**, *130*, 115964. [CrossRef]
61. Yao, S.F.; Chen, X.T.; Ye, H.M. Investigation of structure and crystallization behavior of poly(butylene succinate) by Fourier transform infrared spectroscopy. *J. Phys. Chem. B* **2017**, *121*, 9476–9485. [CrossRef] [PubMed]
62. De Matos Costa, A.R.; Crocitti, A.; de Carvalho, L.H.; Carroccio, S.C.; Cerruti, P.; Santagata, G. Properties of biodegradable films based on poly(butylene succinate) (PBS) and poly(butylene adipate-co-terephthalate) (PBAT) blends. *Polymers* **2020**, *12*, 2317. [CrossRef]
63. Yamadera, R.; Murano, M. The determination of randomness in copolyesters by high resolution nuclear magnetic resonance. *J. Polym. Sci. A-1 Polym. Chem.* **1967**, *5*, 2259–2268. [CrossRef]
64. Panayiotou, C.G. Glass transition temperatures in polymer mixtures. *Polym. J.* **1986**, *18*, 895–902. [CrossRef]
65. Wurm, A.; Ismail, M.; Kretzschmar, B.; Pospiech, D.; Schick, C. Retarded crystallization in polyamide/layered silicates nanocomposites caused by an immobilized interphase. *Macromolecules* **2010**, *43*, 1480–1487. [CrossRef]
66. Sargsyan, A.; Tonoyan, A.; Davtyan, S.; Schick, C. The amount of immobilized polymer in PMMA SiO_2 nanocomposites determined from calorimetric data. *Eur. Polym. J.* **2007**, *43*, 3113–3127. [CrossRef]
67. Soccio, M.; Nogales, A.; Lotti, N.; Munari, A.; Ezquerra, T.A. Evidence of early stage precursors of polymer crystals by dielectric spectroscopy. *Phys. Rev. Lett.* **2007**, *98*, 037801. [CrossRef] [PubMed]
68. Klonos, P.A.; Papadopoulos, L.; Kasimatis, M.; Iatrou, H.; Kyritsis, A.; Bikiaris, D.N. Synthesis, crystallization, structure memory effects, and molecular dynamics of biobased and renewable poly(*n*-alkylene succinate)s with *n* from 2 to 10. *Macromolecules* **2021**, *54*, 1106–1119. [CrossRef]
69. Tai, H.J. Dielectric spectroscopy of poly(butylene succinate) films. *Polymer* **2007**, *48*, 4558–4566. [CrossRef]
70. Schönhals, A.; Szymoniak, P. *Dynamics of Composite Materials*, 1st ed.; Springer: Cham, Switzerland, 2022. [CrossRef]
71. Kremer, F.; Schönhals, A. (Eds.) *Broadband Dielectric Spectroscopy*; Springer: New York, NY, USA, 2003. [CrossRef]
72. Havriliak, S.; Negami, S. A complex plane representation of dielectric and mechanical relaxation processes in some polymers. *Polymer* **1967**, *8*, 161–210. [CrossRef]
73. Cole, K.S.; Cole, R.H. Dispersion and absorption in dielectrics II. Direct current characteristics. *J. Chem. Phys.* **1942**, *10*, 98–105. [CrossRef]
74. Richert, R.; Agapov, A.; Sokolov, A.P. Appearance of a Debye process at the conductivity relaxation frequency of a viscous liquid. *J. Chem. Phys.* **2011**, *134*, 104508. [CrossRef]
75. Tammann, G.; Hesse, W. Die abhängigkeit der viscosität von der temperatur bie unterkühlten flüssigkeiten. *Z. Anorg. Allg. Chem.* **1926**, *156*, 245–257. [CrossRef]

76. Cerveny, S.; Schwartz, G.A.; Otegui, J.; Colmenero, J.; Loichen, J.; Westermann, S. Dielectric study of hydration water in silica nanoparticles. *J. Phys. Chem. C* **2012**, *116*, 24340–24349. [CrossRef]
77. Cerveny, S.; Colmenero, J.; Alegría, A. Dielectric investigation of the low-temperature water dynamics in the poly(vinyl methyl ether)/H_2O system. *Macromolecules* **2005**, *38*, 7056–7063. [CrossRef]
78. Klonos, P.A. Dynamics of molecules physically adsorbed onto metal oxide nanoparticles: Similarities between water and a flexible polymer. *J. Phys. Chem. C* **2018**, *122*, 28825–28829. [CrossRef]
79. Böhmer, R.; Ngai, K.; Angell, C.A.; Plazek, D.J. Nonexponential relaxations in strong and fragile glass formers. *J. Chem. Phys.* **1993**, *99*, 4201–4209. [CrossRef]
80. Delpouve, N.; Saiter, A.; Dargent, E. Cooperativity length evolution during crystallization of poly(lactic acid). *Eur. Polym. J.* **2011**, *47*, 2414–2423. [CrossRef]
81. Aniya, M.; Ikeda, M. A Study on the Correlation between fragility and cooperativity in wide class of glass-forming substances. *Phys. Procedia* **2013**, *48*, 113–119. [CrossRef]
82. Bartos, J.; Kristiak, J. Free volume aspects of the strong–fragile classification of polymer liquids. *J. Non-Cryst. Solids* **1998**, *235–237*, 293–295. [CrossRef]
83. Bikiaris, R.D.; Ainali, N.M.; Christodoulou, E.; Nikolaidis, N.; Lambropoulou, D.; Papageorgiou, G.Z. Thermal stability and decomposition mechanism of poly(alkylene succinate)s. *Macromol* **2021**, *2*, 58–77. [CrossRef]
84. Chrysafi, I.; Ainali, N.M.; Xanthopoulou, E.; Zamboulis, A.; Bikiaris, D.N. Thermal degradation mechanism and decomposition kinetic studies of poly(ethylene succinate)/hemp fiber composites. *J. Compos. Sci.* **2023**, *7*, 216. [CrossRef]
85. Papadopoulou, K.; Tarani, E.; Ainali, N.M.; Chrissafis, K.; Wurzer, C.; Masek, O.; Bikiaris, D.N. The effect of biochar addition on thermal stability and decomposition mechanism of poly(butylene succinate) bionanocomposites. *Molecules* **2023**, *28*, 5330. [CrossRef]
86. Rizzarelli, P.; Carroccio, S. Thermo-oxidative processes in biodegradable poly(butylene succinate). *Polym. Degr. Stabil.* **2009**, *94*, 1825–1838. [CrossRef]

polymers MDPI

Article

Spectroscopic and Thermal Characterisation of Interpenetrating Hydrogel Networks (IHNs) Based on Polymethacrylates and Pluronics, and Their Physicochemical Stability under Aqueous Conditions

David S. Jones *, Marion Westwood, Shu Li and Gavin P. Andrews

School of Pharmacy, Queen's University of Belfast, 97, Lisburn Road, Belfast BT9 7BL, UK; s.li@qub.ac.uk (S.L.); g.andrews@qub.ac.uk (G.P.A.)
* Correspondence: d.jones@qub.ac.uk

Abstract: This study describes the physicochemical characterisation of interpenetrating hydrogel networks (IHNs) composed of either poly(hydroxyethylmethacrylate, p(HEMA)) or poly(methacrylic acid, p(MAA)), and Pluronic block copolymers (grades F127, P123 and L121). IHNs were prepared by mixing the acrylate monomer with Pluronic block copolymers followed by free radical polymerisation. p(HEMA)–Pluronic blends were immiscible, evident from a lack of interaction between the two components (Raman spectroscopy) and the presence of the glass transitions (differential scanning calorimetry, DSC) of the two components. Conversely, IHNs of p(MAA) and each Pluronic were miscible, displaying a single glass transition and secondary bonding between the carbonyl group of p(MAA) and the ether groups in the Pluronic block copolymers (Raman and ATR-FTIR spectroscopy). The effect of storage of the IHNs in Tris buffer on the physical state of each Pluronic and on the loss of Pluronic from the IHNs were studied using DSC and gravimetric analysis, respectively. Pluronic loss from the IHNs was dependent on the grade of Pluronic, time of immersion in Tris buffer, and the nature of the IHN (p(HEMA) or p(MAA)). At equilibrium, the loss was greater from p(HEMA) than from p(MAA) IHNs, whereas increasing ratio of poly(propylene oxide) to poly(ethylene oxide) decreased Pluronic loss. The retention of each Pluronic grade was shown to be primarily due to its micellization; however, hydrogen bonding between Pluronic and p(MAA) (but not p(HEMA)) IHNs contributed to their retention.

Keywords: interpenetrating hydrogel networks (IHNs); biomaterials; spectroscopy; pluronic block copolymers; glass transition; micellization; retention

1. Introduction

Hydrogels are three-dimensional crosslinked hydrophilic polymers that can absorb considerable masses of aqueous fluids through polymer chain expansion and yet maintain their mechanical properties [1]. They can be homopolymers or copolymer networks and are frequently classed according to the nature of their side groups, as either neutral or ionic [2]. The selective choice of acrylate monomers and polymerisation by free radical polymerisation enables the production of polymers that can demonstrate a wide range of physicochemical properties, notably mechanical properties [3], lubricity [4], swelling [5], drug release and degradation [2,6].

Medical-device-associated infections are, unfortunately, the most common nosocomial infection within the primary care setting [7,8], with the incidence of infection being dependent on both the site of implantation and whether the device is external to the body. For example, it has been reported that catheter-associated urinary tract infections account for 9% of all hospital-acquired infections and that approximately two-thirds of these could have been prevented [9]. The annual costs of healthcare-associated infections have been reported to be circa USD 33 billion in the U.S. [10]. Due to their versatility, hydrogels have found application as matrices/coatings for medical devices and as drug delivery systems. Hydrogels are the mainstay of lubricious coatings of urinary catheters, e.g., Lofric and Speedicath [11]. Furthermore, several hydrogel-based approaches have shown promise for the prevention/treatment of medical-device-related infections. These include silver-containing hydrogels [12–14], antimicrobial (antibiotics, antimicrobial polymers/peptides) hydrogels [15,16], photosensitisers [17], polyelectrolyte and polyzwitterionic hydrogels [18], and double network chitosan hydrogels [19].

Despite their widespread usage in urology, several limitations are associated with hydrogels as medical device materials. Firstly, their mechanical properties are often inappropriate for fabricating devices solely using these materials, and whilst this may be addressed by, e.g., changing the monomer composition or the degree of crosslinking, these interventions may compromise other properties, e.g., release kinetics of the bioactive component [20], contact angle [20], and catheter lubricity [21]. An approach that may be used to modify the properties of mono-polymeric systems, including hydrogels, is interpenetrating polymer networks, which contain two or more interlaced polymers and may be designed to retain the key functionalities of each component [22]. In this study, we sought to design biomaterial platforms that may be used as medical device biomaterials and/or drug delivery platforms that provide greater functionality. Specifically, we describe interpenetrating hydrogel networks (IHNs) composed of either poly(hydroxyethylmethacrylate, p(HEMA)) or poly(methacrylic acid, p(MAA)) and block copolymers of poly(ethylene oxide) (PEO) and poly(propylene oxide) (PPO) (Pluronics), arranged in a PEO-PPO-PEO triblock structure. The choice of Pluronic block copolymers as the second network component was informed by the reported use of Pluronic F127 as a brush coating for medical devices to reduce subsequent microbial colonisation [23,24] and the adhesion of blood components [25]. Normally, chemical grafting techniques are required to attach the Pluronic block copolymers to the polymeric device; however, the network structures described in this study may be of potential value in this respect, thus avoiding the need for polymer grafting. Furthermore, the networks described herein may offer advantages for improved delivery of bioactive agents, again designed to control medical-device-related infection.

In this study, we describe the physicochemical (spectroscopic and thermal) properties of IHNs to provide a greater understanding of the physical state of and interactions between the two polymer components and how these interactions affect retention within the network when immersed within an aqueous fluid. This provides an invaluable key step in understanding the practical uses of this IHN as medical device biomaterials/coatings.

2. Materials and Methods

2.1. Chemicals

The materials used were of analytical grade or higher and were used as supplied. 2-Hydroxyethylmethacrlyate 97% (HEMA), methacrylic acid 99% (MAA), benzoyl peroxide, ethylene glycol dimethacrylate 98% (EGDMA), and Pluronic F127 (F127) were purchased from Aldrich, Poole, Dorset, UK. Pluronic P123 (P123) and Pluronic L121 (L121) were supplied by BTC UK, Cheshire, UK. To facilitate sample preparation, medical-grade silicone tubing was purchased from S.F. Medical, Hudson, MA, USA, Poly(ester) film, one-sided siliconised release liner (FL2000TM PET 75 μ 1S) was purchased from Rexam Release B.V., Apeldoorn, The Netherlands, and glass plates (100 mm × 150 mm × 3 mm) were purchased from J.E. Harrison Ltd. Glaziers, Belfast, Northern Ireland.

2.2. Synthesis of Network Homopolymers and IHNs

p(HEMA) homopolymer and p(HEMA)/Pluronic interpenetrating hydrogel networks (IHNs) were prepared by initially dissolving the crosslinking agent EGDMA (2% *w*/*w*) and the free radical initiator benzoyl peroxide (1% *w*/*w*) in the required amount of HEMA in a flat-bottomed beaker. The mixture was stirred mechanically until benzoyl peroxide had fully dissolved. The resultant solution was then injected into a mould comprising two vertical glass plates lined with a siliconised release liner, separated by medical-grade tubing and clamped together by bulldog clips. The mould was placed in a fan-assisted oven at 90 °C for two hours to produce the p(HEMA) control. To produce the IHNs, Pluronic block copolymers (L121, P123 or F127) were dissolved in this monomer mixture at defined concentrations (10%, 20%, 30%, 40% *w*/*w*). The maximum concentration of Pluronic block copolymers, which could be dissolved in HEMA, was 40% *w*/*w*. The resultant solution was then polymerised under the same conditions as p(HEMA).

p(MAA)/Pluronic IHNs were prepared similarly; however, a further concentration of Pluronic (50% *w*/*w*) was added to the monomer solution.

2.3. Modulated Temperature Differential Scanning Calorimetry (MTDSC) of IHNs

MTDSC experiments were performed using a T.A. Instruments Q100 Modulated Differential Scanning Calorimeter (T.A. Instruments, West Sussex, UK) with a refrigerated cooling system. Small sections (7–10 mg) of each material were cut using a scalpel and placed in an oven at 60 °C for 72 h to remove any unreacted monomer. Heat flow was calibrated using an indium standard, and heat capacity with a sapphire standard. Nitrogen was used as the purge gas, with a flow rate of 50 mLmin^{-1} through the DSC cell. Standard aluminium pans and lids from T.A. Instruments were used throughout the experiments. The pans were sealed using a T.A. Instrument press.

Samples were exposed to an initial 5 min isothermal period at −100 °C followed by a modulation amplitude of ±0.7 °C every 50 s with a 2 °C min^{-1} underlying heating rate to 160 °C for p(HEMA) IHNs and to 250 °C for p(MAA) IHNs. Total heat flow, reversing heat flow and non-reversing heat flow were calculated automatically by the Universal Analysis 2000 software as a function of temperature. Experiments were performed at least in quadruplicate for each candidate material.

The glass transition (T_g) was taken as the midpoint of the slope change of the heat capacity. The melting temperature (T_m) was accepted as the minimum of the endothermic peak, whereas the crystallisation temperature (T_c) was determined as the maximum of the exothermic peak. The heat of melting (ΔH_m) and the heat of crystallisation (ΔH_c) were evaluated from the areas of the melting and crystallisation peaks, respectively. The percentage crystallinity, χ_c, was calculated from Equation (1) [26,27]:

$$\chi_{c(PEO)} = \left(\frac{\Delta H_{mPEO}}{w_{cryst}\Delta H^0_{mPEO}}\right) \times 100 \quad (1)$$

where $\Delta H_{m,PEO}$ is the apparent melting enthalpies of PEO present in the network, w_{cryst} is the weight fraction of the crystalline component in the blend and $\Delta H_{mPEO}{}^0$ is the heat of melting per gram of 100% crystalline PEO. It has been reported to be 188 kJ Kg^{-1} [28].

2.4. Dynamic Mechanical Analysis (DMA) of IHNs

DMA was performed using a Dynamic Mechanical Analyser model 2980, (T.A. Instruments). The machine was calibrated for position, force, and electronics every month. Samples of approximate dimensions 30 mm × 10 mm were cut from the films using a scalpel. Using single-cantilever clamps, the sample was subjected to a sinusoidal oscillation at a frequency of 1 Hz while the sample was heated at a rate of 2 °C min^{-1} from 30 °C to 200 °C. The storage modulus and loss tangent (tan δ) were calculated automatically on the Thermal Analysis Universal Analysis 2000 software as a function of temperature. Experiments were performed at least in quadruplicate for each candidate material.

2.5. Raman Spectroscopy of IHNs

Raman spectroscopy of the various IHNs was performed using an Avalon Instruments Raman Station R3 (PerkinElmer Life and Analytical Sciences Inc., Wellesley, MA, USA), with a Class 3B laser emitting at 785 nm. Spectra were recorded from 200 to 3200 cm^{-1}, with a resolution of 2 cm^{-1} [29].

2.6. *Attenuated Total Reflectance–Fourier-Transform Infrared Spectroscopy of IHNs*

Samples were scanned on a PerkinElmer Spectrum One FT-IR spectrometer (PerkinElmer Life and Analytical Sciences Inc., Wellesley, MA, USA) with a diamond attenuated total reflectance accessory, from 4000 cm^{-1} 750 cm^{-1}, with a scan resolution of 4 cm^{-1}. A minimum of 32 sub-scans were signal averaged.

2.7. *Differential Scanning Calorimetry (DSC) of Pluronic Solutions and Hydrated IHNs*

DSC was used to investigate the micellization behaviour of the Pluronic block copolymers in solution and to identify micellization in the hydrated IHNs. Solutions for each Pluronic block copolymer (5, 10, and 20% w/v) were prepared in deionised water and 150 mL of Tris buffer (10% w/w Pluronic only) and stored at 5 °C. Samples of these solutions (approximately 7 mg) were carefully placed in DSC hermetic aluminium pans and sealed prior to analysis. To investigate micellization in the IHNs, the IHNs were soaked for one week in Tris buffer (pH 7.2), after which samples (approximately 9 mg) removed, and, after excess water was carefully removed from the surface, these were placed in an aluminium hermetic pan and sealed. In both cases, DSC experiments were performed using a TA Instruments Q100 Modulated Differential Scanning Calorimeter (TA Instruments Ltd., Crawley, West Sussex, UK) with a refrigerated cooling system attached. Heat flow was calibrated using an indium standard, and heat capacity with a sapphire standard. Nitrogen was used as the purge gas, flowing at a rate of 50 $mLmin^{-1}$ through the DSC cell. The samples were initially cooled to −10 °C, held isothermally for 10 min and then heated at a rate of 2 °C min^{-1}. Micellization was observed in the total heat flow signal as a broad endothermic transition, and the critical micellization temperature (CMT) was accepted as the minimum of the endothermic peak, as determined using Universal Analysis software.

2.8. *Retention of Pluronic Polymers within IHNs*

The IHNs were immersed in 150 mL of Tris buffer (pH 7.2) and at 37 °C with shaking (50 oscillations/min). The samples were removed and placed into a fresh solution at fixed periods, the solvent removed by evaporation, and the residual mass recorded. The mass of the buffer was subtracted from the total mass to enable the mass of Pluronic lost from the IHNs to be determined. The fraction of Pluronic block copolymers released from the IHNs as a function of the theoretical mass of Pluronic block copolymers present was then plotted as a function of time. In all cases, measurements were made on quadruplicate replicate samples.

2.9. *Statistics*

The effects of Pluronic concentration and type on the thermal properties of the IHNs, the critical micelle temperature and the retention of the Pluronic block copolymers within the network were statistically analysed using a two-way analysis of variance (ANOVA) (Prism 10, GraphPad Software). Individual differences between treatments were examined using Dunnett's test with a probability less than 0.05, denoting significance.

3. Results and Discussion

A key challenge to the performance of urinary medical devices is resistance to medical-device-related infection [30]. For example, it has been reported that in 50% of patients who receive urinary catheterisation, infection occurs within 7–10 days [31]. The consequences of catheter-associated infections are considerable, in terms of morbidity/mortality and related healthcare costs [32]. It has been reported (in 2011) that approximately 70% of patients who presented with a urinary tract infection had been catheterised; this occurrence increased to 95% of patients in the intensive care unit setting [33]). Similarly, infection is also a problem for ureteral stents [34]; however, the incidence of infection associated with ureteral stents is lower than with urethral catheters [35]. There have been several proposed strategies to reduce the incidence of urological device-related infection, including changes to device designs (as summarised by [36–38]), the materials from which the device is manufactured (polymeric and non-polymeric [37–39]), and the use of biodegradable stents [40] and novel coatings, e.g., hydrogels, often containing antimicrobial agents [41–45]. To perform optimally, urological devices/coatings of urological devices must offer a wide range of physicochemical properties, notably mechanical properties and the ability to release a microbicidal concentration of drug at the device–fluid interface for a prolonged period; these properties are rarely achieved using a simple, single polymeric system.

Polymer coatings, e.g., brush coatings, have been shown to offer microbial anti-adherent properties and, as a result, have been proposed to lower the incidence of medical-device-related infection [23]. Several studies have used Pluronic F127, either as the brush coating material or as a brush layer onto which other polymers may be grafted/entrapped, leading to enhanced resistance to microbial colonisation of the device surface [23,24] and blood protein repulsion [25]. These systems most often require a chemical grafting process to locate the Pluronic onto the medical device substrate and then, if required, a further grafting of a second polymer using the Pluronic block copolymer as a substrate. We are currently interested in the entrapment and retention of Pluronics within biomaterials as the inclusion of this polymer may offer modified surface properties, properties that are advantageous as medical device coatings. Accordingly, this study examined the formation of interpenetrating hydrogel networks (IHNs) between three candidate Pluronic block copolymers, offering different physicochemical properties with two model hydrogels, poly(hydroxyethylmethacrylate) and poly(methacrylic acid) [46]. Given the importance of the interactions of the components within the polymer network and the retention of the Pluronic component in aqueous conditions for device performance, this study provides a comprehensive investigation of these key properties, thereby enabling polymer candidates to be identified for further consideration as medical device and/or drug delivery biomaterial.

3.1. Thermal and Spectroscopic Properties of Pluronic Block Copolymers

To understand the thermal and spectroscopic properties of IHNs, it was necessary to initially examine the properties of each Pluronic block copolymer. The three grades had the same molecular weight of PPO as the central block (approximately 4000 g mol^{-1}). The PEO molecular weight was dependent on the grade of Pluronic tested and consisted of 70% of the total molecular weight for F127, 30% of P123, and only 10% of the total weight for L121. The thermal properties of Pluronic block copolymers used in this study are presented in Table 1. In addition to differences in the melting properties, differences in the degree of crystallinity, associated with the PEO block component of the polymer, were observed, with F127 exhibiting the greatest crystallinity (due to the largest PEO block length). Thus, the observed melting point for F127 was greater than for P123 [47]. Table 1 additionally shows the glass transition temperatures for the three Pluronic block copolymers. The presence of a common glass transition (as opposed to two separate transitions for the PEO and PPO components) located between the glass transition components for the two polymers (−73 °C and −50 °C for PPO and PEO, respectively [48]) indicates that the amorphous phases of PPO and PEO are considerably mixed [49,50]. Reducing the PEO block length significantly reduced the observed glass transition temperature (Tg).

Table 1. Glass transition (T_g), crystallisation (T_c), and melting (T_m) temperatures; enthalpies of melting ($\Delta H_{m\ PEO}$) and crystallisation ($\Delta H_{c\ PEO}$) and degree of crystallinity ($\chi_{c\ PEO}$) of Pluronic block copolymers (F127, P123, L121).

Pluronic Grade	Mean (±sd) Glass Transition, Melting and Crystallisation Temperatures			Mean (±sd) Enthalpies of Melting and Crystallinity		Degree of Crystallinity
	T_g (°C)	T_m (°C)	T_c (°C)	$\Delta H_{m,PEO}$ (J g^{-1})	$\Delta H_{c,PEO}$ (J g^{-1})	$\chi_{c(PEO)}$ (%)
P F127	−65.8 ± 0.4	56.5 ± 0.3	Not Measured	116.9 ± 1.4	Not Measured	62.2 ± 0.8
P P123	−68.8 ± 0.4	36.8 ± 0.3	Not Measured	42.8 ± 0.8	Not Measured	22.7 ± 0.4
P L121	−71.4 ± 0.8	Not Measured	−56 ± 0.8	Not Measured	6.6 ± 0.4	Not Measured

The Raman spectra of Pluronic F127, P123, and L121 over two wavenumber ranges are shown in Figure 1. The region 250–650 cm^{-1} contained peaks attributed to PEO backbone vibrations [51,52], with F127 showing the strongest (highest intensities) and most well-defined peaks of the three block copolymers (Figure 1a). Differences were observed within the spectral region 1000–1200 cm^{-1} as the PEO/PPO ratio decreased (from F127 to P123 to L121, Figure 1b). The bands in the spectrum of F127 (at 1142 cm^{-1} (C–O stretch) and 1064 cm^{-1} (C–O stretch, CH_2 rock)) were well-defined, whereas the absorbance bands for P123 and L121 were much less distinct and poorly resolved. Strong peaks at 280, 364, 538 and 582 cm^{-1} were detected and assigned to C–C–O, C–O–C bends, and C–C, C–O internal rotations for F127. P123 also showed peaks at these frequencies but were of much lower intensity, whereas peaks for P L121 were not detected at these frequencies. The Raman spectra for Pluronic F127, P123 and L121 are shown in Figure 1a (250–650 cm^{-1}) and Figure 1b (1000–1200 cm^{-1}).

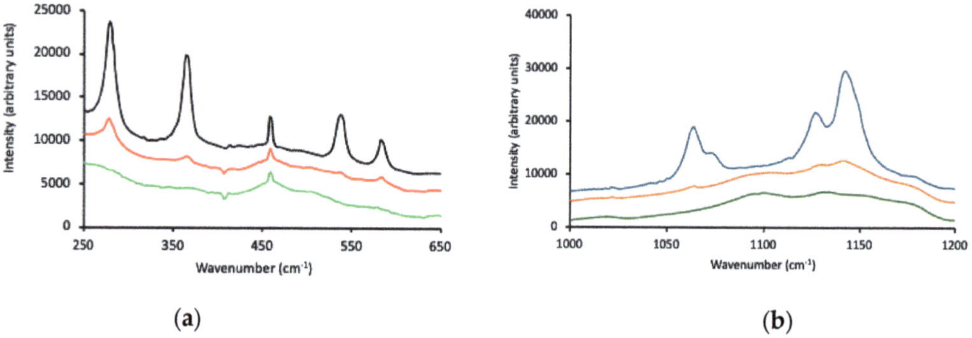

Figure 1. Raman spectra for Pluronic F127 (black line), Pluronic P123 (red line), and Pluronic L121 (green line) within two wavenumber ranges, 250-650 cm^{-1} (**a**) and 1000–1200 cm^{-1} (**b**).

3.2. Thermal and Spectroscopic Properties of IHNs of p(HEMA) and Pluronic Block Co-Polymers

Understanding the miscibility (or immiscibility) of the IHNs may be readily performed by consideration of the thermal properties (Table 2) in conjunction with spectroscopic properties (Figure 2). The glass transition temperatures of the IHNs were examined using modulated temperature differential scanning calorimetry and dynamic mechanical analysis, the latter method offering greater sensitivity. Whilst the values for the glass transition of the polymers were greater when measured using DMA (due to the high frequency employed by this method) [53], good correlations were observed between the two methods. The glass transition temperature of p(HEMA) was 114.2 ± 0.6 °C and 126.6 ± 0.5 °C when measured using MTDSC and DMA, respectively, and agrees with previous reports, e.g., [53]. This transition may be attributed to the onset of large-scale, cooperative displacements of the methylene chains of p(HEMA) in the macromolecular network [54].

Incorporation of Pluronic F127, P123, or L121 sequentially from 10% *w/w* to 40% *w/w* did not change the temperature at which this primary glass transition occurred. Furthermore, a second glass transition was observed with the p(HEMA)/Pluronic IHNs at approximately −67 to −71 °C that was statistically independent of Pluronic concentration, and which was statistically similar to the glass transition temperature of the individual Pluronic components. Accordingly, it may be concluded that each Pluronic is immiscible in the p(HEMA) network; therefore, the p(HEMA)–Pluronic IHNs comprise two distinct phases [54].

Table 2. Glass transition temperatures (T_g), melting points (T_m), melting enthalpies (ΔH_m) and degrees of crystallinity (χ_c) for p(HEMA) and p(HEMA)/Pluronic interpenetrating hydrogel networks (IHNs).

Pluronic Type	Pluronic Conc. (% *w/w*)	Mean ± sd Glass Transition Temperature(°C)			Melt Temperature (°C, T_m)	$\Delta H_{m\ PEO}$ ($J\ g^{-1}$)
		T_g 1	T_g 2	T_g 2*		
None	0, p(HEMA)	Not present	114.2± 0.6	126.6 ± 0.5	Not present	Not present
F127	10	Not present	111.3 ± 0.5	126.7 ± 1.5	43.2 ± 1.3	66.0 ± 3.1
	20	−67.0 ± 0.5	112.0 ± 1.1	124.0 ± 1.9	47.5 ± 1.0	85.7 ± 3.4
	30	−67.0 ± 0.5	113.6 ± 0.6	128.1 ± 1.7	50.3 ± 0.9	99.6 ± 3.3
	40	−66.9 ± 0.5	113.9 ± 0.4	128.9 ± 1.1	53.4 ± 0.6	108.9 ± 3.2
	100	−65.8 ± 0.4	Not present	Not present	56.5 ± 0.3	116.9 ± 1.4
P123	10	−67.1 ± 0.4	113.8 ± 0.6	125.8 ± 0.5	24.8 ± 0.7	2.1 ± 0.6
	20	−67.1 ± 0.3	112.7 ± 0.5	124.8 ± 1.4	28.1 ± 0.3	4.8 ± 0.5
	30	−67.4 ± 0.2	112.0 ± 0.7	126.8 ± 0.9	31.7 ± 0.9	8.8 ± 0.8
	40	−67.6 ± 0.3	112.0 ± 0.4	127.5 ± 0.5	33.9 ± 0.8	13.1 ± 0.7
	100	−68.8 ± 0.4	Not present	Not present	36.8 ± 0.3	42.8 ± 0.8
L121	10	−67.9 ± 0.8	114.2 ± 0.6	126.6 ± 0.5	Not present	Not present
	20	−67.9 ± 0.4	113.4 ± 0.5	124.8 ± 2.4	−41.4 ± 0.4	0.2 ± 0.0
	30	−69.0 ± 1.4	113.7 ± 0.8	121.5 ± 1.6	−42.8 ± 0.7	1.0 ± 0.2
	40	−69.0 ± 1.8	113.6 ± 0.9	125.8 ± 1.9	−48.5 ± 0.9	2.4 ± 0.4
	100	−71.3 ± 0.8	113.0 ± 0.9	127.3 ± 1.6	−51.4 ± 0.6	4.5 ± 0.4

Tg 1: Glass transition temperature of the Pluronic component determined using modulated temperature differential scanning calorimetry. Tg 2: Glass transition temperature of the methacrylate component determined using modulated temperature differential scanning calorimetry. Tg 2* Glass transition temperature of the methacrylate component determined using dynamic mechanical analysis.

The spectroscopic properties of IHNs composed of p(HEMA) and each of the three test Pluronic block copolymers are shown in Figure 2, split into separate wavenumber ranges to enhance clarity. The spectra of p(HEMA)/Pluronic F127 IHNs within the wavenumber ranges of 200–650 cm^{-1}, 1000–1200 cm^{-1}, and 1600–1900 cm^{-1} are shown in Figure 2a–c, whereas the spectra of p(HEMA)/Pluronic P123 IHNs within the wavenumber ranges of 200–700 cm^{-1} and 1600–1900 cm^{-1} are shown in Figure 2d,e. Finally, the spectra of p(HEMA)/Pluronic L121 IHNs within the wavenumber ranges of 200–700 cm^{-1} and 1600–1900 cm^{-1} are shown in Figure 2f,g. To facilitate analysis, the focus was directed to wavenumbers associated with peaks in p(HEMA) and the Pluronic block copolymers and how these change within the IHNs, notably the carbonyl peak for p(HEMA) (1720 cm^{-1}) and the ether peak of PEO (1142 cm^{-1} (C–O stretch) and 1064 cm^{-1} (C–O stretch, CH_2 rock)). Due to the lower intensity of these vibrations in P123 and L121, greater focus was placed on F127 as potential polymer interactions would be easier to observe. Importantly, it was noted for all three IHNs that no changes in the key wavenumbers associated with the molecular groups of the polymeric components of the IHNs occurred. Similarly, within the fingerprint region of the spectrum (200–700 cm^{-1}), the peak at 602 cm^{-1} (associated with C-C-O bond vibration) was present in each of the IHNs at the same wavenumber and the peaks associated with the crystalline regions of F127 (364 and 538 cm^{-1}) were detected in the IHNs without any shift in wavenumber. Similarly, the lack of change in the spectra of individual network components has been reported to be evidence of immiscibility in other studies [55–57], Accordingly, spectroscopy has provided further evidence of the immiscibility of the two polymers in these IHNs.

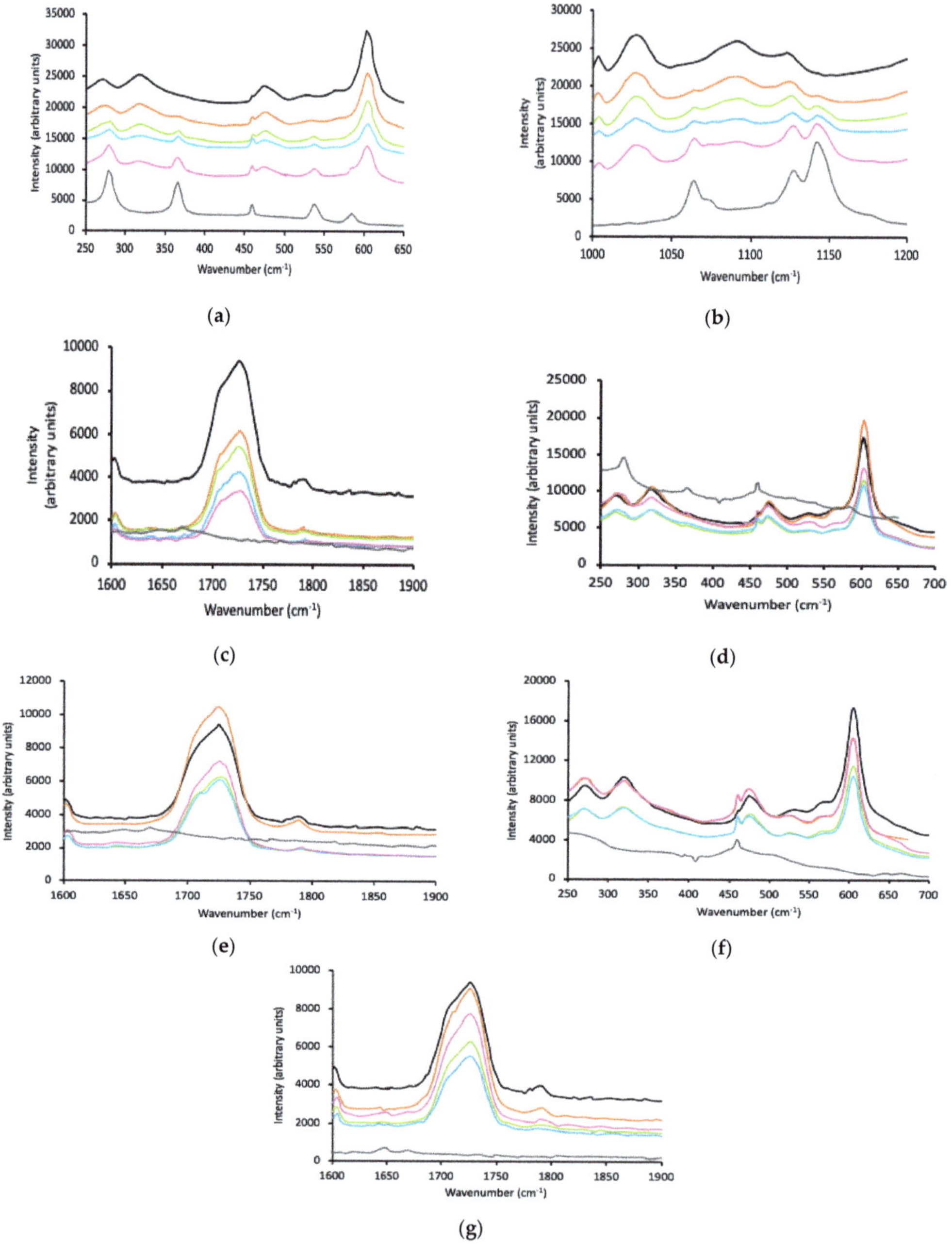

Figure 2. Raman spectra of p(HEMA) (black line), Pluronic (grey line), and p(HEMA)/Pluronic IHNs (10% *w*/*w* red line, 20% *w*/*w*, green line, 30% *w*/*w* blue line, 40% *w*/*w* purple line) within three wavenumber ranges: 250-650 cm^{-1}, 1000-1200 cm^{-1} and 1600-1900 cm^{-1}. (**a**–**c**) refer to F127, (**d**,**e**) P123 and (**f**,**g**) L121.

3.3. Thermal and Spectroscopic Properties of IHNs of p(MAA) and Pluronic Block Co-Polymers

The reported applications of p(MAA) as a medical device biomaterial are less than p(HEMA); however, p(MAA) has been used as brushes as biomaterials and for pH-responsive drug delivery applications [58–60]. This study aims to understand the interaction between p(MAA) and Pluronic copolymers and their stability within an aqueous environment. Previous studies have confirmed an interaction between poly(acrylic acid, p(AA)) and poly(ethylene oxide). For example, Li et al. [61] described a complexation between p(AA) and PEO that was pH-dependent, and similarly, Monir et al. described an interaction between these two chemical components [62]. As in Section 3.2, the interaction between p(MAA) and Pluronic block copolymers was examined using thermal and spectroscopic methods, presented in Table 3 and Figure 3, respectively.

Table 3. Glass transition temperatures (T_g), melting points (T_m), melting enthalpies (ΔH_m), and degrees of crystallinity (χ_c) for p(MAA), P F127, and p(MAA)/Pluronic inter-penetrating hydrogel networks IHNs.

Pluronic Type	Pluronic Conc. (% *w/w*)	Mean ± sd Glass Transition (°C, T_g)			Melt Temperature (°C, T_m)	$\Delta H_{m\ PEO}$ ($J\ g^{-1}$)
		T_g 1	T_g 2	T_g 2*		
None	0, p(MAA)	Not Present	205.5 ± 0.8	211.5 ± 0.4	Not Present	Not Present
F127	10	Not Present	203.2 ± 1.0	206.8 ± 0.6	Not Present	Not Present
	20		194.8 ± 2.2	195.5 ± 0.8		
	30		171.8 ± 2.0	185.7 ± 0.8		
	40		158.5 ± 2.6	173.1 ± 0.7		
	50		142.3 ± 2.5	168.7 ± 1.5		
	100	−65.8 ± 0.4	Not Present	Not Present	56.5 ± 0.3	116.9 ± 1.4
P123	10	Not Present	208.3 ± 1.7	213.7 ± 0.2	Not Present	Not Present
	20		207.0 ± 0.3	205.3 ± 1.0		
	30		198.5 ± 0.7	199.0 ± 0.5		
	40		193.5 ± 1.0	197.8 ± 0.2		
	50		194.1 ± 1.2	197.3 ± 1.1		
	100	−68.8 ± 0.4	Not Present	Not Present	36.8 ± 0.3	42.8 ± 0.8
L121	10	Not Present	210.0 ± 2.8	213.3 ± 0.5	Not Present	Not Present
	20		207.8 ± 1.0	207.3 ± 0.6		
	30		198.4 ± 1.5	205.9 ± 0.4		
	40		191.9 ± 1.4	202.2 ± 0.2		
	50	−68.4 ± 0.6	189.0 ± 1.5	202.0 ± 0.2		
	100	−71.3 ± 0.8	Not Present	Not Present		4.5 ± 0.4

Tg 1: Glass transition temperature of the Pluronic component determined using modulated temperature differential scanning calorimetry. Tg 2: Glass transition temperature of the methacrylate component determined using modulated temperature differential scanning calorimetry. Tg 2*: Glass transition temperature of the methacrylate component determined using dynamic mechanical analysis.

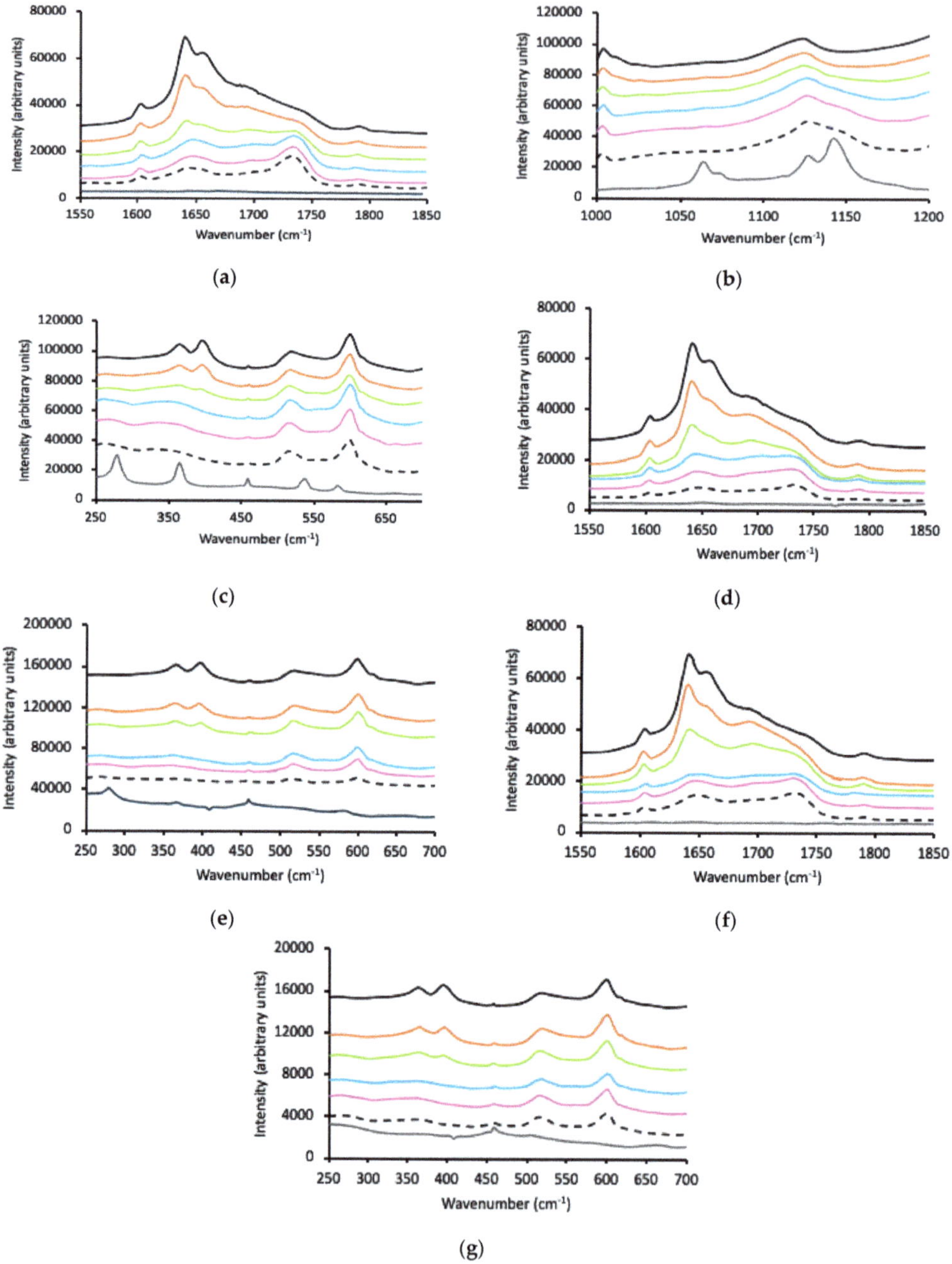

Figure 3. Raman spectra of p(MAA) (black line), Pluronic (grey line), and p(MAA)/Pluronic IHNs (10% *w*/*w* red line, 20% *w*/*w* green line, 30% *w*/*w* blue line, 40% *w*/*w* purple line, 50% *w*/*w* dashed line) within three wavenumber ranges, 250–650 cm^{-1}, 1000–1200 cm^{-1}, and 1600–1900 cm^{-1}; (**a**–**c**) refer to F127, (**d**,**e**) P123 and (**f**,**g**) L121.

The glass transition temperature for p(MAA) was 205.5 ± 0.8 °C and 211.5 ± 0.4 °C when measured using MTDSC and DMA, respectively. Sequential addition of Pluronic

F127 (0–50% w/w) resulted in significant successive reductions in the glass transition temperature of p(MAA), and a disappearance of both the glass transition and the melt transition of F127. Reductions in the glass transition temperature of p(MAA) were noted only upon the addition of 30% w/w of P123 and 30% w/w L121, with reductions in the main glass transition temperature noted following the addition of 40% w/w of these block copolymers (although the addition of 50% did not result in a further decrease). There was no statistical difference in the primary glass transitions of p(MAA) containing equivalent concentrations of P123 and L121. At a 50% w/w loading, the polymer network containing 50% w/w Pluronic L121 also displayed a secondary glass transition temperature associated with the PPO blocks of the Pluronic component, indicative of phase separation at this concentration. Therefore, it may be concluded that, based on thermal analysis, the IHNs of p(MAA) and Pluronic systems form miscible blends [63–65]. The loss of the crystallinity of PEO (and reduction in primary glass transition) suggests an interaction between the PEO component of the Pluronic block copolymers and the carbonyl group of p(MAA) [27,66]. The greater affinity of F127 within the IHNs may be explained by the greater number of PEO blocks.

As before, the spectra of the p(MAA)/Pluronic IHNs have been individually presented in Figure 3 (Figure 3a–c F127, Figure 3d,e P123, and Figure 3f,g L121). p(MAA) typically exhibits a characteristic peak at 1700 cm^{-1} due to the existence of the associated carboxylic acid dimer and a much lower intensity peak at 1730 cm^{-1} due to the presence of the free carbonyl group [67–69]. However, these peaks were partially masked by a second peak at 1640 cm^{-1} with a shoulder at 1656 cm^{-1}, attributed to the C=C bond of residual methacrylic acid. The Raman spectra of Pluronic block copolymers within the 1500–1850 cm^{-1} region were devoid of peaks. As the concentration of Pluronic F127 was increased within the IHNs, the ratio of the intensity of the shoulder present at 1700 cm^{-1} (associated with the p(MAA) dimer) decreased as the intensity of the shoulder at 1730 cm^{-1} (free carbonyl) increased and resolved into a clear peak. It has been proposed that the miscibility of Pluronic block copolymers within the p(MAA) network is due to a hydrogen bonding interaction between the ether oxygen of PEO and the carboxyl group of p(MAA). Therefore, Figure 3b focuses on the spectral region 1200–1000 cm^{-1}, associated with the C-O vibrations in F127. The absence of these peaks in the IHNs, the absence of peaks associated with the the crystallinity of PEO (observed in the wavenumber region 250–650 cm^{-1}), and the shift in the wavenumber of the peak due to the carbonyl group of p(MAA) from the dimer to monomer form is evidence of the interaction between the two components in the IHNs. The intensity of the peaks associated with Pluronics P123 and L121 were dramatically lower than that for F127; however, the emergence of the peak at 1730 cm^{-1} was apparent, indicating an interaction with p(MAA).

ATR/FTIR was used to confirm the possible polymer interactions within p(MAA)/P F127 IHNs, with a particular focus on the carbonyl region of the spectra (a region that was partially masked by the C=C vibrational peak in Raman spectroscopy). The spectra for the IHNs with wavenumber ranges of 1550–1800 cm^{-1} and 1000–1200 cm^{-1} are shown in Figure 4a,b, respectively.

A large, broad peak at 1690 cm^{-1} was detected for the p(MAA) homopolymer due to the mainly associated dimeric carbonyl groups. As the concentration of F127 was increased within the network, this peak shifted to a higher wavenumber, indicating the dissociation of the hydrogen bonding between MAA units. With the increasing concentration of F127, a further peak became apparent due to unassociated carbonyl groups at approximately 1730 cm^{-1}. The peak at 1100 cm^{-1} in F127 is due to the C-O-C asymmetric stretch, which, in the presence of p(MAA), shifted to a lower wavenumber (circa 1080 cm^{-1}, for IHNs containing 30% w/w and 40% w/w F127). This indicates the ether group's involvement in hydrogen bonding with the carbonyl group of p(MAA) [51,70].

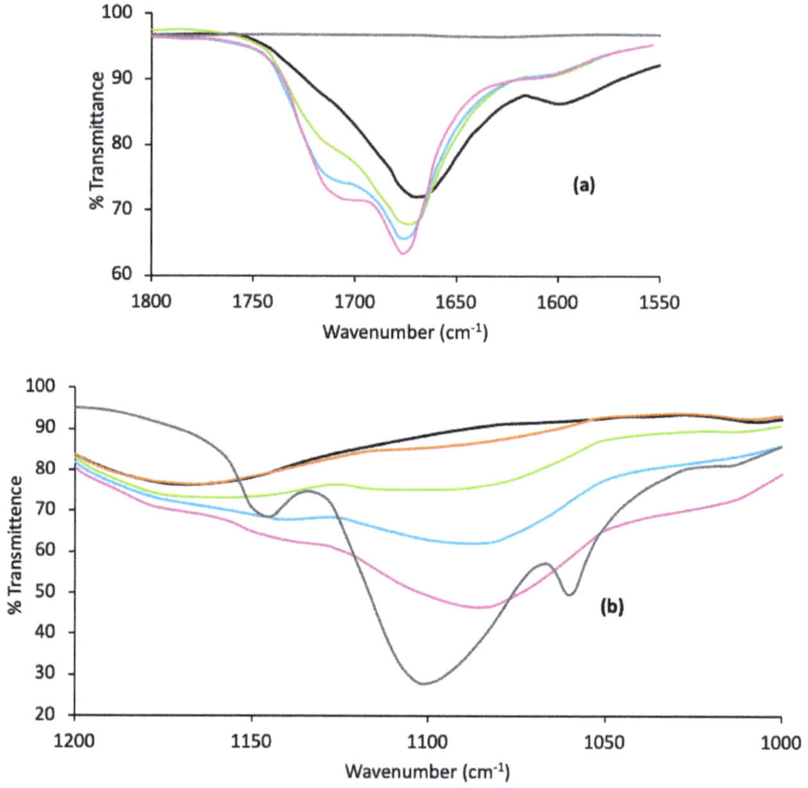

Figure 4. ATR-FTIR spectra of p(MAA) (black line), Pluronic (grey line), and p(MAA)/Pluronic F127 IHNs (10% *w*/*w* red line, 20% *w*/*w* green line, 30% *w*/*w* blue line, 40% *w*/*w* purple line) within two wavenumber ranges, 1550–1800 cm^{-1} (**a**) and 1000–1200 cm^{-1} (**b**).

3.4. Thermal Properties and Loss of Pluronic Block Copolymers from IHNs Following Immersion in Aqueous Fluid

For the IHNs to be clinically successful, either as a medical device biomaterial or as a potential drug delivery platform, the properties of the network should be stable during the period of use. The most potentially catastrophic event would be the rapid loss of the Pluronic block copolymers from the network into the biological fluid, thereby negating any possible advantages of the IHNs. Accordingly, it was important to study the loss of Pluronic block copolymers within the network following the immersion of the IHNs into a simple buffer system (phosphate-buffered saline, pH 7.2) and to identify whether the micellization of the Pluronic block polymers had occurred following immersion in an aqueous environment, the latter phenomenon increasing the effective molecular mass of the Pluronic block copolymers, thereby reducing the diffusion of individual Pluronic monomers from the IHNs. Initially micellization was studied in deionised water and Tris buffer (Table 4).

The temperature at which micellization occurred was dependent on three variables, namely the type and concentration of Pluronic and whether micellization occurred in water or Tris buffer (Table 4). The formation of micelles of the Pluronic copolymers occurs at lower concentrations (CMC) and at lower temperatures (CMT) for Pluronic copolymers with a high PPO:PEO ratio [71]. This is due to the increased hydrophobic interactions in the micelle core compared to Pluronic with a high proportion of PEO. Furthermore, the presence of buffer has been reported to affect the micellization of surfactants.

Table 4. Critical micelle temperatures (CMT) of Pluronic solutions in deionised water and Tris buffer.

Concentration of Pluronic (% *w*/*v*)	Mean (±sd) Critical Micellization Temperature of Pluronic Solutions		
	Pluronic F127	Pluronic P123	Pluronic L121
1	not detected	not detected	16.5 ± 0.1
5	25.1 ± 0.1	19.9 ± 0.1	14.4 ± 0.1
10	22.6 ± 0.1	17.4 ± 0.1	13.5 ± 0.1
10 (Tris buffer)	18.3 ± 0.1	14.1 ± 0.1	11.7 ± 0.1
20	17.1 ± 0.2	14.3 ± 0.1	not tested

Micellization was predominantly dependent on the type of Pluronic and the network composition (p(HEMA) and p(MAA)) of the IHNs (Table 5). IHNs containing Pluronic L121 exhibited the lowest critical micelle temperature. Interestingly, micellization was not observed in p(HEMA) IHNs containing Pluronic F127 that had been stored in Tris Buffer for one week.

Table 5. The effect of Pluronic type and concentration on their critical micelle temperatures within interpenetrating hydrogel networks (IHNs) that had been stored in Tris buffer (pH 7.2) for 1 week.

Network Composition	Pluronic Grade	Mean (±sd) Critical Micelle Temperature (°C) in IHNs Containing Different Concentrations of Pluronic (% *w*/*w*)					
		0	10	20	30	40	50
p(HEMA)	F127	Not detected	Not detected	Not detected	Not detected	Not detected	Not detected
	P123	Not detected	13.3 ± 1.4	12.8 ± 1.3	11.7 ± 1.4	13.8 ± 1.5	Not determined
	L121	Not detected	11.4 ± 1.2	11.3 ± 1.5	10.5 ± 1.8	9.8 ± 2.1	Not determined
p(MAA)	F127	Not detected	Not detected	18.0 ± 1.2	18.5 ± 1.1	18.9 ± 1.3	19.4 ± 0.9
	P123	Not detected	15.2 ± 1.0	15.8 ± 1.0	15.5 ± 0.7	15.9 ± 0.8	16.0 ± 1.1
	L121	Not detected	12.0 ± 1.1	11.8 ± 1.0	12.4 ± 0.8	12.0 ±0.4	11.9 ± 1.0

The losses of each Pluronic block copolymer from p(HEMA) and p(MAA)-based IHNs over an extended period of storage in Tris buffer are shown in Tables 6 and 7, respectively. Pluronic loss was statistically dependent on the hydrogel type (p(HEMA) or p(MAA)), the type of Pluronic (F127, P123, L121), and the time of immersion of the IHNs within the buffer. The loss of F127 from the two networks was significant and markedly greater than P123 and L121 and was significantly lower from the p(MAA) network than the p(HEMA) IHNs. Total loss of F127 occurred between 14 and 21 days from the p(HEMA) network, whereas, at equilibrium, the loss of F127 from p(MAA) IHNs ranged from approximately 2% (10% initial loading) to 25% (50% initial loading).

Table 6. The effect of time of immersion in Tris buffer (pH 7.2) on the % loss of Pluronic block-copolymers within p(HEMA) interpenetrating hydrogel networks (IHNs).

Time (Days)	Mean (±s.d.) % Loss of Pluronic Block Copolymers:											
	Pluronic F127 (% *w*/*w* Initial Loading)				Pluronic P123 (% *w*/*w* Initial Loading)				Pluronic L121 (% *w*/*w* Initial Loading)			
	10	20	30	40	10	20	30	40	10	20	30	40
3	53.4 ± 4.1	73.6 ± 4.9	76.1 ± 5.8	78.1 ± 5.4	0.1 ± 0.0	0.4 ± 0.0	0.8 ± 0.1	2.2 ± 0.0	0.0 ± 0.0	0.9 ± 0.0	1.9 ± 0.2	2.3 ± 0.1
7	87.2 ± 4.0	94.2 ± 7.2	95.2 ± 6.1	96.4 ± 4.9	0.1 ± 0.0	0.5 ± 0.0	2.0 ± 0.3	3.7 ± 0.2	0.1 ± 0.0	1.2 ± 0.1	2.2 ± 0.1	2.7 ± 0.1
14	95.8 ± 5.0	95.0 ± 6.1	94.7 ± 5.4	95.8 ± 6.8	0.1 ± 0.0	1.0 ± 0.0	2.6 ± 0.2	5.5 ± 0.4	0.1 ± 0.0	1.3 ± 0.1	2.2 ± 0.1	3.0 ± 0.2
21	NA	NA	NA	NA	0.2 ± 0.0	1.5 ± 0.2	3.4 ± 0.3	6.4 ± 0.4	0.1 ± 0.0	1.3 ± 0.1	2.3 ± 0.1	3.1 ± 0.3
28	NA	NA	NA	NA	0.3 ± 0.0	2.2 ± 0.2	4.0 ± 0.4	7.5 ± 0.3	0.1 ± 0.0	1.4 ± 0.1	2.4 ± 0.1	3.1 ± 0.2

NA: Not analysed.

Table 7. The effect of time of immersion in Tris buffer (pH 7.2) on the % loss of Pluronic block-copolymers within p(MAA) interpenetrating hydrogel networks (IHNs).

Time (Days)	Mean (±s.d.) % Loss of Pluronic Block Copolymers:														
	Pluronic F127 (% *w*/*w* Initial Loading)					Pluronic P123 (% *w*/*w* Initial Loading)					Pluronic L121 (% *w*/*w* Initial Loading)				
	10	20	30	40	50	10	20	30	40	50	10	20	30	40	50
2	1.5 ± 0.1	2.4 ± 0.1	3.1 ± 0.2	9.1 ± 0.4	17.4 ± 2.4	1.2 ± 0.1	2.0 ± 0.1	2.6 ± 0.2	4.5 ± 0.3	9.0 ± 0.5	ND	ND	0.9 ± 0.1	1.3 ± 0.1	2.4 ± 0.1
4	1.8 ± 0.1	3.0 ± 0.2	3.3 ± 0.1	10.9 ± 6.1	24.5 ± 1.8	1.4 ± 0.1	2.6 ± 0.2	3.0 ± 0.1	5.4 ± 0.2	11.9 ± 1.1	0.1 ± 0.0	0.1 ± 0.0	1.1 ± 0.1	1.5 ± 0.1	3.2 ± 0.2
7	2.0 ± 0.1	3.1 ± 0.1	3.5 ± 0.2	13.1 ± 6.2	25.2 ± 1.8	1.6 ± 0.2	2.6 ± 0.2	2.9 ± 0.2	6.2 ± 0.3	14.0 ± 1.4	0.1 ± 0.0	0.1 ± 0.0	1.4 ± 0.1	1.5 ± 0.2	3.7 ± 0.2
14	2.0 ± 0.1	3.2 ± 0.1	3.5 ± 0.2	13.1 ± 2.4	25.3 ± 2.0	1.7 ± 0.2	2.6 ± 0.2	3.0 ± 0.2	6.2 ± 0.4	14.4 ± 1.3	0.1 ± 0.0	0.1 ± 0.0	1.5 ± 0.1	2.0 ± 0.3	3.7 ± 0.3

ND: Not detected.

In the IHNs, following the uptake of aqueous fluid, the micellization of the different Pluronic block co-polymers and the dissolution of the Pluronic monomers and subsequent loss from the network in the surrounding aqueous fluid occurred. The retention of various Pluronic block copolymers within the IHNs was facilitated by both the stability/size of the micelle and the interaction of the monomers/micelles with the poly(nethacrylates). The micellization of F127, P123, and L121 did not occur within the IHN xerogels, evidenced by the lack of an endothermic event at or close to the micellization temperature of the various Pluronic block copolymers [72]. As the ratio of PPO to PEO of the Pluronic block copolymers increased ((F127 (0.43) to P123 (2.3) to L121 (9.0)), micellization occurred at lower concentrations due to increased hydrophobic interactions in the micelle core. Pluronic loss was investigated at 37 °C, a temperature that exceeded the critical micelle temperature for each Pluronic block copolymer investigated. It is proposed that the greater retention of Pluronic P123 and L121 within the hydrated IHN is due to their existence within a stable micelle structure, one that resists destabilisation in the presence of aqueous fluid and resists monomer diffusion and loss from the network. The loss of Pluronics from p(MAA) IHNs is further reduced through the chemical interaction of these block copolymers with the acrylate, as evidenced by spectroscopy studies; however, the contribution of this phenomenon is relatively limited in comparison to micellization. Interestingly, the loss of Pluronic P127 from the IHNs was markedly greater than the loss of the other Pluronic block copolymers and, moreover, the differences in loss between the p(MAA) and p(HEMA) IHNs were the most pronounced. In p(HEMA) IHNs, the loss of Pluronic F127 was relatively rapid and micelles were not observed following immersion in Tris buffer for 1 week. This suggests that, whilst F127 micelles may have formed, these were not stable and, as a result, F127 monomer was free to diffuse through the network, leading to loss from the IHNs. The loss of F127 from p(MAA) IHNs was less than from the p(HEMA) IHNs with this stabilising at approximately 4 days. We showed that F127 micelles were present in p(MAA) IHNs following storage in Tris buffer for 1 week, at which point stabilisation of F127 loss had occurred. Therefore, the retention of F127 may be accredited to micelle formation; however, based on the previous spectroscopic observations, hydrogen bonding between p(MAA) and F127 may contribute to micelle stabilisation.

4. Conclusions

This manuscript describes novel interpenetrating hydrogel networks (IHNs) composed of p(HEMA) or p(MAA) and different Pluronic block copolymers (F127, P123, and L121) designed for potential use as medical device biomaterials or as platforms for controlled drug delivery. There was minimal interaction with the Pluronic block copolymers in p(HEMA) IHNs, evidenced from Raman spectroscopy and the existence of the individual glass transition temperatures of the components, determined using differential scanning calorimetry (DSC) and dynamic mechanical analysis (DMA). As a result, p(HEMA)-Pluronic block copolymer IHNs may be described as immiscible blends. Conversely, in p(MAA) IHNs, the ether group on the Pluronic block copolymers and the carbonyl group of p(MAA) interacted, as shown from shifts in their wavelength in both Raman spectroscopy and attenuated total reflectance–Fourier–transform spectroscopy. The p(MAA) IHNs exhibited a single glass transition temperature that lay between the two individual network components, the magnitude of which was dependent on the ratio of p(MAA) to Pluronic block polymers. Accordingly, these systems may be described as miscible. The physical state of the Pluronic block polymers within IHN that had been stored in Tris buffer (pH 7.2) for one week was determined using DSC. With the exception of p(HEMA)/F127 IHNs, each Pluronic block copolymer was present as micelles, the critical micelle temperature (CMT) being dependent on the grade of Pluronic block copolymer (decreasing as the ratio of poly(propylene oxide) to poly(ethylene oxide) increased). In the p(HEMA)/F127 IHNs, the micellization of F127 was not observed. Examination of the loss of each Pluronic block copolymer from the two networks following storage in Tris buffer showed that F127 and L121 showed the greatest and lowest loss, respectively. Loss from p(MAA) IHNs was less than from p(HEMA)

IHNs. The retention of the Pluronic block copolymers is accredited to their existence as micelles, the intermolecular bonding and large molecular weight of which suppresses Pluronic monomer diffusion and subsequent loss from the network. However, given the greater retention of Pluronic block copolymer p(MAA) IHNs, it may also be concluded that interactions between each Pluronic block copolymer and p(MAA) significantly contributed to the stability of the Pluronics within this network. Based on the observations in this study, notably the stability of several IHNs under aqueous conditions, we are currently assessing these systems as platforms for medical device biomaterials, particularly for use as urinary biomaterials where, if combined with bioactive agents, enhanced resistance to infection may result.

Author Contributions: Conceptualization, D.S.J. and G.P.A.; Formal analysis, D.S.J., M.W. and S.L.; Investigation, M.W. and S.L.; Writing—original draft, D.S.J.; Writing—review & editing, S.L. and G.P.A.; Supervision, D.S.J. and G.P.A.; Project administration, D.S.J. All authors have read and agreed to the published version of the manuscript.

Funding: A PhD scholarship was provided for M.W. by the Department of Education for Northern Ireland.

Institutional Review Board Statement: No ethical approval was needed.

Data Availability Statement: Data are contained within the article.

Conflicts of Interest: The authors declare no conflict of interest.

References

1. Hameed, H.; Faheem, S.; Paiva-Santos, A.; Sarwar, H.; Jamshaid, M. A Comprehensive Review of Hydrogel-Based Drug Delivery Systems: Classification, Properties, Recent Trends, and Applications. *AAPS PharmSciTech* **2024**, *25*, 64. [CrossRef] [PubMed]
2. Bashir, S.; Hina, M.; Iqbal, J.; Rajpar, A.H.; Mujtaba, M.A.; Alghamdi, N.A.; Wageh, S.; Ramesh, K.; Ramesh, S. Fundamental Concepts of Hydrogels: Synthesis, Properties, and Their Applications. *Polymers* **2020**, *12*, 2702. [CrossRef] [PubMed]
3. Lin, X.; Zhao, X.; Xu, C.; Wang, L.; Xia, Y. Progress in the mechanical enhancement of hydrogels: Fabrication strategies and underlying mechanisms. *J. Polym. Sci.* **2022**, *60*, 2525–2542. [CrossRef]
4. Dai, S.; Gao, Y.; Duan, L. Recent advances in hydrogel coatings for urinary catheters. *J. Appl. Polym. Sci.* **2023**, *140*, e53701. [CrossRef]
5. Feng, W.; Wang, Z. Tailoring the Swelling-Shrinkable Behavior of Hydrogels for Biomedical Applications. *Adv. Sci.* **2023**, *10*, 2303326. [CrossRef]
6. Xu, F.; Dawson, C.; Lamb, M.; Mueller, E.; Stefanek, E.; Akbari, M.; Hoare, T. Hydrogels for Tissue Engineering: Addressing Key Design Needs Toward Clinical Translation. *Front. Bioeng. Biotechnol.* **2022**, *10*, 849831. [CrossRef]
7. Saint, S.; Greene, T.; Krein, S.L.; Rogers, M.A.M.; Ratz, D.; Fowler, K.E.; Edson, B.S.; Watson, S.R.; Meyer-Lucas, B.; Masuga, M.; et al. A Program to Prevent Catheter-Associated Urinary Tract Infection in Acute Care. *N. Engl. J. Med.* **2016**, *374*, 2111–2119. [CrossRef]
8. Meddings, J.; Manojlovich, M.; Ameling, J.M.; Olmsted, R.N.; Rolle, A.J.; Greene, M.T.; Ratz, D.; Snyder, A.; Saint, S. Quantitative Results of a National Intervention to Prevent Hospital-Acquired Catheter-Associated Urinary Tract Infection A Pre-Post Observational Study. *Ann. Intern. Med.* **2019**, *171*, S38–S44. [CrossRef]
9. Van Decker, S.G.; Bosch, N.; Murphy, J. Catheter-associated urinary tract infection reduction in critical care units: A bundled care model. *BMJ Open Qual.* **2021**, *10*, e001534. [CrossRef]
10. Yang, K.R.; Han, Q.; Chen, B.P.; Zheng, Y.H.; Zhang, K.S.; Li, Q.; Wang, J.C. Antimicrobial hydrogels: Promising materials for medical application. *Int. J. Nanomed.* **2018**, *13*, 2217–2263. [CrossRef]
11. Stensballe, J.; Looms, D.; Nielsen, P.N.; Tvede, V. Hydrophilic-coated catheters for intermittent catheterisation reduce urethral microtrauma: A prospective, randomised, participant-blinded, crossover study of three different types of catheters. *Eur. Urol.* **2005**, *48*, 978–983. [CrossRef] [PubMed]
12. Boonkaew, B.; Kempf, M.; Kimble, R.; Supaphol, P.; Cuttle, L. Antimicrobial efficacy of a novel silver hydrogel dressing compared to two common silver burn wound dressings: Acticoat and PolyMem Silver®. *Burns* **2014**, *40*, 89–96. [CrossRef]
13. Sung, A.Y.; Kim, T.-H. Optical application of poly(HEMA-co-MMA) containing silver nanoparticles and N,N-dimethylacrylamide. *Korean J. Chem. Eng.* **2012**, *29*, 686–691. [CrossRef]
14. Obradovic, B.; Stojkovska, J.; Jovanovic, Z.; Miskovic-Stankovic, V. Novel alginate based nanocomposite hydrogels with incorporated silver nanoparticles. *J. Mater. Sci. Mater. Med.* **2012**, *23*, 99–107. [CrossRef] [PubMed]
15. Ng, V.W.L.; Chan, J.M.W.; Sardon, H.; Ono, R.J.; García, J.M.; Yang, Y.Y.; Hedrick, J.L. Antimicrobial hydrogels: A new weapon in the arsenal against multidrug-resistant infections. *Adv. Drug Deliv. Rev.* **2014**, *78*, 46–62. [CrossRef]

16. Copling, A.; Akantibila, M.; Kumaresan, R.; Fleischer, G.; Cortes, D.; Tripathi, R.S.; Carabetta, V.J.; Vega, S.L. Recent Advances in Antimicrobial Peptide Hydrogels. *Int. J. Mol. Sci.* **2023**, *24*, 22. [CrossRef]
17. Craig, R.A.; McCoy, C.P.; Gorman, S.P.; Jones, D.S. Photosensitisers—The progression from photodynamic therapy to anti-infective surfaces. *Expert Opin. Drug Deliv.* **2015**, *12*, 85–101. [CrossRef]
18. Wang, X.W.; Wang, J.; Yu, Y.; Yu, L.; Wang, Y.X.; Ren, K.F.; Ji, J. A polyzwitterion-based antifouling and flexible bilayer hydrogel coating. *Compos. Part B Eng.* **2022**, *244*, 110164. [CrossRef]
19. Carpa, R.; Farkas, A.; Dobrota, C.; Butiuc-Keul, A. Double-Network Chitosan-Based Hydrogels with Improved Mechanical, Conductive, Antimicrobial, and Antibiofouling Properties. *Gels* **2023**, *9*, 278. [CrossRef]
20. Jones, D.S.; Andrews, G.P.; Gorman, S.P. Characterization of crosslinking effects on the physicochemical and drug diffusional properties of cationic hydrogels designed as bioactive urological biomaterials. *J. Pharm. Pharmacol.* **2005**, *57*, 1251–1259. [CrossRef]
21. Kazmierska, K.; Szwast, M.; Ciach, T. Determination of urethral catheter surface lubricity. *J. Mater. Sci. Mater. Med.* **2008**, *19*, 2301–2306. [CrossRef] [PubMed]
22. Dai, Z.; Yang, X.; Wu, F.; Wang, L.; Xiang, K.; Li, P.; Lv, Q.; Tang, J.; Dohlman, A.; Dai, L.; et al. Living fabrication of functional semi-interpenetrating polymeric materials. *Nat. Commun.* **2021**, *12*, 3422. [CrossRef] [PubMed]
23. Nejadnik, M.R.; van der Mei, H.C.; Norde, W.; Busscher, H.J. Bacterial adhesion and growth on a polymer brush-coating. *Biomaterials* **2008**, *29*, 4117–4121. [CrossRef]
24. Stepulane, A.; Rajasekharan, A.K.; Andersson, M. Multifunctional Surface Modification of PDMS for Antibacterial Contact Killing and Drug-Delivery of Polar, Nonpolar, and Amphiphilic Drugs. *ACS Appl. Bio Mater.* **2022**, *5*, 5289–5301. [CrossRef]
25. Auxier, J.A.; Dill, J.K.; Schilke, K.F.; McGuire, J. Blood protein repulsion after peptide entrapment in pendant polyethylene oxide layers. *Biotechnol. Appl. Biochem.* **2014**, *61*, 371–375. [CrossRef] [PubMed]
26. Conti, D.; Yoshida, M.; Pezzin, S.; Coelho, L. Miscibility and crystallinity of poly(3-hydroxybutyrate)/poly(3-hydroxybutyrate-co-3-hydroxyvalerate) blends. *Thermochim. Acta* **2006**, *450*, 61–66. [CrossRef]
27. Rocco, A.; Pereira, R.; Felisberti, M. Miscibility, crystallinity and morphological behavior of binary blends of poly(ethylene oxide) and poly(methyl vinyl ether-maleic acid). *Polymer* **2001**, *42*, 5199–5205. [CrossRef]
28. Cimmino, S.; Dipace, E.; Martuscelli, E.; Silvestre, C. Evaluation of the equilibrium melting temperature and structure analysis of poly(ethylene oxide) poly(methyl methacrylate) blends. *Makromol. Chem. Macromol. Chem. Phys.* **1990**, *191*, 2447–2454. [CrossRef]
29. Andrews, G.; Laverty, T.; Jones, D. Mucoadhesive Polymeric Polyologels Designed for the Treatment of Periodontal and Related Diseases of the Oral Cavity. *Polymers* **2024**, *16*, 589. [CrossRef]
30. Gilmore, B.F.; Hamill, T.M.; Jones, D.S.; Gorman, S.P. Validation of the CDC Biofilm Reactor as a Dynamic Model for Assessment of Encrustation Formation on Urological Device Materials. *J. Biomed. Mater. Res. Part B Appl. Biomater.* **2010**, *93*, 128–140. [CrossRef]
31. Andersen, M.J.; Flores-Mireles, A.L. Urinary Catheter Coating Modifications: The Race against Catheter-Associated Infections. *Coatings* **2020**, *10*, 23. [CrossRef]
32. Smith, D.R.M.; Pouwels, K.B.; Hopkins, S.; Naylor, N.R.; Smieszek, T.; Robotham, J.V. Epidemiology and health-economic burden of urinary-catheter-associated infection in English NHS hospitals: A probabilistic modelling study. *J. Hosp. Infect.* **2019**, *103*, 44–54. [CrossRef] [PubMed]
33. Shuman, E.K.; Chenoweth, C.E. Urinary Catheter-Associated Infections. *Infect. Dis. Clin. N. Am.* **2018**, *32*, 885–897. [CrossRef] [PubMed]
34. Chew, B.H.; Lange, D. Ureteral stent symptoms and associated infections: A biomaterials perspective. *Nat. Rev. Urol.* **2009**, *6*, 440–448. [CrossRef] [PubMed]
35. Ilie, V.G.; Ilie, V.I. Ureteric Stent Use—Part of the Solution and Part of the Problem. *Curr. Urol.* **2018**, *11*, 126–130. [CrossRef]
36. Zhang, Y.; Chen, C.; Hu, W.; Li, S.; Li, Y.; Zhu, R. Antireflux Ureteral Stent Improves Stent-Related Symptoms and Quality of Life: A Prospective Randomized Controlled Trial. *J. Endourol.* **2023**, *37*, 761–767. [CrossRef]
37. Grazia, A.D.; Somani, B.K.; Soria, F.; Carugo, D.; Mosayyebi, A. Latest advancements in ureteral stent technology. *Transl. Androl. Urol.* **2019**, *8*, S436–S441. [CrossRef]
38. Mosayyebi, A.; Manes, C.; Carugo, D.; Somani, B.K. Advances in Ureteral Stent Design and Materials. *Curr. Urol. Rep.* **2018**, *19*, 35. [CrossRef]
39. Wierzbicka, A.; Krakos, M.; Wilczek, P.; Bociaga, D. A comprehensive review on hydrogel materials in urology: Problems, methods, and new opportunities. *J. Biomed. Mater. Res. Part B Appl. Biomater.* **2023**, *111*, 730–756. [CrossRef]
40. Jin, L.; Yao, L.; Yuan, F.; Dai, G.C.; Xue, B.X. Evaluation of a novel biodegradable ureteral stent produced from polyurethane and magnesium alloys. *J. Biomed. Mater. Res. Part B Appl. Biomater.* **2021**, *109*, 665–672. [CrossRef]
41. Barros, A.A.; Duarte, A.R.C.; Pires, R.A.; Sampaio-Marques, B.; Ludovico, P.; Lima, E.; Mano, J.F.; Reis, R.L. Bioresorbable ureteral stents from natural origin polymers. *J. Biomed. Mater. Res. Part B Appl. Biomater.* **2015**, *103*, 608–617. [CrossRef] [PubMed]
42. Barros, A.A.; Oliveira, C.; Lima, E.; Duarte, A.R.C.; Reis, R.L. Gelatin-based biodegradable ureteral stents with enhanced mechanical properties. *Appl. Mater. Today* **2016**, *5*, 9–18. [CrossRef]
43. Zmejkoski, D.Z.; Markovic, Z.M.; Zdravkovic, N.M.; Trisic, D.D.; Budimir, M.D.; Kuzman, S.B.; Kozyrovska, N.O.; Orlovska, I.V.; Bugárová, N.; Petrovic, D.Z.; et al. Bactericidal and antioxidant bacterial cellulose hydrogels doped with chitosan as potential urinary tract infection biomedical agent. *RSC Adv.* **2021**, *11*, 8559–8568. [CrossRef] [PubMed]

44. Park, J.H.; Cho, Y.W.; Cho, Y.-H.; Choi, J.M.; Shin, H.J.; Bae, Y.H.; Chung, H.; Jeong, S.Y.; Kwon, I.C. Norfloxacin-releasing urethral catheter for long-term catheterization. *J. Biomater. Sci. Polym. Ed.* **2003**, *14*, 951–962. [CrossRef] [PubMed]
45. Cho, Y.W.; Park, J.H.; Kim, S.H.; Cho, Y.H.; Choi, J.M.; Shin, H.J.; Bae, Y.H.; Chung, H.; Jeong, S.Y.; Kwon, I.C. Gentamicin-releasing urethral catheter for short-term catheterization. *J. Biomater. Sci. Polym. Ed.* **2003**, *14*, 963–972. [CrossRef] [PubMed]
46. Dragan, E.S. Design and applications of interpenetrating polymer network hydrogels. A review. *Chem. Eng. J.* **2014**, *243*, 572–590. [CrossRef]
47. Lü, H.; Zheng, S.X.; Tian, G.H. Poly(hydroxyether sulfone) and its blends with poly(ethylene oxide): Miscibility, phase behavior and hydrogen bonding interactions. *Polymer* **2004**, *45*, 2897–2909. [CrossRef]
48. Andrews, R.J.; Grulke, E.A. *Glass Transition Temperatures of Polymers*, 4th ed.; John Wiley and Sons: New York, NY, USA, 2004.
49. Mothé, C.G.; Drumond, W.S.; Wang, S.H. Phase behavior of biodegradable amphiphilic poly(L,L-lactide)-b-poly(ethylene glycol)-b-poly(L,L-lactide). *Thermochim. Acta* **2006**, *445*, 61–66. [CrossRef]
50. Kuo, S.W.; Huang, C.F.; Tung, P.H.; Huang, W.J.; Huang, J.M.; Chang, F.C. Synthesis, thermal properties, and specific interactions of high Tg increase in poly(2,6-dimethyl-1,4-phenylene oxide)-block-polystyrene copolymers. *Polymer* **2005**, *46*, 9348–9361. [CrossRef]
51. Fuller, C.S.; MacRae, R.J.; Walther, M.; Cameron, R.E. Interactions in poly(ethylene oxide)-hydroxypropyl methylcellulose blends. *Polymer* **2001**, *42*, 9583–9592. [CrossRef]
52. Rinaldi, A.; Matos, R.; Rubira, A.; Ferreira, O.; Girotto, E. Electrical, spectroscopic, and thermal properties of blends formed by PEDOT, PVC, and PEO. *J. Appl. Polym. Sci.* **2005**, *96*, 1710–1715. [CrossRef]
53. Siddiqui, M.N.; Redhwi, H.H.; Tsagkalias, I.; Softas, C.; Ioannidou, M.D.; Achilias, D.S. Synthesis and characterization of poly(2-hydroxyethyl methacrylate)/silver hydrogel nanocomposites prepared via in situ radical polymerization. *Thermochim. Acta* **2016**, *643*, 53–64. [CrossRef]
54. Fekete, E.; Földes, E.; Pukánszky, M. Effect of molecular interactions on the miscibility and structure of polymer blends. *Eur. Polym. J.* **2005**, *41*, 727–736. [CrossRef]
55. Marwat, Z.K.; Baloch, M.K. Solvent induced miscibility between polymers, and its influence on the morphology, and mechanical properties of their blends. *Eur. Polym. J.* **2015**, *66*, 520–527. [CrossRef]
56. Buzarovska, A.; Grozdanov, A. Crystallization kinetics of poly(hydroxybutyrate-*co*-hydroxyvalerate) and poly(dicyclohexylitaconate) PHBV/PDCHI blends: Thermal properties and hydrolytic degradation. *J. Mater. Sci.* **2009**, *44*, 1844–1850. [CrossRef]
57. Aquino, K.A.D.; da Silva, F.F.; Araújo, E.S. Investigation of Poly(vinyl chloride)/Polystyrene Mixture Miscibility: Comparison of Viscometry Criteria with Fourier Transform Infrared Spectroscopy and Refractive Index Measurements. *J. Appl. Polym. Sci.* **2011**, *119*, 2770–2777. [CrossRef]
58. Qu, Z.; Xu, H.; Gu, H. Synthesis and Biomedical Applications of Poly((meth)acrylic acid) Brushes. *ACS Appl. Mater. Interfaces* **2015**, *7*, 14537–14551. [CrossRef]
59. Hong, C.Y.; Li, X.; Pan, C.Y. Smart Core-Shell Nanostructure with a Mesoporous Core and a Stimuli-Responsive Nanoshell Synthesized Via Surface Reversible Addition-Fragmentation Chain Transfer Polymerization. *J. Phys. Chem. C* **2008**, *112*, 15320. [CrossRef]
60. Sun, J.T.; Yu, Z.Q.; Hong, C.Y.; Pan, C.Y. Biocompatible Zwitterionic Sulfobetaine Copolymer-Coated Mesoporous Silica Nanoparticles for Temperature-Responsive Drug Release, Macromol. *Rapid Commun.* **2012**, *33*, 811. [CrossRef]
61. Li, Y.; Li, H.; Oigiviy, L.Y.; Zhu, F.; Yang, J.; Zhu, Y. Complexation of poly(acrylic acid) and poly(ethylene oxide) investigated by enhanced Rayleigh scattering method. *J. Polym. Sci. Part B Polym. Phys.* **2010**, *48*, 1847–1852. [CrossRef]
62. Monir, T.S.B.; Afroz, S.; Khan, R.A.; Miah, M.Y.; Takafuji, M.; Alam, M.A. pH-Sensitive Hydrogel from Polyethylene Oxide and Acrylic acid by Gamma Radiation. *J. Compos. Sci.* **2019**, *3*, 58. [CrossRef]
63. Kaczmarek, H.; Szalla, A. Photochemical transformation in poly(acrylic acid)/poly(ethylene oxide) complexes. *J. Photochem. Photobiol. A Chem.* **2006**, *180*, 46–53. [CrossRef]
64. Zhang, F.; Meng, F.; Lubach, J.; Koleng, J.; Watson, N.A. Properties and mechanisms of drug release from matrix tablets containing poly(ethylene oxide) and poly(acrylic acid) as release retardants. *Eur. J. Pharm. Biopharm.* **2016**, *105*, 97–105. [CrossRef] [PubMed]
65. Zou, S.F.; Lv, R.H.; Tong, Z.Z.; Na, B.; Fu, K.; Liu, H.S. In situ hydrogen-bonding complex mediated shape memory behavior of PAA/PEO blends. *Polymer* **2019**, *183*, 121878. [CrossRef]
66. Rocco, A.M.; Bielschowsky, C.E.; Pereira, R.P. Blends of poly(ethylene oxide) and poly(4-vinylphenol-co-2-hydroxyethyl methacrylate): Thermal analysis, morphological behaviour and specific interactions. *Polymer* **2003**, *44*, 361–368. [CrossRef]
67. Akiba, I.; Ohba, Y.; Akiyama, S. Phase structure in blends of poly(ethylene glycol) and poly(styrene-co-methacrylic acid). *Macromolecules* **1999**, *32*, 1175–1179. [CrossRef]
68. Akiyama, S.; Ishikawa, K.; Kawahara, S.; Akiba, I. Interactions in blends of bisphenol-A polycarbonate and poly(styrene-co-methacrylic acid). *Polymer* **2001**, *42*, 6657–6660. [CrossRef]
69. Lee, J.; Painter, P.; Coleman, M. Hydrogen bonding in polymer blends. 3. Blends involving polymers containing methacrylic acid and ether groups. *Macromolecules* **1988**, *21*, 346–354. [CrossRef]
70. Abu-Diak, O.A.; Jones, D.S.; Andrews, G.P. Understanding the performance of melt-extruded poly(ethylene oxide)-bicalutamide solid dispersions: Characterisation of microstructural properties using thermal, spectroscopic and drug release methods. *J. Pharm. Sci.* **2012**, *101*, 200–213. [CrossRef]

71. Alexandiridis, P.; Hatton, T.A. Poly(ethylene oxide)-poly(propylene oxide)-poly(ethylene oxide) block copolymer surfactants in aqueous solutions and interfaces: Thermodynamics, structure, dynamics and modelling. *Colloids Surf. A* **1995**, *96*, 1–46. [CrossRef]
72. Hecht, E.; Hoffmann, H. Kinetic and calorimetric investigations on micelle formation of block co-polymers of the Poloxamer type. *Colloids Surf. A Physicochem. Eng. Asp.* **1995**, *96*, 181–197. [CrossRef]

polymers

MDPI

Article

On the Properties of Styrene–Maleic Acid Copolymer–Lipid Nanoparticles: A Solution NMR Perspective

Vladislav V. Motov [1,2], Erik F. Kot [1,3], Svetlana O. Kislova [4], Eduard V. Bocharov [1], Alexander S. Arseniev [1], Ivan A. Boldyrev [4], Sergey A. Goncharuk [1,*] and Konstantin S. Mineev [1,*,†]

[1] Shemyakin-Ovchinnikov Institute of Bioorganic Chemistry RAS, 117997 Moscow, Russia; motov.vladislave@gmail.com (V.V.M.); erick.kot@gmail.com (E.F.K.); bon@nmr.ru (E.V.B.); aars@nmr.ru (A.S.A.)
[2] Moscow Center for Advanced Studies, Kulakova Str. 20, 140829 Moscow, Russia
[3] Faculty of Biology, Shenzhen MSU-BIT University, Shenzhen 518172, China
[4] Frumkin Institute of Physical Chemistry and Electrochemistry, Russian Academy of Sciences, Leninsky Prospect 31, 119071 Moscow, Russia; s.o.kislova@phyche.ac.ru (S.O.K.); i_boldyrev@mail.ru (I.A.B.)
* Correspondence: ms.goncharuk@gmail.com (S.A.G.); mineev@nmr.uni-frankfurt.de (K.S.M.)
† Current address: Institute of Organic Chemistry and Chemical Biology, Goethe University Frankfurt, 60438 Frankfurt am Main, Germany.

Abstract: The production of functionally active membrane proteins (MPs) in an adequate membrane environment is a key step in structural biology. Polymer–lipid particles based on styrene and maleic acid (SMA) represent a promising type of membrane mimic, as they can extract properly folded MPs directly from their native lipid environment. However, the original SMA polymer is sensitive to acidic pH levels, which has led to the development of several modifications: SMA-EA, SMA-QA, and others. Here, we introduce a novel SMA derivative with a negatively charged taurine moiety, SMA-tau, and investigate the formation and characteristics of lipid–SMA-EA and lipid–SMA-tau membrane-mimicking particles. Our findings demonstrate that both polymers can form nanodiscs with a patch of lipid bilayer that can undergo phase transitions at temperatures close to those of the lipid bilayer membranes. Finally, we discuss the potential applications of these SMAs for NMR spectroscopy.

Keywords: membrane proteins; membrane mimetics; NMR spectroscopy; SMALP

1. Introduction

Membrane proteins attract the most interest in structural biology, as they may serve as drug targets for a wide range of severe diseases. However, studies on membrane proteins are hampered due to the need to place these proteins into an environment closely resembling the cell membrane. None of the existing high-resolution approaches can work efficiently in the presence of liposomes, the most adequate membrane mimetic. Therefore, it is necessary to utilize lipid-containing nanoparticles of relatively small size (a radius of 2–6 nm is desirable). Several types of such membrane mimetics have previously been suggested for structural studies, namely, detergent micelles, bicelles, amphipols, and lipid–protein nanodiscs (LPNs) [1–12], which all have their advantages and drawbacks. Among the most recent advances in the field, one can highlight so-called SMA–lipid particles, SMALPs, or lipodiscs [13].

SMA is an amphipathic copolymer of styrene and maleic acid. When added to liposomes or cell membranes, SMA can extract lipids and form discoidal particles, which are believed to contain a patch of lipid bilayer [14]. This is an essential advantage of SMA in comparison to bicelles and LPNs, because SMA can extract the properly folded proteins directly in their native lipid environment, and no detergent is used [14–22]. At the same time, SMALPs represent an "open system", similar to bicelles, and SMALPs can exchange their contents upon collision [23]. Additionally, SMA can form nanodiscs of various sizes in the range of 10–60 nm, and the largest particles can align spontaneously in a strong magnetic field, which is beneficial for solid-state NMR applications [24,25]. The original SMA polymer cannot tolerate an acidic pH or the presence of divalent cations [14,26]; therefore, several modifications to SMA have been introduced (SMA-QA, SMA-EA, SMA-ED, and many others [27–29]). SMA nanodiscs have been successfully applied in structural studies using cryo-electron microscopy [4,30–32], and their principal applicability for the NMR investigation of membrane-anchored soluble domains has been demonstrated [16,33,34]. On the other hand, the possibility of the observation of transmembrane domains has not yet been shown.

In addition to the general applicability of lipodiscs for structural studies, an essential question is whether the state of lipids in particles resembles the interior of the cell membrane. The same question stands for other membrane mimetics, including lipid–protein nanodiscs [35] and phospholipid bicelles [36,37]. One of the major properties of lipid bilayers is that they undergo phase transitions, and in the case of SMALPs, this property was investigated for particles formed by the original SMA polymer [38]. It appeared that polymers with different ratios between styrene and maleic acid affected lipid behavior to various extents, and the effect was additionally dependent on the contents of the polymer and lipids [26,27,34]. However, the properties of lipodiscs formed by the modified SMA analogs were not studied in much detail. Here, we utilize the NMR-based approach that we previously proposed to investigate the structure of lipid bicelles [36,37,39] and study the phase transitions, size, and applicability limits of lipodiscs based on the modified SMA polymers.

2. Materials and Methods

2.1. SMA-EA and SMA-Tau Synthesis

The SMA modifications used in this work were synthesized using previously characterized commercially available SMAnh (Sigma-Aldrich, Burlington, MA, USA, Mw = 1900 g/mol, ratio of styrene to maleic units of 3:1) [40].

To synthesize SMA-EA, SMAnh (1 g, 2.55 mmol) was dissolved in 50 mL of Dimethylformamide (DMF) with stirring. Ethanolamine (311.5 mg, 5.1 mmol) was dissolved in 2 mL of DMF. The solutions were mixed and stirred with magnetic stirrer (5drops-HS-19, 5drops.ru, Moscow, Russia) for 1 h (until the reaction mixture turned light yellow). A total of 50 mL $CHCl_3$ was added (until the reaction mixture became cloudy and heated). Then, 30 mL of petroleum ether was added and left to stir for 2 days, during which a white crystalline precipitate formed. It was filtered under vacuum conditions and freeze-dried (INEJ-6, Institute for Biological Instrumentation of RAS, Pushchino, Russia). The yield was 30% [41].

To synthesize SMA-tau, 15 mL of distilled DMF was added to SMAnh (300 mg, 0.765 mmol) and stirred for an hour. Taurine (191 mg, 1.53 mmol) and KOH (86 mg, 1.53 mmol) were dissolved in 5 mL distilled water. The solutions were drained and stirred for two hours. The resulting solution was evaporated on a rotary evaporator (Rotavapor R-300, Buchi, Flawil, Switzerland). The resulting white crystals were lyophilized. Then, they were dissolved in 10 mL of water, and dialysis was carried out (the dialysis bag used allows substances weighing up to 1 kDa to pass through). The bag was placed in a 1 L glass container and stirred overnight. The yield was 58%.

2.2. Protein Expression and Purification

p75-TM-ICD was expressed in *E. coli* strain BL21(DE3) in M9 medium. To obtain the ^{15}N-labeled protein, $^{15}NH_4Cl$ was used. The cells were grown overnight at 28 °C in a shaking incubator (New Brunswick Innova 44R, Eppendorf SE, Hamburg, Germany) at 250 rpm. Protein expression was induced at OD600 ~ 0.6 by adding isopropyl β-d-1-thiogalactopyranoside (IPTG) to a final concentration of 0.1 mM, and cultivation was continued at 13 °C for 36 h. Cells were harvested by centrifugation (H1750R, Cence, Changsha, China) at 7000× *g* for 10 min at 4 °C and stored at − 20 °C for later use.

The protein purification protocol is described in detail in [42]. Briefly, the cells were lysed in Lys buffer (20 mM Tris, pH 8.0, 250 mM NaCl, 0.2 mM phenylmethylsulfonyl fluoride, 1 mM EDTA, 10 mM β-mercaptoethanol) by sonication (Bandelin SONOPULS HD 2200, Bandelin, Berlin, Germany). The lysate was clarified by centrifugation (H1750R, Cence, Changsha, China) at 14,000× *g* for 30 min at 4 °C, and the membrane fraction was harvested by centrifugation at 150,000× *g* for 1 h at 4 °C. The precipitate was resuspended in IMAC buffer (20 mM Tris, pH 8.0, 250 mM NaCl, 10 mM Imidazole, 10 mM β-mercaptoethanol, 0.5 % lauryl sarcosine) and purified by immobilized metal affinity chromatography using Ni-Sepharose HP resin (GE HealthCare, Chicago, IL, USA). The fractions containing the target protein were combined, and the protein was further purified by size-exclusion chromatography (SEC) in SEC buffer (10 mM Tris, pH 8.0, 100 mM NaCl, 10 mM β-mercaptoethanol, 0.5 % lauryl sarcosine). To prepare liposomes with p75-TM-ICD, the lipid was added to the protein sample, and detergent was removed using BioBeads SM2 resin (BioRad, Hercules, CA, USA) [42]. Detergent removal was monitored by ^{1}H-NMR.

2.3. Lipodisc Preparation

To prepare the lipodisc samples, stock mixtures of lipid MLVs and SMA-EA or SMA-tau in water were used. Lipid and SMA stocks contained 20 mM phosphate buffer (pH 8) and 5% D_2O. The final concentration of SMA was 1% w/w, and the final concentration of lipid depended on the desired mass ratio of lipid to SMA, $q = mass_{lipid}/mass_{SMA}$. Samples were incubated in a thermomixer (Thermomixer comfort, Eppendorf SE, Germany, Hamburg) at 30 °C overnight.

To prepare the DMPC/DMPG/SMA samples, lipids were mixed in chloroform and dried using a centrifugal vacuum concentrator (Savant SPD1010, Thermo Scientific, Waltham, MA, USA). Next, 100 mM PBS buffer (pH 8) was added to the lipids, and the resulting mixture was vortexed and supplied with the SMA stock. The final concentration of SMA was 0.75% (w/w), and the final concentration of lipids was 1.5% (w/w) (individual lipid concentrations depended on the desired DMPC/DMPG ratio). Samples were incubated in the thermomixer at 30 °C overnight.

To prepare SMALPs containing an *E. Coli* membrane fraction with p75-TM-ICD, 20 mM phosphate buffer (pH 8) was added to the membrane fraction, the resulting mixture was vortexed, and the SMA stock was added afterwards. The final concentration of *E. coli* lipids was 0.55% (w/w), and the final concentration of SMA depended on the desired q ratio. Samples were incubated in the thermomixer at 303 K overnight and centrifuged (Centrifuge 5417 R, Eppendorf SE, Germany, Hamburg) at 25,000× *g* at 4 °C for 120 min.

2.4. Size-Exclusion Chromatography

Size-exclusion chromatography (SEC) experiments were performed on a Superdex 200 Increase 10/300 GL column (GE HealthCare, Chicago, IL, USA). An amount of 500 µL of the 2% w/w SMA-EA/tau sample or 500 µL of the 2% w/w DMPC/SMA-EA/tau sample with a lipid/SMA mass ratio of q = 1 was injected into the column. The working buffer was PBS pH 7.5. The flow rate was 0.5 mL/min.

2.5. Dynamic Light Scattering

Dynamic light scattering (DLS) experiments were performed on a DynaPro instrument (Wyatt Technology, Santa Barbara, CA, USA) using a 12 µL cuvette. Data were recorded in 25 scans of 5 s each. DLS data were processed using Wyatt Dynamics 6.7.7.9 software. Each measurement was repeated 3 times, and the mean size distribution was calculated.

2.6. NMR Spectroscopy

^{1}H diffusion spectra for the DMPC/SMA-EA and DMPC/SMA-tau samples were acquired on a Bruker Avance 700 MHz spectrometer equipped with a room temperature triple resonance probe (Bruker, Billerica, MA, USA) using a double-stimulated echo pulse sequence with the suppression of convection effects and the solvent signal [43]. The diffusion delay was set to 400 ms.

^{31}P NMR spectra were recorded on a Bruker Avance III 600 MHz spectrometer equipped with a broadband double resonance probe (Bruker, Billerica, MA, USA). A single pulse experiment was used, with a $\pi/6$ pulse of 4 µs, a relaxation delay of 3 s, and an acquisition time of 2.7 s, and with broadband WALTZ16 proton decoupling at a field strength of 1.42 kHz. A pre-acquisition delay of 10 min was used before each experiment to ensure the equilibrium state of the system.

^{1}H-^{15}N-HSQC spectra of p75-TM-ICD were recorded on Bruker Avance III 800 MHz and Bruker Avance 700 MHz spectrometers equipped with cryogenic triple-resonance probes (Bruker, Billerica, MA, USA).

2.7. NMR Data Processing

Diffusion NMR spectra were processed using Wolfram Mathematica software. The diffusion of DMPC was measured using signals at 0.9 and 1.3 ppm. The dependencies of DMPC/SMA particle radii on the q ratio were approximated by the ideal bicelle model with a cylindrical rim:

$$R(q) = r\left(1 + \frac{q}{\lambda} + \frac{\sqrt{q(q+\lambda)}}{\lambda}\right) \quad (1)$$

where R is the SMALP radius, r is the width of the rim, formed by SMA molecules, and λ is the ratio between the lipid and SMA densities [36].

Based on the ideal bicelle model approximation, the average amount of DMPC molecules in the NMR-observable particles was estimated using the following formula:

$$N_{lipids} = \frac{2\pi(R-r)^2}{S_0}, \quad (2)$$

where R is the SMALP radius, r is the width of the rim, formed by SMA molecules, and S_0 is the surface area of a single DMPC polar headgroup (was taken equal to 0.6 nm^2 [44]).

To determine the chemical shifts, signals in the ^{31}P spectra were deconvoluted using the Lorentzian lineshapes and nonlinear regression in Wolfram Mathematica software. The temperature dependence of the chemical shift was approximated by the following function:

$$\delta(T) = (a_1T + b_1)\frac{Exp[c(d-T)]}{1 + Exp[c(d-T)]} + (a_2T + b_2)\frac{Exp[c(T-d)]}{1 + Exp[c(T-d)]} \quad (3)$$

the inflection point of this function was calculated to determine the temperature of the phase transition.

3. Results

Our approach is based on three major techniques. First, by studying the dependence of particle size on the lipid/polymer ratio, q, we tested different models of the SMALP structure [36]. Second, we used ^{31}P NMR spectroscopy to monitor the lipid phase transitions in the SMALPs [37]. Finally, using a model membrane protein, we analyzed the prospects of lipodiscs for solution NMR [42,45–47]. These three steps were taken to study the nanodiscs composed of two SMA-based polymers, modified with ethanolamine (SMA-EA) and taurine moieties (SMA-tau) (Figure 1). The ability of the SMA-EA copolymer to form lipodiscs of defined sizes, which can be controlled by adjusting the q ratio, makes it one of the most versatile SMA polymers [41]. The SMA-tau modification, a novel analog of SMA-EA in which the hydroxyl group is replaced with sulfonic acid, has not been proposed previously and is introduced in this study for the first time. We hypothesized that this modification could increase the solubility and monodispersity of SMA and SMA-based lipodiscs due to the double-negative charge on the modified maleic acid groups.

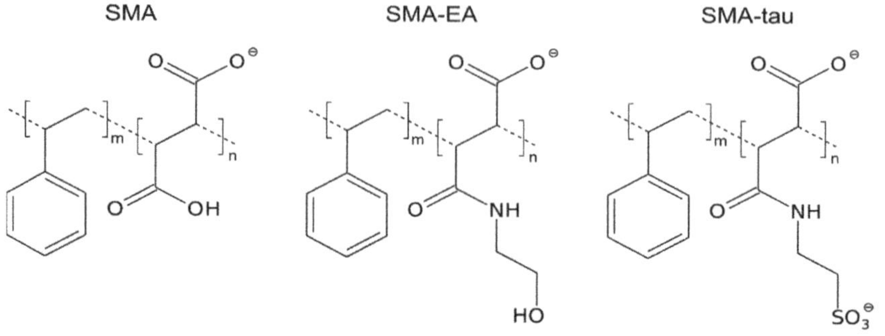

Figure 1. The chemical structure of SMA and its modifications SMA-EA and SMA-tau.

3.1. SMA-EA and SMA-Tau Form SMA–Lipid Particles

As previously reported, SMA forms particles with lipids [48]. Therefore, the first step of our work was to estimate the size distribution of lipodisc particles and to find out if the particles are suitable for solution NMR studies. We used size-exclusion chromatography (SEC) and compared the profiles of pure SMA-EA and SMA-tau (derivatives hereafter referred to as SMA) and their aggregates with DMPC lipids (Figure 2a). First, the elution volumes for the main peaks of SMA-EA and SMA-tau correspond to particles with a mass of about 40 kDa. At the same time, the SMA-EA profile is highly asymmetric and has a wide dispersion. This indicates an oligomerization of SMA-EA in solution, which is in agreement with previously reported data, where the formation of micellar species was observed [48]. Secondly, the profiles of the DMPC/SMA samples are shifted to the left and show a narrower, distinct peak in contrast to pure SMA. This indicates the formation of mixed massive DMPC/SMA particles with almost all SMA molecules being in the lipid-bound state. Based on the column calibration (Figure 2a), the molecular weight of the DMPC/SMA particles (1:1, w/w) exceeds 150 kDa, which corresponds to the expected size of lipodiscs (~5 nm in radius).

We analyzed the size of DMPC/SMA particles using dynamic light scattering (DLS) and NMR spectroscopy (Figure 2b–d). The results are consistent with the data provided by SEC. The peak fractions of the DMPC/SMA-EA and DMPC/SMA-tau mixtures consist of particles with average radii of 5.22 ± 0.05 and 4.33 ± 0.04 nm measured by DLS and of 4.36 ± 0.03 and 4.12 ± 0.06 nm measured by NMR. At the same time, it should be noted that there are some discrepancies in the values obtained by the different methods. We believe that this is explained by the fact that in NMR, larger particles contribute less to the signal, whereas in DLS, larger particles contribute more. Thus, there is most likely a particle size distribution in the sample, with the center lying between the DLS and NMR values.

In conclusion, SMA-EA and SMA-tau are able to interact with lipids and form particles with a size of ~10 nm, which corresponds to conventional lipid–protein nanodiscs used in solution NMR spectroscopy [49].

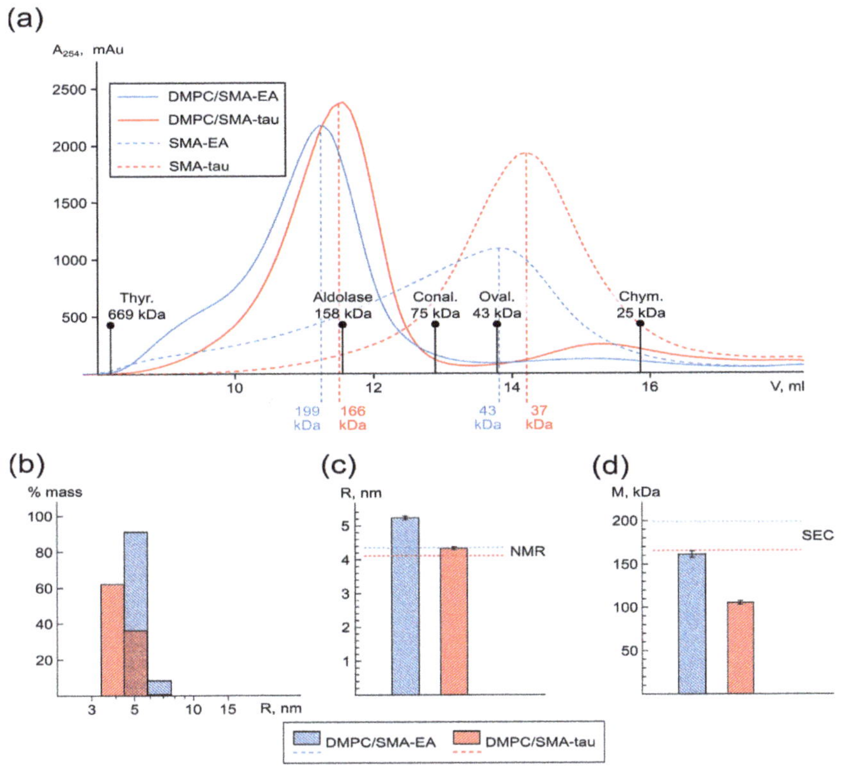

Figure 2. Size of SMA and SMA–lipid particles. (**a**) SEC profiles of SMA-EA (dashed blue line), DMPC/SMA-EA (blue line), SMA-tau (dashed red line), and DMPC/SMA-tau (red line) samples. (**b**) Distribution of radii for DMPC/SMA-EA (blue bars) and DMPC/SMA-tau (red bars) samples at 303 K obtained by DLS. The DMPC/SMA ratio in all cases was 1:1 (w:w). (**c**) Average radii of DMPC/SMA-EA (blue bar) and DMPC/SMA-tau (red bar) particles measured by DLS and NMR (blue and red dashed lines, respectively). (**d**) Average mass of DMPC/SMA-EA (blue bar) and DMPC/SMA-tau (red bar) measured by DLS compared to SEC data (blue and red dashed lines, respectively).

3.2. SMA-EA and SMA-tau Nanoparticles Reveal Bicelle-Like Behavior

To control the size and evaluate the morphology of the particles at typical working temperatures, we examined the sizes of DMPC/SMA-EA and DMPC/SMA-tau particles as a function of the DMPC/SMA w/w ratio, q, using the NMR diffusion measurements as described previously [43]. Since NMR diffusion measurements are less demanding in terms of sample preparation, and are more accurate and reproducible than the DLS, we continued our study with NMR as the main approach. The dependence of the average SMALP hydrodynamic radii on q for the DMPC/SMA-EA and DMPC/SMA-tau samples at 303 K and 313 K is shown in Figure 3.

For both DMPC/SMA-EA and DMPC/SMA-tau, the average lipodisc radius grows with increasing lipid content in the samples, similar to phospholipid bicelles. With this in mind, we applied the ideal bicelle model to describe the obtained dependence [36,50]. The ideal bicelle model is an equation that relates the size of the particle to the composition of the mixture, assuming the complete separation of the components within the particle, and can generally be applied to any type of discoidal aggregate. For SMA mixtures, we used the model that assumes a cylindrical shape of the lipodisc rim, which has two parameters: rim thickness and the ratio of the volumes that are occupied by the disc and rim molecules [36].

According to this analysis, SMA-EA forms a rim as thick as 2.8 nm at 303 K and 2.5 nm at 313 K, and SMA-tau forms rims of 3.0 and 2.7 nm at 303/313 K, respectively (Table S1).

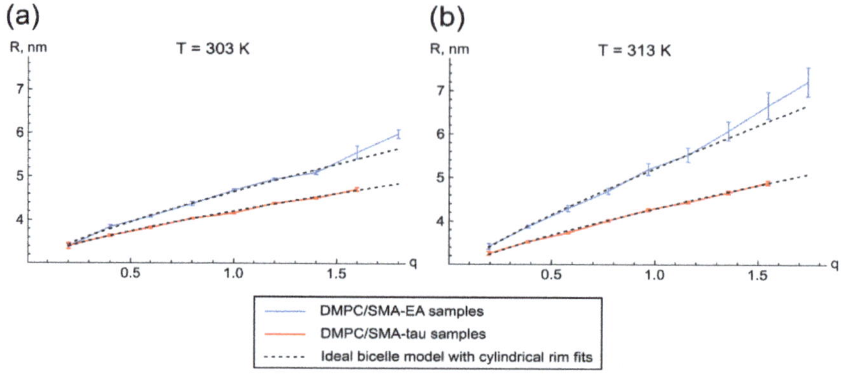

Figure 3. Dependencies of the lipid–SMA particle radii on q (**a**) obtained at 303 K and (**b**) at 313 K. Blue and red lines correspond to the data for DMPC/SMA-EA and DMPC/SMA-tau samples, respectively. Dashed lines correspond to their approximation by the ideal bicelle model with a cylindrical rim. Sample buffer contained 20 mM NaPi, pH 8.

Based on the ideal bicelle model approximation, the calculated average number of lipid molecules per SMA-EA particle increased from 9.0 to 96.3 in the q range of 0.2 to 1.2 at 313 K. The average number of lipid molecules per SMA-tau particle increased from 3.3 to 50.7 at 313 K in the q range of 0.2 to 1.6 (Figure S1). Therefore, the DMPC/SMA-EA samples demonstrate a more rapid increase in lipid number compared to DMPC/SMA-tau, with a ratio of over twofold at q = 1.0 at both 303 K and 313 K, for instance.

It is noteworthy that, in contrast to bicelles, the DMPC/SMA-EA samples exhibit pronounced temperature-dependent behavior. The SMA-EA–lipid particles are substantially larger at higher temperatures, which could be an undesired effect for the NMR applications. Like bicelles, DMPC/SMA-EA particles deviate from the ideal bicelle model starting from a specific SMA/lipid ratio, a phenomenon not observed in DMPC/SMA-tau mixtures [36].

We also tested how pH, ionic strength, and volume fraction of SMALPs affect lipodisc radius. Changes in pH moderately affected the DMPC/SMA-EA particles. An increase in pH from 7.0 to 8.0 resulted in an expansion of the particles, reaching 14% at 303 K and 20% at 313 K (Figure S2a). These changes can be attributed to the differing charges of the polymers present in the rim. Interestingly, the radii of the DMPC/SMA-tau particles remained almost unaffected. The addition of NaCl up to 100 mM had no pronounced effect on the DMPC/SMA-EA particle radii (Figure S2b). These particles showed moderate growth (~8% at 303 K and ~12% at 313 K), with an increase in volume fraction from 1.5% to 3% (Figure S2c).

To summarize, the radii of DMPC/SMA-EA and DMPC/SMA-tau particles behave similarly to bicelles and can be analyzed using the ideal bicelle model [36]. This model reveals that the SMA molecules form a thick rim, and the fraction of the volume occupied by the SMA in the SMALP particle is greater than the fraction occupied by the lipid for the q values considered. One might even wonder whether the lipid packing in SMALPs is disturbed and whether the properties of the lipids are similar to those in lipid bilayer membranes.

3.3. Phase Transitions in SMA Nanoparticles Reveal the Presence of Lipid Bilayer

Phase transition is a fundamental property of a lipid bilayer. Therefore, the study of such transitions in lipid–SMA samples would allow us to conclude whether lipodiscs mimic the properties of lipid bilayers. For this purpose, we analyzed the temperature-dependent behavior in the ^{31}P NMR spectra of the lipid–SMA samples. The presence of inflection in the temperature–chemical shift dependence plots indicates the process of lipid phase transition [37].

Temperature dependencies of the DMPC ^{31}P chemical shift for the DMPC/SMA-EA samples with q 0.3, 0.6, and 1.0 and DMPC/SMA-tau samples with q 0.6, 1.0, and 1.6 are shown in Figure 4a,b. No inflection is observed in the plots until the q value is higher than 0.6 in the case of the DMPC/SMA-EA samples and 1.0 in the case of the DMPC/SMA-tau samples, and this inflection is more pronounced at higher q values, which allows us to determine the critical temperatures of the phase transitions (Table S2, Figure 4d, Figure S3). Furthermore, the DMPC/SMA plots are similar to those observed for the DMPC/DHPC bicelles and lipid–protein nanodiscs [37] (Figure 4c).

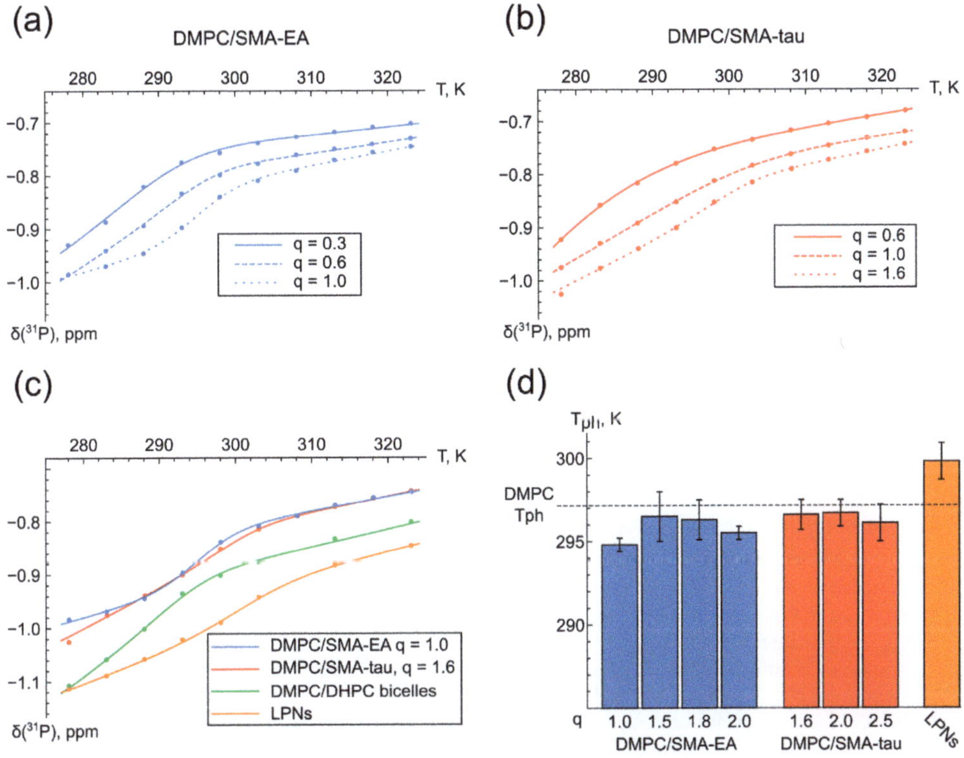

Figure 4. Analysis of chemical shifts in ^{31}P spectra at different temperatures. Temperature dependence of chemical shifts for (**a**) DMPC/SMA-EA samples and (**b**) for DMPC/SMA-tau samples (**c**). Temperature dependence of chemical shifts for DMPC/SMA samples with DMPC/DHPC bicelles and lipid-protein nanodiscs (LPNs); (**d**) Phase transition temperatures for DMPC/SMA-EA, DMPC/SMA-tau, and LPNs compared with available data for DMPC in a lipid bilayer (dashed line) [51].

The next step of the work was to study lipodiscs with alternative lipid compositions in order to simulate various properties of the lipid bilayer, such as thickness or charge (Figure 5). For this purpose, ^{31}P NMR spectra were obtained for POPC/SMA-EA (q = 1.5), DPPC/SMA-EA (q = 1.5), DMPC/DMPG/SMA-EA, and DMPC/DMPG/SMA-tau (q = 2) at various DMPC/DMPG ratios (60/40, 70/30, and 80/20).

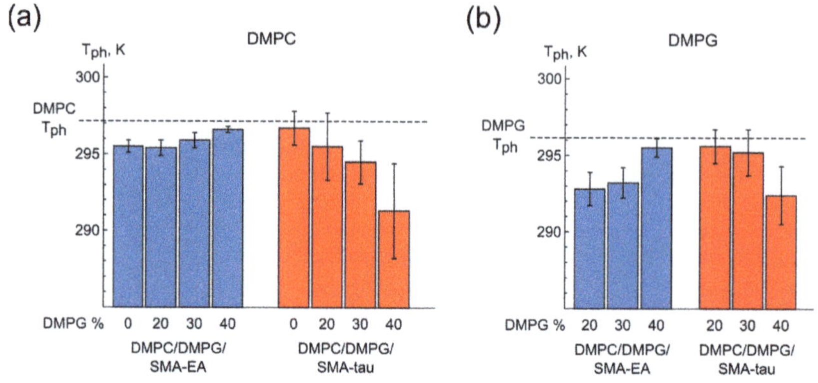

Figure 5. Phase transition temperatures of (**a**) DMPC and (**b**) DMPG in DMPC/DMPG/SMA-EA and DMPC/DMPG/SMA-tau samples compared to literature data for DMPC and DMPG bilayers (dashed lines) [51].

First, no phase transition was observed in the POPC/SMA-EA particles (Figure S4a). This is consistent with the fact that the phase transition temperature of POPC is below the water melting point [52]. For the DPPC/SMA-EA samples, the phase transition was observed at 311.3 $\pm$ 1.2 K, which is also in agreement with the higher phase transition temperature reported for DPPC bilayers [53] (Figure S4a, Table S2).

For the DMPG-containing samples, changes in the DMPC/DMPG ratio as well as the choice of SMA modification did not reveal a significant effect on the phase transition temperature (Figures 5 and S5b, Table S2).

Thus, phase transitions reveal the presence of a lipid bilayer in the particles with a sufficiently high lipid/SMA ratio. The phase behavior of different lipids in SMALPs is similar to the behavior of these lipids in bilayer membranes; therefore, the properties of the lipids are likely to be retained to a certain extent in the lipid–SMA particles under investigation.

3.4. *Effect of the SMA Rim on Membrane Protein*

So far, we have demonstrated that the tested SMALPs exhibit bicelle-like behavior and contain a patch of lipid bilayer that undergoes phase transitions [36,37,39], which is observed in the q range of 0.6–1.4 for SMA-EA and 1.0–1.6 for SMA-tau.

As a final step, we investigated the applicability of SMA-EA and SMA-tau lipodiscs as membrane mimetics for membrane proteins. As a model protein, we selected a fragment of the human p75 receptor, comprising the transmembrane and intracellular domains (p75-TM-ICD). NMR spectra of this protein have been previously obtained in various mimetics: bicelles, LPNs, and liposomes [42,45–47]. Furthermore, this protein is trafficked to the *E. coli* membrane during expression, allowing us to test the ability of SMA to extract the protein from the membrane.

We prepared protein-containing SMALP samples by extracting p75-TM-ICD from either POPC liposomes or the *E. coli* membrane fraction (MF), recorded the ^{1}H-^{15}N-HSQC NMR spectra (Figures 6 and S5), and compared them with spectra of the same protein obtained directly in liposomes or the membrane fraction.

In the initial spectra of p75-TM-ICD (Figure 6a,c), the signals of the mobile ICD domain are observed. The TMD signals are broadened and invisible because TMD is embedded in the large lipid particles (liposomes or membrane vesicles). The addition of SMA-EA to the protein in POPC at q = 1.3 resulted in a dramatic decrease in signal intensity (Figure 6b). The addition of SMA-EA to the protein in MF at q = 2.0 left the intensity unchanged (Figure 6c,d), but further addition of SMA to q = 1.5 also resulted in a significant decrease in intensity (Figure 6e). Unlike SMA-EA, the addition of SMA-tau resulted in signal loss at both q = 2.5 and q = 1.5 (Figure S5d,e).

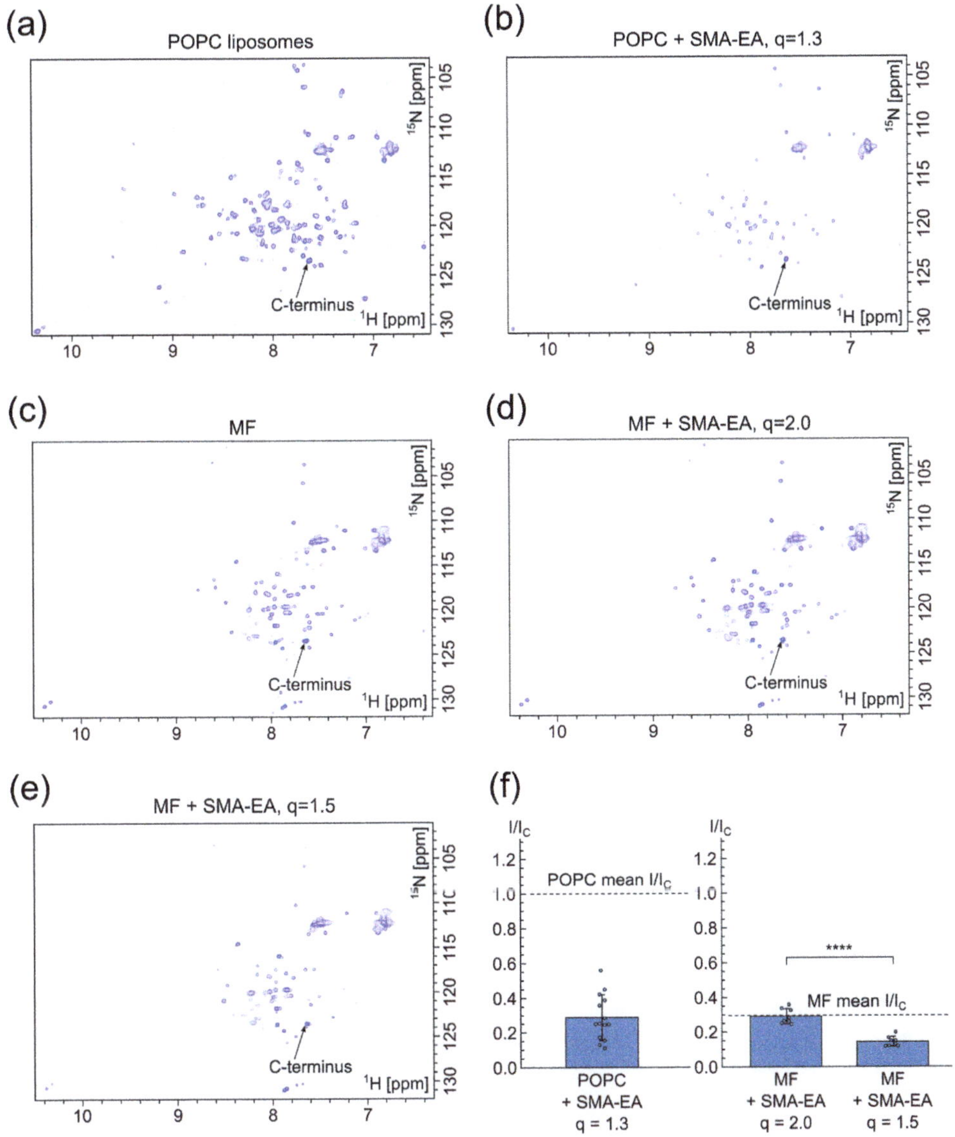

Figure 6. ^{1}H-^{15}N-HSQC spectra of p75-TM-ICD in different membrane mimetics: (**a**) in POPC liposomes; (**b**) in POPC/SMA-EA at q = 1.3; (**c**) in *E. coli* membrane fraction (MF); (**d**) in MF/SMA-EA at q = 2.0; (**e**) in MF/SMA-EA at q = 1.5; (**f**) relative change in cross-peak intensities in the ^{1}H-^{15}N-HSQC spectra of p75-TM-ICD after the SMA-EA addition.

The observed effect of intensity decrease upon SMA addition could be caused either by protein precipitation (since the effect is proportional to the amount of SMA added) or by the interaction between the ICD domain and SMA molecules. To test the first hypothesis, we compared the intensities of the signal corresponding to the amino group of the C-terminal residue before and after SMA addition. The intensities of this signal remained weakly affected (Figure S5a), ruling out protein precipitation. Therefore, we assume that there are transient interactions between the ICD domain and SMA molecules, leading to a decrease in ICD mobility.

To quantitatively assess the immobilization of the ICD, we calculated the average decrease in NMR cross-peak intensity of the protein amide groups. The values were normalized to the intensity of the C-terminal amide group signal and plotted in Figure 6f. In the case of the MF/SMA-EA sample, at q = 2.0, almost no intensity decrease is observed after SMA addition compared to q = 1.5, where signal intensities decreased twofold. In contrast, in the case of MF/SMA-tau, there is no significant difference in spectral quality for either q value (Figure S5d,e).

To summarize, we show here that both SMA-EA and SMA-tau can efficiently extract MPs from liposomes and cell membranes; however, the quality of the NMR spectra could be strongly affected by the protein–SMA interaction, as previously shown for SMA derivatives [34].

4. Discussion

The search for membrane mimetics that sufficiently reproduce the native membrane environment of MPs and at the same time are compatible with NMR, X-ray crystallography, and CryoEM remains a relevant task of modern structural biology. Bilayer-containing discoidal particles, bicelles and LPNs, have proven to be a suitable choice in several cases [1–12], but they have a number of disadvantages and limitations, including non-specific affinity, e.g., for water-soluble proteins. Therefore, the expansion of the existing set of available membrane mimetics is essential for the structural biology of membrane proteins. Analysis of the physical and chemical properties characteristic of hydrophobic particles formed in membrane mimetic solutions is also important to understand the limits of their applicability and to rationally select the mimetic for a particular membrane protein. In the current work, we applied the methodology previously developed in our group to investigate small isotropic bicelles [36,37,39] to study the structure and physical properties of SMALPs, a recently introduced and actively developed type of membrane mimetic [13,14,26,27,54–56].

First, we proposed the novel SMA derivative, SMA-tau. The polymer carries a negatively charged taurine moiety, an addition to the carboxyl group of SMA, which makes it a unique dianion in the set of available SMA variants [27–29]. We investigated the formation and properties of lipid–SMA-EA and lipid–SMA-tau membrane-mimicking particles. These SMALPs form particles with molecular weights greater than 150 kDa, which is similar to conventional lipid–protein nanodiscs. Using diffusion and ^{31}P NMR in solution, we have shown that the lipid–SMA particles exhibit bicelle-like behavior, and their structural parameters can be obtained from the ideal bicelle model [36]. Both SMAs form a rim that is at least two times thicker than the detergent rim of commonly used bicelles and three times thicker than the rim thickness of lipid–protein nanodiscs [57]. At the same time, the data obtained here differ from the values presented in previous works [58–62]. According to the measurements obtained by SANS/SAXS methods, the rim thickness varies in the range of 0.6–1.8 nm depending on the SMA modification. This discrepancy can be explained by one of the following possibilities: (1) the SMA derivatives obtained here form a special rim that is different from the previously studied SMA; (2) the packing is highly dependent on the q parameter and the ideal bicelle model we used does not describe this system correctly; (3) the double-cylinder model used to process the SAXS data does not allow obtaining the correct rim thickness values due to the smooth transition between the disk and rim areas of the particles. Considering that the SMA-EA modification has been characterized previously, the problem is most likely explained by the inaccuracy of one of the models. Nevertheless, the studied SMALPs contain a lipid bilayer patch that can undergo phase transitions at temperatures close to those of lipid bilayer membranes [51–53].

Next, we employed SMA-EA and SMA-tau to extract a fragment of the p75 receptor from POPC liposomes and the membrane fraction of *E. coli* cells. The extraction was successful to some extent; however, we observed undesired interactions between the water-soluble domains of p75 and the SMA rim of the lipodiscs, and similar effects were previously found for the modified SMA lipodiscs and cytochrome p450 [34]. These interactions were stronger in the case of SMA-tau, making the protein almost unobservable in the NMR spectra. We believe that there are some non-specific interactions between the soluble domain of p75 and SMA. Higher values of q correspond to larger bilayer sizes, resulting in a reduced probability of the contact between the p75 intracellular domain and the SMA rim. This explains why we observed a decrease in signal intensity as the q value decreased. Thus, we can provide additional evidence that the properties of the SMA rim are extremely important for the solubilization of the membrane proteins under investigation. In this regard, the introduction of SMA-tau could be a valuable development. For the particular case of p75NTR, SMA-tau does not provide a significant improvement over SMA-EA; however, the situation may be different for the other proteins. Therefore, the introduction of SMA-tau, with its unique physical and chemical properties, expands the existing palette of SMA derivatives and may be useful for structural studies of other membrane proteins with different molecular surface properties. In addition, the overall smaller size of SMA-tau–lipid particles compared to SMA-EA particles formed at the same q is potentially advantageous for NMR applications.

SMALPs have numerous advantages over micelles, bicelles, and LPNs. They allow avoiding the use of detergents that can unfold the target protein [14–22]. Moreover, SMA polymers can extract the protein directly from lipid membranes, preserving the native lipid environment. The properties of SMA polymers can be altered by various covalent modifications to adapt to different pH values. SMALPs have demonstrated their applicability in a number of studies of membrane proteins by CryoEM [30–32]. However, as with other membrane mimetics, SMALPs have their drawbacks and limitations, especially when it comes to their application for the needs of solution NMR spectroscopy. As we have found here, the size of SMA-EA SMALPs depends on the temperature, which is an undesirable phenomenon, considering that the object size is an important parameter that defines the quality of NMR spectra. Furthermore, for the low q values (<0.6 for SMA-EA and <1.0 for SMA-tau), where the relatively small particles are formed, we did not observe any phase transitions, indicating that the environment provided by these particles is far from native. Conversely, at high q values (>1 for SMA-EA and SMA-tau), excessively large particles were found, which are not applicable for solution NMR. Furthermore, the experiments with the p75 protein revealed that the ICD domain can interact non-specifically with the rim. Thus, although these SMA polymers have all the properties of beneficial membrane mimetics, their use in solution NMR is limited and varies with the system under investigation. Screening experiments are necessary for each particular protein.

Supplementary Materials: The following supporting information can be downloaded at https://www.mdpi.com/article/10.3390/polym16213009/s1, Table S1: Parameters of approximation DMPC/SMA R(q) dependencies using ideal bicelle model with cylindrical rim; Table S2: Phase transition temperatures in studied lipid–SMA samples; Figure S1: Number of lipid molecules at different q values in observed DMPC/SMA-EA and DMPC/SMA-tau particles estimated with ideal bicelle model; Figure S2: (a) Radii dependence on pH for DMPC/SMA-EA at q = 1.0 (w/v = 3%) and DMPC/SMA-tau at q = 1.0 (w/v = 3%) samples at 303 K and 313 K, (b) radii dependence on NaCl concentration for DMPC/SMA-EA q = 1.0, pH = 7.0 samples at 303 K and 313 K, (c) radii values for DMPC/SMA-EA q = 1.0, pH = 7.0, w/w 3% and 1.5% samples at 303 K and 313 K; Figure S3: Temperature dependence of chemical shifts in ^{31}P spectra for DMPC/SMA-EA (a) and DMPC/SMA-tau samples (b); Figure S4: Temperature dependence of chemical shifts for (a) POPC/SMA-EA and DPPC/SMA-EA (q = 1.5) and (b) for DMPC/DMPG/SMA-EA and DMPC/DMPG/SMA-tau samples (q = 2) at DMPC/DMPG ratios of 60/40, 70/30, and 80/20; Figure S5: ^{1}H-^{15}N-HSQC spectra of p75-TM-ICD in different membrane mimetics.

Author Contributions: Experiments: V.V.M. and S.O.K. Analyzing and interpreting: V.V.M. and E.F.K. Supervising: I.A.B., S.A.G. and K.S.M. Funding and resources: A.S.A. and E.V.B. Writing (original draft): V.V.M. and E.F.K. Writing (editing): all the authors. All authors have read and agreed to the published version of the manuscript.

Funding: This work was supported by the Russian Science Foundation grant No 22-14-00020.

Institutional Review Board Statement: Not applicable.

Data Availability Statement: All the data are available on request to the authors.

Conflicts of Interest: The authors declare no conflicts of interest.

References

1. Warschawski, D.E.; Arnold, A.A.; Beaugrand, M.; Gravel, A.; Chartrand; Marcotte, I. Choosing membrane mimetics for NMR structural studies of transmembrane proteins. *Biochim. Biophys. Acta (BBA)-Biomembr.* **2011**, *1808*, 1957–1974. [CrossRef]
2. Mineev, K.S.; Nadezhdin, K.D. Membrane mimetics for solution NMR studies of membrane proteins. *Nanotechnol. Rev.* **2016**, *6*, 15–32. [CrossRef]
3. Majeed, S.; Ahmad, A.B.; Sehar, U.; Georgieva, E.R. Lipid Membrane Mimetics in Functional and Structural Studies of Integral Membrane Proteins. *Membranes.* **2021**, *11*, 685. [CrossRef]
4. E Autzen, H.; Julius, D.; Cheng, Y. Membrane mimetic systems in CryoEM: Keeping membrane proteins in their native environment. *Curr. Opin. Struct. Biol.* **2019**, *58*, 259–268. [CrossRef]
5. Denisov, I.G.; Sligar, S.G. Nanodiscs for structural and functional studies of membrane proteins. *Nat. Struct. Mol. Biol.* **2016**, *23*, 481–486. [CrossRef]
6. Denisov, I.G.; Sligar, S.G. Nanodiscs in Membrane Biochemistry and Biophysics. *Chem. Rev.* **2017**, *117*, 4669–4713. [CrossRef]
7. Hagn, F.; Etzkorn, M.; Raschle, T.; Wagner, G. Optimized Phospholipid Bilayer Nanodiscs Facilitate High-Resolution Structure Determination of Membrane Proteins. *J. Am. Chem. Soc.* **2013**, *135*, 1919–1925. [CrossRef]
8. Hagn, F.; Nasr, M.L.; Wagner, G. Assembly of phospholipid nanodiscs of controlled size for structural studies of membrane proteins by NMR. *Nat. Protoc.* **2017**, *13*, 79–98. [CrossRef]
9. Sanders, C.R.; Prosser, R.S. Bicelles: A model membrane system for all seasons. *Structure* **1998**, *6*, 1227–1234. [CrossRef] [PubMed]
10. Sanders, C.R.; Landis, G.C. Reconstitution of Membrane Proteins into Lipid-Rich Bilayered Mixed Micelles for NMR Studies. *Biochemistry* **1995**, *34*, 4030–4040. [CrossRef] [PubMed]
11. Czerski, L.; Sanders, C.R. Functionality of a Membrane Protein in Bicelles. *Anal. Biochem.* **2000**, *284*, 327–333. [CrossRef] [PubMed]
12. Glover, K.J.; Whiles, J.A.; Wu, G.; Yu, N.-J.; Deems, R.; Struppe, J.O.; Stark, R.E.; Komives, E.A.; Vold, R.R. Structural Evaluation of Phospholipid Bicelles for Solution-State Studies of Membrane-Associated Biomolecules. *Biophys. J.* **2001**, *81*, 2163–2171. [CrossRef] [PubMed]
13. Tonge, S.; Tighe, B. Responsive hydrophobically associating polymers: A review of structure and properties. *Adv. Drug Deliv. Rev.* **2001**, *53*, 109–122. [CrossRef] [PubMed]
14. Knowles, T.J.; Finka, R.; Smith, C.; Lin, Y.-P.; Dafforn, T.; Overduin, M. Membrane Proteins Solubilized Intact in Lipid Containing Nanoparticles Bounded by Styrene Maleic Acid Copolymer. *J. Am. Chem. Soc.* **2009**, *131*, 7484–7485. [CrossRef]
15. Swainsbury, D.J.K.; Scheidelaar, S.; van Grondelle, R.; Killian, J.A.; Jones, M.R. Bacterial Reaction Centers Purified with Styrene Maleic Acid Copolymer Retain Native Membrane Functional Properties and Display Enhanced Stability. *Angew. Chem. Int. Ed.* **2014**, *53*, 11803–11807. [CrossRef]
16. Krishnarjuna, B.; Ravula, T.; Ramamoorthy, A. Detergent-free extraction, reconstitution and characterization of membrane-anchored cytochrome-b5 in native lipids. *Chem. Commun.* **2020**, *56*, 6511–6514. [CrossRef]
17. Stroud, Z.; Hall, S.C.; Dafforn, T.R. Purification of membrane proteins free from conventional detergents: SMA, new polymers, new opportunities and new insights. *Methods* **2018**, *147*, 106–117. [CrossRef]
18. Jamshad, M.; Lin, Y.-P.; Knowles, T.J.; Parslow, R.A.; Harris, C.; Wheatley, M.; Poyner, D.R.; Bill, R.M.; Thomas, O.R.; Overduin, M.; et al. Surfactant-free purification of membrane proteins with intact native membrane environment. *Biochem. Soc. Trans.* **2011**, *39*, 813–818. [CrossRef]
19. Hardy, D.; Bill, R.M.; Rothnie, A.J.; Jawhari, A. Stabilization of Human Multidrug Resistance Protein 4 (MRP4/ABCC4) Using Novel Solubilization Agents. *SLAS Discov. Adv. Sci. Drug Discov.* **2019**, *24*, 1009–1017. [CrossRef]
20. Grime, R.L.; Goulding, J.; Uddin, R.; Stoddart, L.A.; Hill, S.J.; Poyner, D.R.; Briddon, S.J.; Wheatley, M. Single molecule binding of a ligand to a G-protein-coupled receptor in real time using fluorescence correlation spectroscopy, rendered possible by nano-encapsulation in styrene maleic acid lipid particles. *Nanoscale* **2020**, *12*, 11518–11525. [CrossRef]
21. Dörr, J.M.; Koorengevel, M.C.; Schäfer, M.; Prokofyev, A.V.; Scheidelaar, S.; van der Cruijsen, E.A.W.; Dafforn, T.R.; Baldus, M.; Killian, J.A. Detergent-free isolation, characterization, and functional reconstitution of a tetrameric K^+ channel: The power of native nanodiscs. *Proc. Natl. Acad. Sci. USA* **2014**, *111*, 18607–18612. [CrossRef] [PubMed]
22. Brady, N.G.; Li, M.; Ma, Y.; Gumbart, J.C.; Bruce, B.D. Non-detergent isolation of a cyanobacterial photosystem I using styrene maleic acid alternating copolymers. *RSC Adv.* **2019**, *9*, 31781–31796. [CrossRef]
23. Arenas, R.C.; Danielczak, B.; Martel, A.; Porcar, L.; Breyton, C.; Ebel, C.; Keller, S. Fast Collisional Lipid Transfer Among Polymer-Bounded Nanodiscs. *Sci. Rep.* **2017**, *7*. [CrossRef]
24. Yamamoto, K.; Gildenberg, M.; Ahuja, S.; Im, S.-C.; Pearcy, P.; Waskell, L.; Ramamoorthy, A. Probing the Transmembrane Structure and Topology of Microsomal Cytochrome-P450 by Solid-State NMR on Temperature-Resistant Bicelles. *Sci. Rep.* **2013**, *3*, srep0255. [CrossRef]

25. Radoicic, J.; Park, S.H.; Opella, S.J. Macrodiscs Comprising SMALPs for Oriented Sample Solid-State NMR Spectroscopy of Membrane Proteins. *Biophys. J.* **2018**, *115*, 22–25. [CrossRef]
26. Scheidelaar, S.; Koorengevel, M.C.; van Walree, C.A.; Dominguez, J.J.; Dörr, J.M.; Killian, J.A. Effect of Polymer Composition and pH on Membrane Solubilization by Styrene-Maleic Acid Copolymers. *Biophys. J.* **2016**, *111*, 1974–1986. [CrossRef]
27. Ravula, T.; Hardin, N.Z.; Di Mauro, G.M.; Ramamoorthy, A. Styrene maleic acid derivates to enhance the applications of bio-inspired polymer based lipid-nanodiscs. *Eur. Polym. J.* **2018**, *108*, 597–602. [CrossRef]
28. Lindhoud, S.; Carvalho, V.; Pronk, J.W.; Aubin-Tam, M.-E. SMA-SH: Modified Styrene–Maleic Acid Copolymer for Functionalization of Lipid Nanodiscs. *J. Biol. Macromol.* **2016**, *17*, 1516–1522. [CrossRef]
29. Zhang, H.; Shou, C.; Li, C.-X.; Pan, J.; Xu, J.-H. Synthesis and modification of alternative copolymers for activating membrane enzymes in vitro. *Eur. Polym. J.* **2023**, *200*. [CrossRef]
30. Parmar, M.; Rawson, S.; Scarff, C.A.; Goldman, A.; Dafforn, T.R.; Muench, S.P.; Postis, V.L. Using a SMALP platform to determine a sub-nm single particle cryo-EM membrane protein structure. *Biochim. Biophys. Acta (BBA)-Biomembr.* **2017**, *1860*, 378–383. [CrossRef]
31. Qiu, W.; Fu, Z.; Xu, G.G.; Grassucci, R.A.; Zhang, Y.; Frank, J.; Hendrickson, W.A.; Guo, Y. Structure and activity of lipid bilayer within a membrane-protein transporter. *Proc. Natl. Acad. Sci. USA* **2018**, *115*, 12985–12990. [CrossRef]
32. Sun, C.; Benlekbir, S.; Venkatakrishnan, P.; Wang, Y.; Hong, S.; Hosler, J.; Tajkhorshid, E.; Rubinstein, J.L.; Gennis, R.B. Structure of the alternative complex III in a supercomplex with cytochrome oxidase. *Nature* **2018**, *557*, 123–126. [CrossRef]
33. Drewniak, P.; Al-Abdul-Wahid, S.; Ladizhansky, V.; Brown, L.S. Investigation into aquaporin 1's conserved structural features and the C-terminal tail. *Biophys. J.* **2023**, *122*, 194a. [CrossRef]
34. Ravula, T.; Hardin, N.Z.; Bai, J.; Im, S.-C.; Waskell, L.; Ramamoorthy, A. Effect of polymer charge on functional reconstitution of membrane proteins in polymer nanodiscs. *Chem. Commun.* **2018**, *54*, 9615–9618. [CrossRef]
35. Denisov, I.G.; McLean, M.A.; Shaw, A.W.; Grinkova, Y.V.; Sligar, S.G. Thermotropic Phase Transition in Soluble Nanoscale Lipid Bilayers. *J. Phys. Chem. B* **2005**, *109*, 15580–15588. [CrossRef]
36. Mineev, K.S.; Nadezhdin, K.D.; Goncharuk, S.A.; Arseniev, A.S. Characterization of Small Isotropic Bicelles with Various Compositions. *Langmuir* **2016**, *32*, 6624–6637. [CrossRef]
37. Kot, E.F.; Goncharuk, S.A.; Arseniev, A.S.; Mineev, K.S. Phase Transitions in Small Isotropic Bicelles. *Langmuir* **2018**, *34*, 3426–3437. [CrossRef]
38. Pardo, J.D.; Dörr, J.; Renne, M.; Ould-Braham, T.; Koorengevel, M.; van Steenbergen, M.; Killian, J. Thermotropic properties of phosphatidylcholine nanodiscs bounded by styrene-maleic acid copolymers. *Chem. Phys. Lipids* **2017**, *208*, 58–64. [CrossRef]
39. Kot, E.F.; Arseniev, A.S.; Mineev, K.S. Behavior of Most Widely Spread Lipids in Isotropic Bicelles. *Langmuir* **2018**, *34*, 8302–8313. [CrossRef] [PubMed]
40. Kislova, S.; Motov, V.; Myasnyanko, I.; Pytskii, I.; Goncharuk, S.; Boldyrev, I. Conformational transitions of maleic acid segment drive pH induced changes in SMA polymer structure and solubility. *J. Mol. Liq.* **2024**, *398*. [CrossRef]
41. Ravula, T.; Ramadugu, S.K.; Di Mauro, G.; Ramamoorthy, A. Bioinspired, Size-Tunable Self-Assembly of Polymer–Lipid Bilayer Nanodiscs. *Angew. Chem. Int. Ed.* **2017**, *56*, 11466–11470. [CrossRef] [PubMed]
42. Goncharuk, S.A.; Artemieva, L.E.; Nadezhdin, K.D.; Arseniev, A.S.; Mineev, K.S. Revising the mechanism of p75NTR activation: Intrinsically monomeric state of death domains invokes the "helper" hypothesis. *Sci. Rep.* **2020**, *10*, 1–15. [CrossRef]
43. Zheng, G.; Price, W.S. Simultaneous convection compensation and solvent suppression in biomolecular NMR diffusion experiments. *J. Biomol. NMR* **2009**, *45*, 295–299. [CrossRef]
44. Bragin, P.; Kuznetsov, A.; Bocharova, O.; Volynsky, P.; Arseniev, A.; Efremov, R.; Mineev, K. Probing the effect of membrane contents on transmembrane protein-protein interaction using solution NMR and computer simulations. *Biochim. Biophys. Acta (BBA)-Biomembr.* **2018**, *1860*, 2486–2498. [CrossRef]
45. Mineev, K.S.; Goncharuk, S.A.; Kuzmichev, P.K.; Vilar, M.; Arseniev, A.S. NMR Dynamics of Transmembrane and Intracellular Domains of p75NTR in Lipid-Protein Nanodiscs. *Biophys. J.* **2015**, *109*, 772–782. [CrossRef]
46. Nadezhdin, K.D.; García-Carpio, I.; Goncharuk, S.A.; Mineev, K.S.; Arseniev, A.S.; Vilar, M. Structural Basis of p75 Transmembrane Domain Dimerization. *J. Biol. Chem.* **2016**, *291*, 12346–12357. [CrossRef]
47. Franco, M.L.; Nadezhdin, K.D.; Light, T.P.; Goncharuk, S.A.; Soler-Lopez, A.; Ahmed, F.; Mineev, K.S.; Hristova, K.; Arseniev, A.S.; Vilar, M. Interaction between the transmembrane domains of neurotrophin receptors p75 and TrkA mediates their reciprocal activation. *J. Biol. Chem.* **2021**, *297*, 1009260. [CrossRef]
48. Di Mauro, G.M.; La Rosa, C.; Condorelli, M.; Ramamoorthy, A. Benchmarks of SMA-Copolymer Derivatives and Nanodisc Integrity. *Langmuir* **2021**, *37*, 3113–3121. [CrossRef]
49. Nakano, M.; Fukuda, M.; Kudo, T.; Miyazaki, M.; Wada, Y.; Matsuzaki, N.; Endo, H.; Handa, T. Static and Dynamic Properties of Phospholipid Bilayer Nanodiscs. *J. Am. Chem. Soc.* **2009**, *131*, 8308–8312. [CrossRef] [PubMed]
50. Triba, M.N.; Warschawski, D.E.; Devaux, P.F. Reinvestigation by Phosphorus NMR of Lipid Distribution in Bicelles. *Biophys. J.* **2005**, *88*, 1887–1901. [CrossRef] [PubMed]
51. Lewis, R.N.; Zhang, Y.-P.; McElhaney, R.N. Calorimetric and spectroscopic studies of the phase behavior and organization of lipid bilayer model membranes composed of binary mixtures of dimyristoylphosphatidylcholine and dimyristoylphosphatidylglycerol. *Biochim. Biophys. Acta (BBA)-Biomembr.* **2005**, *1668*, 203–214. [CrossRef] [PubMed]

52. Jiménez-Monreal, A.M.; Villalaín, J.; Aranda, F.J.; Gómez-Fernández, J.C. The phase behavior of aqueous dispersions of unsaturated mixtures of diacylglycerols and phospholipids. *Biochim. Biophys. Acta (BBA)-Biomembr.* **1998**, *1373*, 209–219. [CrossRef]
53. Biltonen, R.L.; Lichtenberg, D. The use of differential scanning calorimetry as a tool to characterize liposome preparations. *Chem. Phys. Lipids* **1993**, *64*, 129–142. [CrossRef]
54. Dörr, J.M.; Scheidelaar, S.; Koorengevel, M.C.; Dominguez, J.J.; Schäfer, M.; van Walree, C.A.; Killian, J.A. The styrene–maleic acid copolymer: A versatile tool in membrane research. *Eur. Biophys. J.* **2015**, *45*, 3–21. [CrossRef]
55. Juarez, J.F.B.; Harper, A.J.; Judge, P.J.; Tonge, S.R.; Watts, A. From polymer chemistry to structural biology: The development of SMA and related amphipathic polymers for membrane protein extraction and solubilisation. *Chem. Phys. Lipids* **2019**, *221*, 167–175. [CrossRef]
56. Chen, A.; Majdinasab, E.J.; Fiori, M.C.; Liang, H.; Altenberg, G.A. Polymer-Encased Nanodiscs and Polymer Nanodiscs: New Platforms for Membrane Protein Research and Applications. *Front. Bioeng. Biotechnol.* **2020**, *8*. [CrossRef]
57. Skar-Gislinge, N.; Simonsen, J.B.; Mortensen, K.; Feidenhans'l, R.; Sligar, S.G.; Møller, B.L.; Bjørnholm, T.; Arleth, L. Elliptical Structure of Phospholipid Bilayer Nanodiscs Encapsulated by Scaffold Proteins: Casting the Roles of the Lipids and the Protein. *J. Am. Chem. Soc.* **2010**, *132*, 13713–13722. [CrossRef]
58. Jamshad, M.; Grimard, V.; Idini, I.; Knowles, T.J.; Dowle, M.R.; Schofield, N.; Sridhar, P.; Lin, Y.; Finka, R.; Wheatley, M.; et al. Structural analysis of a nanoparticle containing a lipid bilayer used for detergent-free extraction of membrane proteins. *Nano Res.* **2014**, *8*, 774–789. [CrossRef]
59. Hall, S.C.L.; Tognoloni, C.; Charlton, J.; Bragginton, C.; Rothnie, A.J.; Sridhar, P.; Wheatley, M.; Knowles, T.J.; Arnold, T.; Edler, K.J.; et al. An acid-compatible co-polymer for the solubilization of membranes and proteins into lipid bilayer-containing nanoparticles. *Nanoscale* **2018**, *10*, 10609–10619. [CrossRef]
60. Bjørnestad, V.A.; Orwick-Rydmark, M.; Lund, R. Understanding the Structural Pathways for Lipid Nanodisc Formation: How Styrene Maleic Acid Copolymers Induce Membrane Fracture and Disc Formation. *Langmuir* **2021**, *37*, 6178–6188. [CrossRef]
61. Maier, R.; Arenas, R.C.; Zhang, F.; García-Sáez, A.; Schreiber, F. Structural Insights into Polymer-Bounded Lipid Nanodiscs. *Langmuir* **2023**, *39*, 2450–2459. [CrossRef]
62. Neville, G.M.; Morrison, K.A.; Shilliday, E.R.; Doutch, J.; Dalgliesh, R.M.; Price, G.J.; Edler, K.J. The effect of polymer end-group on the formation of styrene – maleic acid lipid particles (SMALPs). *Soft Matter* **2023**, *19*, 8507–8518. [CrossRef]

polymers

MDPI

Article

Preparation and Characterization of Melamine Aniline Formaldehyde-Organo Clay Nanocomposite Foams (MAFOCF) as a Novel Thermal Insulation Material

Ahmet Gürses [1,*] and Elif Şahin [2]

[1] Department of Chemistry Education, K.K. Education Faculty, Ataturk University, Erzurum 25240, Turkey
[2] Nano Science and Nano Engineering Department, Ataturk University, Erzurum 25240, Turkey; sahine@atauni.edu.tr
* Correspondence: ahmetgu@yahoo.com or agurses@atauni.edu.tr

Abstract: The main purpose of this study is to prepare a melamine aniline formaldehyde foam, an MAF copolymer, with lower water sensitivity and non-flammability properties obtained by the condensation reaction of melamine, aniline, and formaldehyde. In addition, the preparation of MAFF composites with organoclay reinforcement was determined as a secondary target in order to obtain better mechanical strength, heat, and sound insulation properties. For the synthesis of foams, the microwave irradiation technique, which offers advantages such as faster reactions, high yields and purities, and reduced curing times, was used together with the heating technique and the effect of organoclay content on the structural and textural properties of foams and both heat insulation and mechanical stability was investigated. Virgin melamine formaldehyde foam, MFF, melamine aniline formaldehyde foam, MAFFF, and melamine aniline formaldehyde–organoclay nanocomposite foams prepared with various organoclay contents, MAFOCFs, were characterized by HRTEM, FTIR, SEM, and XRD techniques. From spectroscopic and microscopic analyses, it was observed that organoclay flakes could be exfoliated without much change in the resin matrix with increasing clay content. In addition, it was determined that aniline formaldehyde, which is thought to enter the main polymer network as a bridge, caused textural changes in the polymeric matrix, and organoclay reinforcement also affected these changes. Although the highest compressive strength was obtained in MAFOCF5 foam with high organoclay content (0.40 MPa), it was determined that the compressive strengths in the nanocomposites were generally quite high despite their low bulk densities. In the prepared nanocomposite with 0.30% organoclay content (MAFOCF2), 0.33 MPa compressive strength and 0.051 thermal conductivity coefficient were measured. For virgin polymers and composites, bulk density, thermal conductivity, and compressive strength values were determined in the order of magnitude as MFF > MAFOCF1 > MAFOCF5 > MAFOCF6 > MAFF > MAFOCF3 > MAFOCF2 > MAFOCF4; MFF > MAFF > MAFOCF6 > MAFOCF5 > MAFOCF1 > MAFOCF4 > MAFOCF3 > MAFOCF2 and MAFOCF5 > MAFOCF4 > MAFOCF2 > MAFF > MAFOCF6 > MFF > MAFOCF1 > MAFOCF3. As a result, both compressive strength and thermal conductivity values indicate that nanocomposite foam with 0.20 wt% organoclay content can be a promising new insulation material.

Keywords: melamine formaldehyde; aniline formaldehyde; melamine aniline formaldehyde; organoclay; polymer nanocomposite; foam; thermal insulation; compressive strength

1. Introduction

Properties such as crack resistance, ductility, and thermal stability of thermoset resins, which exhibit hardness, brittleness, and low toughness due to excessive crosslink density, can be improved by adding various reinforcements such as nanofillers and -fibers and using the copolymerization process [1–4]. However, nanomaterials and -fibers in particular can easily agglomerate due to their high specific surface area fibrous structures, and preventing their agglomeration is extremely important for their effective adhesive interactions with polymer molecules [4–8]. Melamine formaldehyde (MF) resins, which have excellent thermal stability and high aging resistance, have advantages such as low cost and versatile usability as leather tanning and coating material, as well as high flame retardancy, advanced sound and heat insulation, low density, and high void volume [9,10]. They have higher application potential compared to traditional organic foams in fields such as construction, transportation, and aviation. Melamine formaldehyde prepolymers need to be cured by heating to gradually connect the existing oligomers with methylene and ether bridges to form a three-dimensional network structure. On the other hand, reinforcements such as nanoparticles and -fibers and different polymers have the ability to form a steric barrier against the proximity of triazine rings in the melamine formaldehyde polymer network. In particular, nanoparticles added at very low rates and uniformly dispersed in the polymeric matrix can absorb some of the impact energy generated by an external force and thus provide increased strength. In order to provide superior properties such as high strength, high thermal insulation performance, good wettability, high thermal stability, toughness, flame retardancy to melamine formaldehyde resins, clay minerals, various oxides, carbon nanotubes, graphite nanofillers, and cellulose nanofibers, oxide nanoparticles such as SiO_2, ZnO, Al_2O_3, and TiO_2 and metals such as Fe and Ni are widely used as reinforcements. In particular, organo-montmorillonite produced from montmorillonite clay, which has the advantages of low cost, abundance and prevalence, high aspect ratio, and excellent particle–surface interaction potential, can significantly improve the thermal and chemical stability and processability of thermosets [7,11–20]. On the other hand, the addition of flexible short-chain polymers such as starch, carbon fiber, polyvinyl alcohol, and polyurethane to melamine formaldehyde resin, which is highly brittle due to its molecular structure containing a large number of triazine rings and short flexible carbon chains, is also a common modification process. It has been observed that modified rigid closed-cell resins, whose flexibility is increased by increasing the flexible chain length between triazine rings, can exhibit better compressive strength, flexural strength, cellular structure integrity, thermal insulation performance, and flame resistance compared to unmodified melamine formaldehyde foams. Modifiers, which generally contain various functional groups that can react with hydroxy methyl groups or amino groups, can create less crosslink density than methylene linkage by forming links between triazine rings and thus increase the flexibility of the resin network [21,22]. There are many studies on various copolymers synthesized by condensing a mixture of phenol or various amines and formaldehyde [23,24]. These modified and copolymer resins have practical applications in many fields such as electronics, insulation materials, protective adhesives, and aerospace industries due to their high thermal stability, heat and chemical resistance, and electrical insulation properties. Melamine and aniline formaldehyde resins are well-known thermoset resins, and the most obvious advantage of aniline formaldehyde over melamine formaldehyde is its high resistance to water. Therefore, the presented study aims to characterize the surface morphological, structural, and textural properties of a new foam, both a terpolymer foam synthesized from melamine, formaldehyde, and aniline (MAFF) and its organoclay nanocomposite foams (MAFOCFs) with different organoclay ratios, and to investigate their thermal insulation behaviors and mechanical stability.

2. Materials and Methods

2.1. Materials

Raw montmorillonite (MMT), supplied from Karakayalar A.Ş. in Çankırı province, Turkey, with a specific surface area of 64.2 m^2/g and X-ray Fluorescence (XRF) composition given in Table 1, was used for the synthesis of organo-montmorillonite (OMMT) to be used in the preparation of nanocomposites [25].

Table 1. X-ray Fluorescence (XRF) chemical compositions of montmorillonite (MMT).

Component (%)									
SiO_2	Al_2O_3	Fe_2O_3	MgO	CaO	Na_2O	K_2O	TiO_2	SO_3	Other
59.32	17.19	5.95	3.63	2.21	1.68	0.97	0.74	0.51	7.81

Melamine, aniline (99.5% wt.), formaldehyde (37% wt.), NaOH, acetic acid, Tween 80, a nonionic surfactant (polyoxyethylene (20) sorbitan monooleate), and analytical-grade glycerin (all supplied by Merck KGaA, Darmstadt, Germany) were used to prepare melamine formaldehyde (MFF), melamine aniline formaldehyde (MAFF) and composite foams (MAFOCFs). In addition, gasoline, a mixture of isooctane, butane, and 3-ethyltoluene used for foaming, was also supplied by a local gas station.

2.2. Method

2.2.1. Organoclay Synthesis

Montmorillonite (MMT), a cationic surfactant, cetyltrimethylammonium bromide, and CTAB (Merck Co.) suspensions, a hydrocarbon material that is a product of petroleum refining and some of the properties of which are given in Table 2, were used to prepare organoclay using the solution intercalation method [25]. The procedure applied for the synthesis of organo-montmorillonite (OMMT) is the same as in our previous work.

Table 2. Some characteristics of the hydrocarbon material [25].

Density (15 °C), kg/m^3	Calorific Value MJ/kg	Flash Point °C	Water by Distillation, wt. %	C	H	N	S	Ash
990.70	42.74	105.80	0.10	83.40	11.90	0.80	1.50	0.03

2.2.2. Preparation of Melamine Formaldehyde Prepolymer (MF), Melamine Aniline Formaldehyde Prepolymer (MAF), and MAF–Organoclay Nanocomposites

A mixture of melamine, aniline (99.5 wt%), and formaldehyde (37 wt%) (mass ratio: 0.67:0.02:0.31) was placed in a flat-bottomed flask equipped with a thermometer and cooling equipment and heated with stirring at approximately 60 °C until a clear solution was obtained. Then, the pH of the mixture was adjusted to 8.5 with a concentrated NaOH solution and refluxed at approximately 95 °C for 1 h. After neutralization using concentrated acetic acid, the prepolymer containing 1.0 wt% glycerin, 1.0 wt% Tween 80, and 6.0 wt% gasoline was mechanically stirred for homogenization. Virgin melamine formaldehyde resin (MF) was also prepared by the same procedure using melamine and formaldehyde (37 wt%) in a mass ratio of 0.68:0.32. In the preparation of resin–organoclay nanocomposites by in situ synthesis, organoclay was added to the flask containing other components at weight ratios of 0.10, 0.15, 0.20, 0.30, 0.40, and 0.45%. For the foaming process, mixtures with and without organoclay reinforcement were exposed to microwave radiation in a microwave oven for 2 min in a suitable container. Finally, the mixtures were placed in a modular aluminum mold of 10.0 cm × 10.0 cm × 1.0 cm and heated in an oven at 140 °C for 1 h. The samples were prepared in duplicate. The compositions and codes of virgin MFF and MAFF and MAFOCFs are presented in Table 3.

Table 3. Codes and compositions of virgin MFF and MAFF and MAFOCF1-6 foams.

Sample No.	Sample Code	Organoclay wt. %
1	Melamine formaldehyde foam MFF	-
2	Melanin aniline formaldehyde foam MAFF	-
3	Melamine aniline formaldehyde–organoclay nanocomposite foam1 MAFOCF1	0.10
4	Melamine aniline formaldehyde–organoclay nanocomposite foam2 MAFOCF2	0.15
5	Melamine aniline formaldehyde–organoclay nanocomposite foam3 MAFOCF3	0.20
6	Melamine aniline formaldehyde–organoclay nanocomposite foam4 MAFOCF4	0.30
7	Melamine aniline formaldehyde–organoclay nanocomposite foam5 MAFOCF5	0.40
8	Melamine aniline formaldehyde–organoclay nanocomposite foam6 MAFOCF6	0.45

2.2.3. Spectroscopic and Microscopic Analyses and Measurements of Compressive Strength and Thermal Conductivity

Structural, crystallographic, and textural characterization of virgin MFF foam, virgin MAF foam, and MAF–organoclay nanocomposite foams were performed using spectroscopic techniques such as XRD and FTIR and microscopic techniques such as HRTEM and SEM.

XRD diffractograms for the prepared samples were obtained using a PANalytical Empyrean X-ray diffractometer with Cu Ka (1540 Å) radiation operating at 5 kV and 40 mA for 2 h in the 9–90° range and a scan rate of 4/min (Malvern Panalytical Ltd., Malvern, UK).

FTIR spectra were obtained using a Vertex 70 V FTIR spectrometer with a mid-IR ceramic source in the range of 4000 to 400 cm^{-1}, an average of 100 scans, and a resolution of 1 cm^{-1} (Bruker Optics Inc., Billerica, MA, USA).

HRTEM images of the samples were taken using a HITACHI HT7700 high-resolution transmission electron microscope (LaB6 filament) operating at 120.0 kV (Hitachi Ltd., Tokyo, Japan).

SEM (FEI-INSPECT S50 model) at 30 kV was used to take SEM patterns of virgin MFF, virgin MAFF, and MAF–organoclay nanocomposite foams.

To measure the compressive strength of virgin MFF and MAFF and MAFOCFs, a universal testing machine (Zwick/Roell) was used according to the DIN ISO 844:2009-10 standard [17,20].

A thermal conductivity meter (Quick Thermal Conductivity Meter QTM-500, Minami-ku, Kyoto, Japan) with a probe consisting of a single heating wire and a thermocouple was used to measure the thermal conductivity coefficients of virgin MFF and MAFF and MAFOCFs. The sample thickness was 8.0 mm, and the measurements were made in duplicate.

3. Results and Discussion

3.1. Textural Structure Analysis of Melamine Formaldehyde Foam (MFF), Melamine Aniline Foam (MAFF), and Melamine Aniline Copolymer Organoclay Nanocomposite Foams (MAFOCFs1-6)

For the textural characterization of melamine formaldehyde foam (MFF), melamine aniline copolymer foam (MAFF) foam, and melamine aniline copolymer organoclay nanocomposite foams (MAFOCFs1-6) prepared with various organoclay contents, HRTEM images are given in Figure 1.

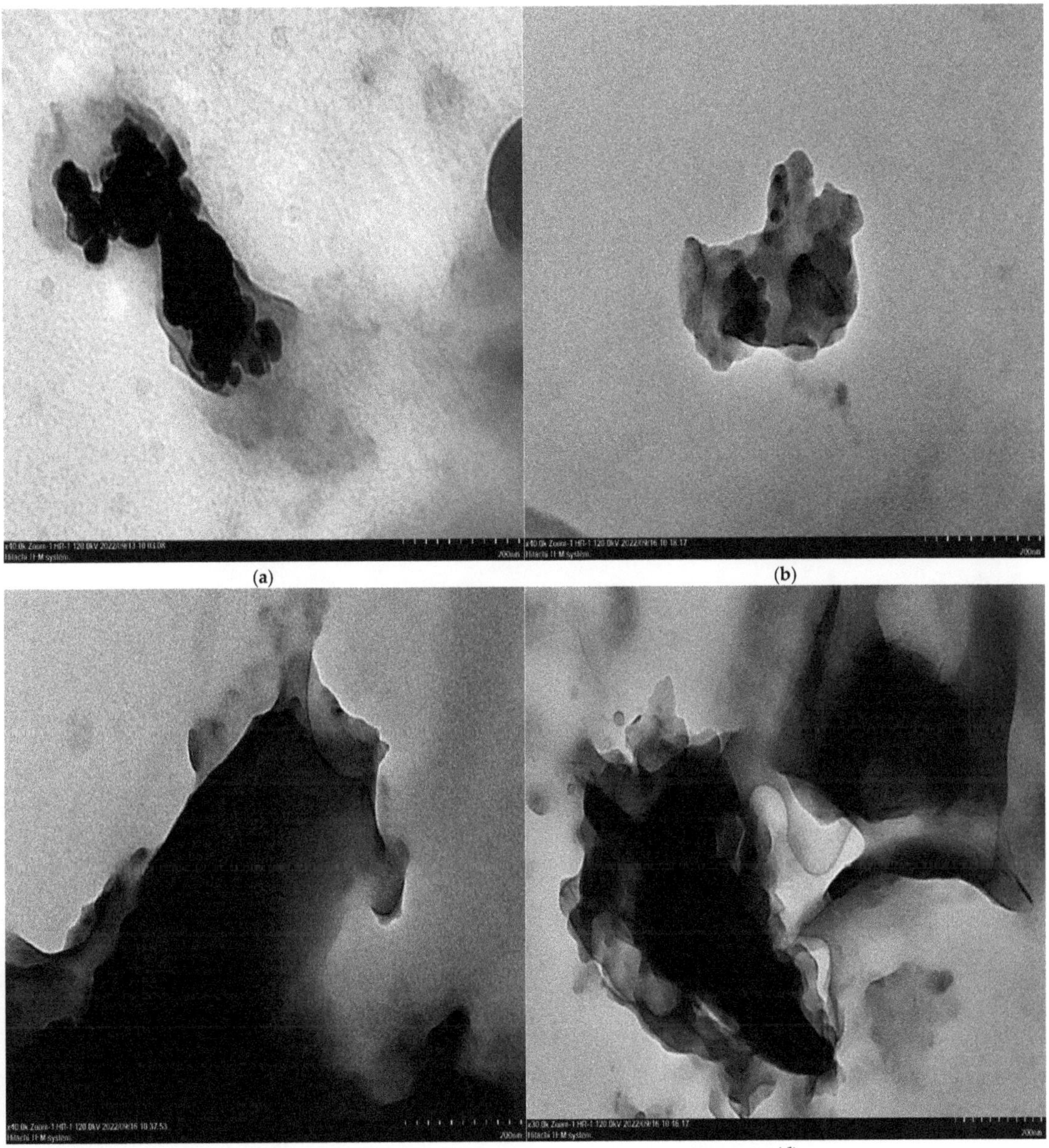

(a) (b) (c) (d)

Figure 1. *Cont.*

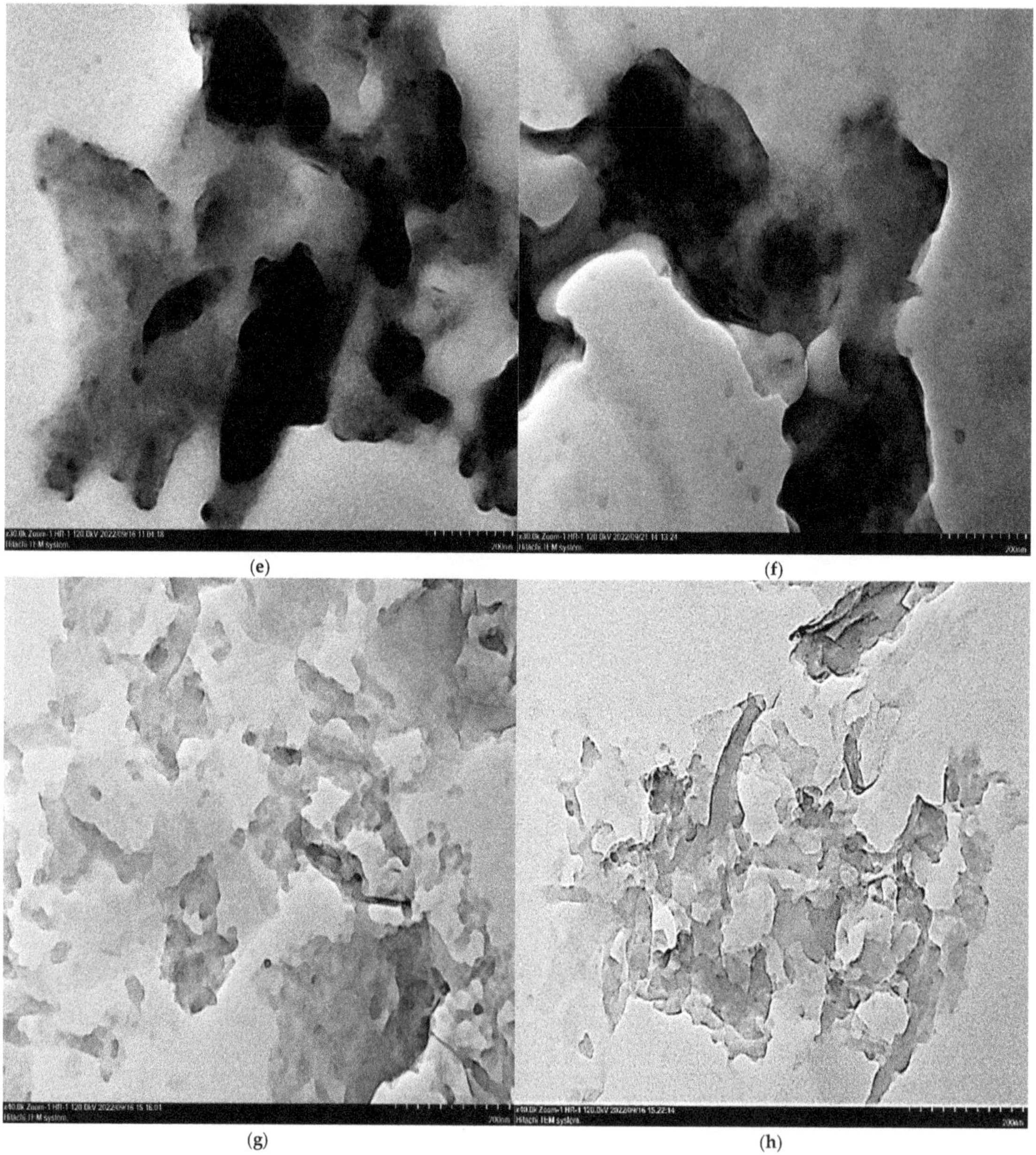

Figure 1. HRTEM images of virgin melamine formaldehyde foam (MFF) (**a**), virgin melamine aniline co-polymer foam (MAFF) (**b**), and melamine aniline copolymer organoclay nanocomposite foams (MAFOCFs1-6) (**c**–**h**).

From Figure 1a, it can be seen that in virgin MFF, the aggregated resin microspheres interact with each other and form a cluster-like growth in three dimensions. This was associated with the emergence of micro water vapor bubbles by the use of microwave irradiation before curing by heating. From Figure 1b, it can be seen that in virgin MAFF, a relatively planar structure emerges, reflecting the formation of aniline–formaldehyde bridges, which probably cause the triazine rings to be more distant from each other, as well as the presence of more effective particle interactions, compared to virgin MFF. The images of the composites prepared with various organoclay contents in Figure 1c–h show that increasing the organoclay content produces more pronounced textural changes compared to virgin MAFF, and the appearance of organoclay platelets, depicted as dark lines. From Figure 1d, it can be seen that despite the organoclay platelets tending to partially aggregate, they are predominantly exfoliated, resulting in a textural structure with different topographic forms. In Figure 1e,f, it is seen that, despite the higher organoclay contents, the organoclay platelets are almost completely exfoliated, which leads to the emergence of a more regular structure and a more planar appearance of the topographic forms. This situation is explained by the fact that aniline formaldehyde groups are included in the network in the form of bridging in the formation of the melamine formaldehyde polymeric network, thus causing the planar structure to dominate, in contrast to the possibility of increasing bending capabilities. In the images of the composites with higher organoclay contents (MAFOCF5 and MAFOCF6), a textural structure is observed in which the organoclay platelets are more homogeneously dispersed and more rectangular pyramidal topographic arrangements are formed in the resin.

3.2. Surface Morphological Analysis of Melamine Formaldehyde Foam (MFF), Melamine Aniline Copolymer Foam (MAFF), and Melamine Aniline Copolymer Organoclay Nanocomposite Foams (MAFOCFs1-6)

The SEM patterns taken for the surface morphological characterization of melamine formaldehyde foam (MFF), melamine aniline copolymer foam (MAFF) foam, and melamine aniline copolymer organoclay nanocomposite foams (MAFOCFs1-6) prepared with various organoclay contents are presented in Figure 2. The pattern of MFF in Figure 2a reflects a surface morphology indicating the resin microsphere clusters, grown in three dimensions and interacting with each other [17,20]. It can be seen from Figure 2b that MAFF exhibits a more regular morphology compared to MFF, and a network structure is formed in MAFF indicating the coherent association of relatively smaller resin microspheres. The surface morphology of MAFOCF with low organoclay content shows the formation of elliptical and probably tubular cavities and a surface morphology reflecting a regular network structure formed by the aggregation of very small resin microspheres with higher compliance, and significant changes compared to that of virgin MAFF (Figure 2c). The surface morphologies of MAFOCFs given in Figure 2d–g show that the aggregation of very small resin microspheres with higher compliance leads to the formation of a more holistic structure and the elliptical cavities gradually become circular. This shows that the organoclay platelets dispersed in the polymeric matrix are significantly effective in both the porosity development and the topographic orientations of the polymer skeleton. From Figure 2h, it can be seen that in the composite coded MAFOCF6, morphological formations in the form of defects with different network structures, probably resulting from the heterogeneous distribution of organoclay platelets, occur [26].

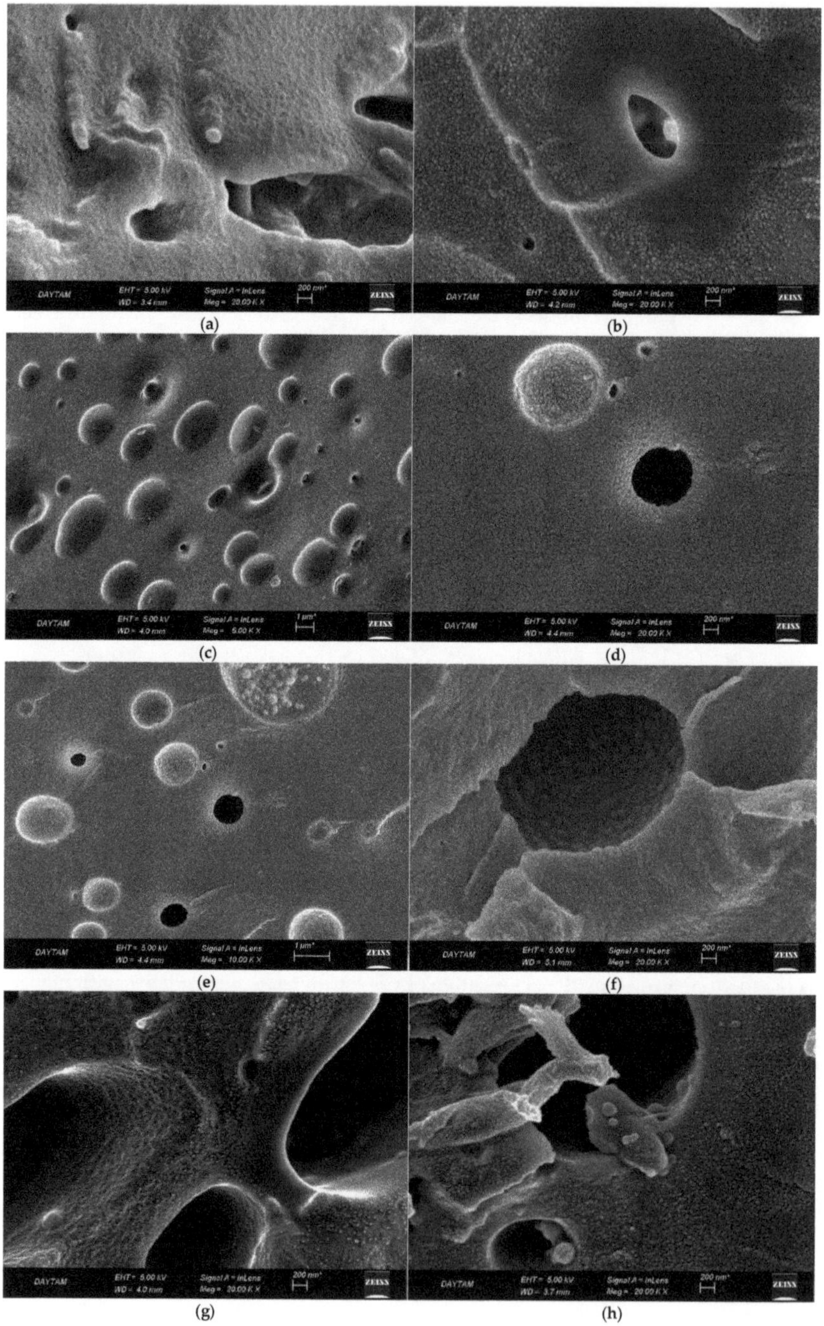

Figure 2. SEM patterns of virgin melamine formaldehyde foam (MFF) (**a**), virgin melamine aniline copolymer foam (MAFF) (**b**), and melamine aniline copolymer organoclay nanocomposite foams (MAFOCFs1-6) (**c**–**h**).

3.3. Structural Analysis of Melamine Formaldehyde Foam (MFF), Melamine Aniline Foam (MAFF), and Melamine Aniline Copolymer Organoclay Nanocomposite Foams (MAFOCFs1-6)

FTIR spectra of virgin MFF and MAFF and resin–organoclay nanocomposite foams (MAFOCF1, MAFOCF2, MAFOCF3, MAFOCF4, MAFOCF5, and MAFOCF6) are given in Figure 3.

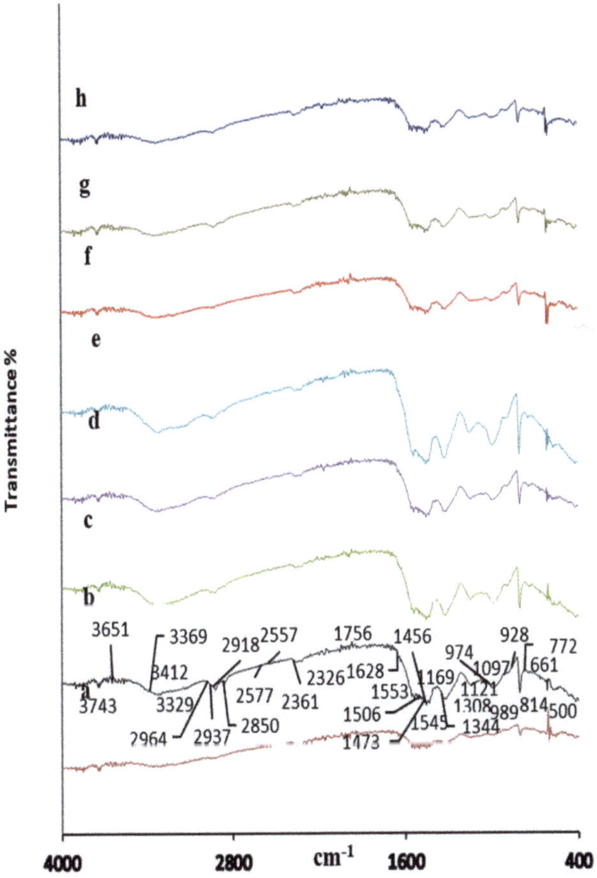

Figure 3. FTIR analysis of virgin melamine formaldehyde foam (MFF) (**a**), virgin melamine aniline formaldehyde foam (MAFF) (**b**), and melamine aniline formaldehyde-organoclay nanocomposite foams (MAFOCFs1-6) (**c**–**h**).

The peaks at 3329 cm^{-1}, 989 cm^{-1}, 1545 cm^{-1}, and 1456 cm^{-1} in the FTIR spectrum of MFF, MAFF, and also MAFOCF correspond to the stretching vibrations of N-H, O-H, C-O-C, and C-N bonds, respectively [27]. From Figure 3, it can be seen that in the spectrum of the MAF–organoclay nanocomposite foams (MAFOCFs), there was a strong peak at 3329 cm^{-1} originating from CTAB as well as the peaks at 2326 cm^{-1}, 2361 cm^{-1}, and 2937 cm^{-1}, which correspond to C–H anti-stretching, C–H stretching, and amine group stretching vibrations. The peaks observed between 999 cm^{-1} and 500 cm^{-1} correspond to the bonding vibrations of the main functional groups Si-O and Al-OH. It is observed that the intensity of these peaks increases with increasing organoclay content. The relatively strong, broad peak at 3329 cm^{-1} and the peak at 3743 cm^{-1} indicate the presence of hydroxyl and N–H bonds, and the peak intensities increased with increasing organoclay content [17,20].

On the other hand, the relatively low intensities of these peaks can be attributed to the intense interactions between the clay layers and CTA^+ ions bound to long-chain hydrocarbons. The peaks at 1545 cm^{-1} and 814 cm^{-1} are characteristic of the triazine ring of melamine. The fact that the spectra of MAFF and MAFOCFs are almost similar may imply that the addition of organoclay does not affect the functional groups on the polymer surface and that they are sterically effective. The peaks at 1558 cm^{-1} and 3400 cm^{-1} observed in the spectra of MAFOCFs correspond to cyclic aromatic C=C vibrations and N-H stretching, indicating the incorporation of aniline into the polymer backbone [28,29]. The peaks at 500 cm^{-1} and 444 cm^{-1} can be associated with Si–O–Al (octahedral Al) and Si-O–Si bending vibrations originating from organoclay. The peaks observed in the range of 2850–2964 cm^{-1} correspond to the antisymmetric and symmetric stretching of CH_2 groups of the CTA^+ ions on the organoclay surface [30].

3.4. XRD Analysis of Melamine Formaldehyde Foam (MFF), Melamine Aniline Copolymer Foam (MAFF), and Melamine Aniline Copolymer Organoclay Nanocomposite Foams (MAFOCFs1-6)

The XRD diffractograms of melamine formaldehyde foam (MFF), melamine aniline copolymer foam (MAFF), and melamine aniline copolymer organoclay nanocomposite foams (MAFONCFs1-6) are given in Figure 4. The two typical broad peaks at 8.0° and 21.9° in the XRD pattern of MFF indicate the amorphous structure of the resin [17,20]. The peaks at 18.4°, 21.1°, 26.3°, 28.0°, and 29.3° in the diffractogram of virgin MAFF indicate the incorporation of aniline and formaldehyde into the polymer network, probably in the form of bridging [31,32]. On the other hand, it can be seen in Figure 4 that the characteristic smectite peak observed at approximately 10° in organoclay-reinforced MAFOCsF1-6 partially overlaps with the smaller peak belonging to MF and this peak also shifts to the left in composites with low organoclay content. The relative shift in the characteristic smectite peak to the left indicates that the organoclay platelets in the polymer matrix are exfoliated. In addition, in the diffractograms of the resin–organoclay composite foams, the peaks that emerged with the addition of aniline formaldehyde to the amorphous MF polymer network are shown to have decreased in intensity, or some of them disappeared with organoclay reinforcement, which reveals the effect and decisive role of organoclay platelets in the change in crystallographic structure.

3.5. Analysis of Thermal Insulation Characteristics of Virgin Melamine Formaldehyde Foam (MF), Virgin Melamine Aniline Formaldehyde Foam (MAFF), and Melamine Aniline Formaldehyde–Organoclay Nanocomposite Foams (MAFOCFs1-6)

Thermal conductivity is an important property used to evaluate the thermal performance of a material under steady-state conditions. The thermal insulation performances of the prepared foams were evaluated by measuring their thermal conductivity coefficients [33–36]. Table 4 shows the bulk density and thermal conductivity values of virgin MFF and MAFF and resin–organoclay composite foams with different organoclay contents, such as MAFOCFs1-6. From this table, it can be seen that the lowest conductivity coefficients are exhibited by composite foams with high organoclay content, such as MAFOCF2, MAFOCF3, and MAFOCF4. This behavior, which may be related to the formation of a very regular and holistic textural structure, indicates the decisive effect of the exfoliated organoclay platelets on the formation of the polymeric network structure reflecting the void textural organization. It can be said that, especially in these composite foams, the transformation of the textural network formed by the aggregation of microspheres from sphericity to a relatively planar or layered structure may have led to the emergence of a more effective barrier system in terms of heat conduction with a more tortuous arrangement (see Figure 1c–e). The calculated bulk density values of 0.16, 0.17, and 0.15 g/cm^3 for these composite foams are significantly lower than those of virgin MFFs and MAFFs, respectively, confirming the significant and void-friendly textural changes and the formation of regular voids by the incorporation of organoclay platelets into the matrix. On the other hand, MAFOCF5 and -6-coded foams with higher organoclay content exhibited relatively higher thermal conductivity values compared to other composite foams, with lower thermal conductivity values compared to virgin MFF foam, which was attributed to possible textural changes leading to partial disruption of the regular void structure due to the increase in organoclay content. In particular, the better thermal insulation behavior of virgin MAFF compared to virgin MFF indicates the effect of a more uniformly packed microsphere arrangement and more elastic network structure including aniline–formaldehyde bridges on the formation of the void structure. The approximately 45% increase in bulk density observed in composites with higher organoclay content may indicate a significant disruption in the structural arrangement producing voids due to textural change.

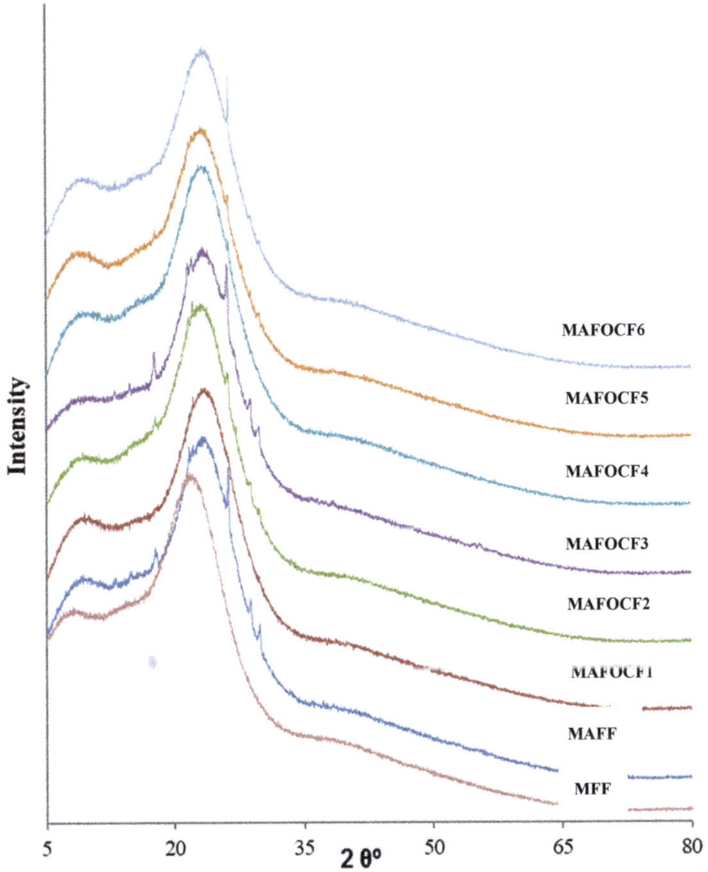

Figure 4. XRD diffractograms of virgin melamine formaldehyde foam (MF), virgin melamine aniline formaldehyde foam (MAFF), and melamine aniline formaldehyde–organoclay nanocomposite foams (MAFOCFs1-6).

3.6. Analysis of Compressive Strengths of Virgin Melamine Formaldehyde Foam (MF), Virgin Melamine Aniline Formaldehyde Foam (MAFF), and Melamine Aniline Formaldehyde–Organoclay Nanocomposite Foams (MAFOCFs1-6)

The compressive strength of materials is a critical design property measured under compressive loads and is also used to characterize the structure of composite materials [37,38]. There are differences in compressive fracture behavior between brittle materials, which are prone to splitting and sudden fracture, and ductile materials, which are prone to flow without fracture. Thermoset resins and their composites are brittle materials. In many materials, mechanical stability depends on porosity, and microcracks can grow until their dimensions are the same as the pores. Even at low porosity and hardness, microcracks can develop between weak interfaces under load, and these interfaces can propagate and form new cracks [39–41]. Table 5 shows the compressive strengths of virgin MFF and MAFF and composite foams prepared with various organoclay ratios, MAFOCFs1-6, obtained from measurements made on two replicates and 8.0 mm thick samples, together with standard deviation values. As can be seen from Table 5, the compressive strength of MAF foam is higher compared to MF foam. This means that the increase in elasticity caused by the inclusion of aniline–formaldehyde bridges in the melamine formaldehyde polymer network positively affects its mechanical stability, thus making the MAF foam more ductile. In addition, it can be seen from the same table that the foam with the highest organoclay content, MAFOCF6, has significantly lower compressive strength compared to other composite foams with both relatively high densities and high thermal conductivity coefficients. It is known that microcracks are the source of mechanical weakness in brittle materials and that slippage along the cracks causes tensile forces along the cracks. Microcracks, which develop for different reasons, usually occur close to the crack tips and can interact with microstructural anomalies, leading to energy accumulation. Therefore, the weakened mechanical strength of MAFOCF6 with high organoclay content can be attributed to microcracks that may have occurred due to anomalies such as planar textural changes and the formation of small polymeric fragments due to the dispersion of organoclay platelets dispersed in the resin (see Figure 2h).

Table 4. Bulk densities and thermal conductivities of virgin MFF and MAFF and composites, such as MAFOCFs1-6, prepared with various organoclay ratios.

Codes of Specimen	Bulk Density g/cm^3	Thermal Conductivity Coefficient λ, W/(m K)	Standard Deviation
MFF	0.26	0.083	0.001
MAFF	0.19	0.069	0.011
MAFOCF1	0.23	0.063	0.008
MAFOCF2	0.16	0.051	0.010
MAFOCF3	0.17	0.052	0.007
MAFOCF4	0.15	0.053	0.008
MAFOCF5	0.23	0.065	0.004
MAFOCF6	0.22	0.076	0.003

Table 5. Compressive strengths of MAFOCF1-6 composites with virgin MFF and MAFF with various organoclay ratios.

Codes of Specimen	Compressive Strength, MPa	Standard Deviation
MFF	0.27	0.020
MAFF	0.30	0.011
MAFOCF1	0.24	0.080
MAFOCF2	0.33	0.010
MAFOCF3	0.19	0.090
MAFOCF4	0.34	0.010
MAFOCF5	0.40	0.090
MAFOCF6	0.29	0.003

4. Conclusions

This study aims to prepare and analyze the structural and textural properties of melamine aniline formaldehyde copolymer organoclay nanocomposite foams and to investigate both their thermal insulation and strength properties.

Melamine aniline formaldehyde foam and melamine aniline formaldehyde–organoclay nanocomposites were successfully prepared. It was determined that the addition of organoclay caused significant changes in the textural structure of the resin.

In the lowest organoclay content nanocomposite foam, MAFOCF1, exfoliation and partial aggregation of organoclay occurred, but it was determined that this nanocomposite was similar to virgin MAFF in terms of textural and morphological features.

From HRTEM and SEM images, except for MAFOCF6, a more regular structural arrangement was observed with increasing organoclay content (MAFOCF2 and -5 nanocomposites), indicating more advanced organoclay exfoliation. From the SEM patterns, it was observed that the surface morphologies of MAFF with high organoclay content reflected the formation of a holistic structure by effective and harmonious aggregation of small microspheres, and the elliptical cavities gradually became circular.

From the XRD spectra of virgin MAFF and MAFOCFs, it was observed that in nanocomposites, the smaller of the two peaks of MAFF partially overlapped with the characteristic smectite peak, and the smectite peak was partially shifted to the left, implying intercalated or exfoliated dispersion of organoclay platelets.

The better thermal insulation behavior of virgin MAFF compared to virgin MFF indicates the effect of a more uniformly packed microsphere arrangement and more elastic network structure including aniline–formaldehyde bridges on the formation of void structure. The thermal insulation values gradually decreased with the increase in organoclay content in resin–organoclay nanocomposites.

The mechanical stability of virgin MAFF and resin–organoclay nanocomposites was relatively higher than that of virgin MFF, indicating that the increase in elasticity caused by the incorporation of aniline–formaldehyde bridges into the polymer network had a positive effect in terms of ductility and mechanical stability. Moreover, MAFOCF2-5-coded resin–organoclay nanocomposites exhibited significantly higher compressive strength compared to virgin resin foams and other composite foams despite having low density and low thermal conductivity coefficients.

Author Contributions: Conceptualization, A.G.; Methodology, E.Ş.; Validation, A.G. and E.Ş.; Investigation, E.Ş.; Resources, E.Ş.; Writing—original draft, A.G.; Funding acquisition, A.G. All authors have read and agreed to the published version of the manuscript.

Funding: This study was supported by Atatürk University Scientific Research Projects Coordination Unit (BAP) (Project Code: FDK-2022-10258). The authors are grateful to Atatürk University for their financial support.

Institutional Review Board Statement: Not applicable for studies not involving humans or animals.

Data Availability Statement: The original contributions presented in this study are included in the article. Further inquiries can be directed to the corresponding author.

Conflicts of Interest: The authors of this study declare that they have no conflict of interest.

References

1. Vaia, R.A.; Maguire, J.F. Polymer nanocomposites with prescribed morphology: Going beyond nanoparticle-filled polymers. *Chem. Mater.* **2007**, *19*, 2736–2751. [CrossRef]
2. Bashar, M.; Sundararaj, U.; Mertiny, P. Study of matrix micro-cracking in nano clay and acrylic tri-block-copolymer modified epoxy/basalt fiber-reinforced pressure-retaining structures. *Express Polym. Lett.* **2011**, *5*, 882e96. [CrossRef]
3. Chandgude, S.; Salunkhe, S. Biofiber-reinforced polymeric hybrid composites: An overview on mechanical and tribological performance. *Polym. Compos.* **2020**, *41*, 3908–3939. [CrossRef]
4. Taguet, A.; Cassagnau, P.; Lopez-Cuesta, J.M. Structuration, selective dispersion and compatibilizing effect of (nano) fillers in polymer blends. *Prog. Polym. Sci.* **2014**, *39*, 1526–1563. [CrossRef]
5. Idumah, C.I.; Obele, C.M. Understanding interfacial influence on properties of polymer nanocomposites. *Surf. Interfaces* **2021**, *22*, 100879. [CrossRef]
6. Guchait, A.; Saxena, A.; Chattopadhyay, S.; Mondal, T. Influence of nanofillers on adhesion properties of polymeric composites. *ACS Omega* **2022**, *7*, 3844–3859. [CrossRef]

7. Rajeshkumar, G.; Seshadri, S.A.; Ramakrishnan, S.; Sanjay, M.R.; Siengchin, S.; Nagaraja, K.C. A comprehensive review on natural fiber/nano-clay reinforced hybrid polymeric composites: Materials and technologies. *Polym. Compos.* **2021**, *42*, 3687–3701. [CrossRef]
8. Cho, J.; Joshi, M.S.; Sun, C.T. Effect of inclusion size on mechanical properties of polymeric composites with micro and nano particles. *Compos. Sci. Technol.* **2006**, *66*, 1941–1952. [CrossRef]
9. Wang, Y.; Chen, Z.; Lu, Y.; Yang, L.; Xu, T.; Wu, H.; Zhang, J.; He, L. A review of application, modification, and prospect of melamine foam. *Nanotechnol. Rev.* **2023**, *12*, 20230137. [CrossRef]
10. Resalati, H.; Hatam, A.; Dehghani Firouzabadi, M.R. Preparation, characterization and abrasion resistance property of melamine formaldehyde/montmorillonite nanocomposite coatings. *Prog. Color Color. Coat.* **2015**, *8*, 267–281.
11. Kotal, M.; Bhowmick, A.K. Polymer nanocomposites from modified clays: Recent advances and challenges. *Prog. Polym. Sci.* **2015**, *51*, 127–187. [CrossRef]
12. Bee, S.L.; Abdullah MA, A.; Bee, S.T.; Sin, L.T.; Rahmat, A.R. Polymer nanocomposites based on silylated-montmorillonite: A review. *Prog. Polym. Sci.* **2018**, *85*, 57–82. [CrossRef]
13. Guo, F.; Aryana, S.; Han, Y.; Jiao, Y. A review of the synthesis and applications of polymer–nanoclay composites. *Appl. Sci.* **2018**, *8*, 1696. [CrossRef]
14. Uddin, M.N.; Hossain, M.T.; Mahmud, N.; Alam, S.; Jobaer, M.; Mahedi, S.I.; Ali, A. Research and applications of nanoclays: A review. *SPE Polym.* **2024**, *5*, 507–535. [CrossRef]
15. Ray, S.S. Recent trends and future outlooks in the field of clay-containing polymer nanocomposites. *Macromol. Chem. Phys.* **2014**, *215*, 1162–1179. [CrossRef]
16. Annappa, A.R.; Basavarajappa, S.; Davim, J.P. Effect of organoclays on mechanical properties of glass fiber-reinforced epoxy nanocomposite. *Polym. Bull.* **2022**, *79*, 5085–5103. [CrossRef]
17. Gürses, A.; Şahin, E. Preparation of Melamine Urea Formaldehyde Organo Clay Nanocomposite Foams Using Thermal Processing and Microwave Irradiation Techniques and Investigation of Their Thermal Insulation and Compressive Strength. *Polym. Sci.* **2024**, *66*, 117–128.
18. Mishra, T.; Mandal, P.; Rout, A.K.; Sahoo, D. A state-of-the-art review on potential applications of natural fiber-reinforced polymer composite filled with inorganic nanoparticle. *Compos. Part C Open Access* **2022**, *9*, 100298. [CrossRef]
19. Omran, A.A.B.; Mohammed, A.A.; Sapuan, S.M.; Ilyas, R.A.; Asyraf, M.R.M.; Rahimian Koloor, S.S.; Petrů, M. Micro-and nanocellulose in polymer composite materials: A review. *Polymers* **2001**, *13*, 231. [CrossRef] [PubMed]
20. Gürses, A.; Şahin, E. Preparation of Melamine Formaldehyde Foam and a Melamine-Formaldehyde-Organo-Clay Nanocomposite and Hybrid Composites. *Minerals* **2023**, *13*, 1407. [CrossRef]
21. Mousavi, S.M.M.; Afra, E.; Tajvidi, M.; Bousfield, D.W.; Dehghani-Firouzabadi, M. Application of cellulose nanofibril (CNF) as coating on paperboard at moderate solids content and high coating speed using blade coater. *Prog. Org. Coat.* **2018**, *122*, 207–218. [CrossRef]
22. Kovacevic, Z.; Flincec Grgac, S.; Bischof, S. Progress in Biodegradable Flame-Retardant Nano-Biocomposites. *Polymers* **2021**, *13*, 741. [CrossRef]
23. Gürses, A.; Güneş, K.; Mindivan, F.; Korucu, M.E.; Açıkyıldız, M.; Doğar, Ç. The investigation of electrokinetic behaviour of micro-particles produced by CTA+ ions and Na-montmorillonite. *Appl. Surf. Sci.* **2014**, *318*, 79–84. [CrossRef]
24. Singla, P.; Mehta, R.; Upadhyay, S.N. Clay modification by the use of organic cations. *Green Sustain. Chem.* **2012**, *2*, 21–25. [CrossRef]
25. Acikyildiz, M.; Gurses, A.; Yolcu, H. Synthesis of super hydrophobic clay by solution intercalation method from aqueous dispersions. *Acta Phys. Pol. A* **2015**, *127*, 1156–1160. [CrossRef]
26. Rashid, M.; Islam, M.M.; Minami, H.; Aftabuzzaman, M.; Rahman, M.A.; Hossain, M.M.; Hoque, S.M.; Alam, M.A.; Ahmad, H. Nickel decorated melamine-formaldehyde resin/polyaniline composites for high specific capacitance. *Mater. Chem. Phys.* **2020**, *249*, 122957. [CrossRef]
27. Azarpour, H.; Moghadam, P.N.; Ehsanimehr, S.; Vahabi, H. Novel electrically conductive nanocomposites based on polyaniline and poly (aniline-co-melamine) copolymers grafted on melamine–formaldehyde resin. *Iran. Polym. J.* **2022**, *31*, 1033–1045. [CrossRef]
28. Gao, Y.; Liu, S.; Wang, Q.; Wang, G. Preparation of melamine–formaldehyde resin grafted by (3-aminopropyl) triethoxy silane for high-performance hydrophobic materials. *J. Appl. Polym. Sci.* **2020**, *137*, 48664. [CrossRef]
29. Ma, Y.; Zhang, W.; Wang, C.; Xu, Y.; Li, S.; Chu, F. Preparation and characterization of melamine modified urea-formaldehyde foam. *Int. Polym. Process.* **2013**, *28*, 188–198. [CrossRef]
30. Adesakin, A.O.; Ajayi, O.O.; Imosili, P.E.; Attahdaniel, B.E.; Olusunle, S.O.O. Characterization and evaluation of mechanical properties of dolomite as filler in polyester. *Chem. Mater. Res.* **2013**, *3*, 36–40.
31. Gürses, A.; Doğar, Ç.; Köktepe, S.; Mindivan, F.; Güneş, K.; Aktürk, S. Investigation of thermal properties of PUF/colored organoclay nanocomposites. *Acta Phys. Pol. A* **2012**, *127*, 979–983. [CrossRef]
32. Li, X.; Luo, J.; Li, J.; Gao, Q. Effects of diatomite inorganic fillers on the properties of a melamine–urea–formaldehyde resin. *J. Appl. Polym. Sci.* **2016**, *133*. [CrossRef]

33. Han, S.; Chen, Y.; Lyu, S.; Chen, Z.; Wang, S.; Fu, F. Effects of processing conditions on the properties of paraffin/melamine-urea-formaldehyde microcapsules prepared by in situ polymerization. *Colloids Surf. A Physicochem. Eng. Asp.* **2020**, *585*, 124046. [CrossRef]
34. Ivanova, N.M.; Visurkhanova, Y.A.; Soboleva, E.A. Structure and electrocatalytic properties of copper-containing aniline–melamine–formaldehyde polymer composites. *Russ. J. Appl. Chem.* **2018**, *91*, 396–40354. [CrossRef]
35. Amaral, C.; Vicente, R.; Ferreira, V.M.; Silva, T. Polyurethane foams with microencapsulated phase change material: Comparative analysis of thermal conductivity characterization approaches. *Energy Build.* **2017**, *153*, 392–402. [CrossRef]
36. Jia, X.; Wang, J.; Hou, C.; Tan, Y.; Zhang, Y. Effective insensitiveness of melamine urea-formaldehyde resin via interfacial polymerization on nitratmine explosives. *Nanoscale Res. Lett.* **2018**, *13*, 402. [CrossRef] [PubMed]
37. Dharkar, K.P.; Khamborkar, A.K.; Kalambe, A.B. Thermal Degradation Analysis of Melamine-Aniline-Formaldehyde Terpolymeric Ligand. *Res. J. Chem. Sci.* **2012**, *2*, 11–16.
38. Hatta, H.; Goto, K.; Aoki, T. Strengths of C/C composites under tensile, shear, and compressive loading: Role of interfacial shear strength. *Compos. Sci. Technol.* **2005**, *65*, 2550–2562. [CrossRef]
39. Liu, J.; Yue, K.; Xu, L.; Wu, J.; Chen, Z.; Wang, L.; Liu, W.; Lu, W. Bonding performance of melamine-urea–formaldehyde and phenol-resorcinol–formaldehyde adhesive glulams at elevated temperatures. *Int. J. Adhes. Adhes.* **2020**, *98*, 102500. [CrossRef]
40. Saba, N.; Jawaid, M.; Sultan, M.T.H. An overview of mechanical and physical testing of composite materials. In *Mechanical and Physical Testing of Bio Composites, Fibre-Reinforced Composites and Hybrid Composites*; Elsevier: Amsterdam, The Netherlands, 2019; pp. 1–12.
41. Jumahat, A.; Soutis, C.; Mahmud, J.; Ahmad, N. Compressive properties of nanoclay/epoxy nanocomposites. *Procedia Eng.* **2012**, *41*, 1607–1613. [CrossRef]

polymers

MDPI

Article

One-Pot Synthesis of Amphiphilic Linear and Hyperbranched Polyelectrolytes and Their Stimuli-Responsive Self-Assembly in Aqueous Solutions

Angelica Maria Gerardos [1,2], Aleksander Forys [3], Barbara Trzebicka [3] and Stergios Pispas [1,*]

[1] Theoretical and Physical Chemistry Institute, National Hellenic Research Foundation, 48 Vassileos Constantinou Avenue, 11635 Athens, Greece
[2] Department of Chemistry, National and Kapodistrian University of Athens, Panepistimiopolis, Zografou, 15771 Athens, Greece
[3] Centre of Polymer and Carbon Materials, Polish Academy of Sciences, 34 ul. M. Curie-Skłodowskiej, 41-819 Zabrze, Poland; aforys@cmpw-pan.pl (A.F.); btrzebicka@cmpw-pan.pl (B.T.)
* Correspondence: pispas@eie.gr

Abstract: Stimuli-responsive polymeric nanostructures are compelling vectors for a wide range of application opportunities. The objective we sought was to broaden the array of self-assembling amphiphilic copolymers with stimuli-responsive characteristics by introducing a hydrophilic tunable monomer, (2-dimethylamino)ethyl methacrylate (DMAEMA), together with a hydrophilic one, lauryl methacrylate (LMA), within linear and branched copolymer topologies. Size exclusion chromatography was used to evaluate the resultant linear and hyperbranched copolymers' molecular weight and dispersity, and FT-IR and ^{1}H-NMR spectroscopy techniques were used to delineate their chemical structure. The structural changes in the obtained self-organized supramolecular structures were thoroughly investigated using aqueous media with varying pH and salinity by dynamic light scattering (DLS), fluorescence spectroscopy (FS), and transmission electron microscopy (TEM). The nanoscale assemblies formed by the amphiphiles indicate significant potential for applications within the field of nanotechnology.

Keywords: amphiphilic copolymer; random copolymer; polyelectrolyte; hyperbranched

1. Introduction

Stimuli-responsive (or "smart") polymers represent a category of nanomaterials capable of undergoing reversible or irreversible alterations in their structure and physicochemical properties in response to various stimuli, both chemical and physical [1]. Numerous factors, including pH, temperature, and ion concentration, can trigger such changes. These environmental parameters can significantly influence the self-assembly processes in these materials [2]. Amphiphilic macromolecules, a significant class of such materials, are differentiated by the presence of both hydrophilic and hydrophobic moieties. When they are introduced into aqueous environments, this duality allows them to self-assemble into various nanostructures [3]. These macromolecules, akin to surfactants, create micelle-like structures, rendering them valuable for various applications, notably in nanomedicine, due to their capacity to facilitate hydrophobic compound encapsulation [4]. This process can be accomplished through non-covalent interactions such as hydrogen bonding and hydrophobic interactions. However, this approach often yields low encapsulation efficiencies, which may prematurely release the contents before they reach their intended environment in biomedical applications [5]. Macromolecules with modifiable properties are needed to overcome these limitations. Cationic macrosurfactants are given particular consideration among the amphiphilic compounds. Diverse cationic surfactants have been successfully developed, featuring various pendant groups, polymeric structures, etc. [6]. To achieve stimuli-responsive characteristics, the former is essential; amine-bearing polymeric structures are a desirable option because of their pH and temperature sensitivity. Striking examples are the (co)polymers based on poly[(2-dimethylamino)ethyl methacrylate] (PDMAEMA) and other alkylamine methacrylates, and poly(N-isopropylacrylamide). Moreover, the ability of DMAEMA's amine group to form complexes through hydrogen bonding and electrostatic interactions makes it a highly suitable candidate for utilization as a functional monomer [7]. The pH-dependency of the tertiary amino group protonation characteristics of PDMAEMA dictates the moiety's effective charge and, consequently, its water solubility. Additionally, it establishes the capacity to form complexes with anionic compounds, including biomolecules [8]. PDMAEMA exhibits thermoresponsiveness, whereby its lower critical solution temperature (LCST) is contingent upon its molecular weight and pH level. The presence of ionizable groups affects the LCST; specifically, it tends to decrease under basic conditions and increase in acidic environments. This phenomenon occurs due to the protonated state, which enhances the hydrophilicity of the polymer [9,10]. PDMAEMA-based copolymers have been effectively complexed with proteins [11], enzymes [12], and DNA [13]. Hydrophobic moieties are also important in producing structured nano-assemblies of amphiphiles. Bearing a long alkyl chain (n-dodecyl side chain), lauryl methacrylate (LMA) is a hydrophobic methacrylate ester-type monomer. The n-dodecyl chain provides ample side chain flexibility and high hydrophobicity compared to other alkyl methacrylates to facilitate spontaneous self-assembly. Moreover, LMA is appealing from an environmental standpoint and suitable for naturally produced copolymers as it can be made from plant oils, a renewable resource [14,15]. Hyperbranched polymers are gaining traction as a viable option for nanotechnology applications, leveraging the unique properties of copolymer topology. These materials are often more cost-effective than dendrimers while still providing similar functionalities, such as a high number of end and in-chain groups that can effectively interact and form complexes with various molecules [16]. Moreover, these polymers can be efficiently produced through a one-pot method. Established in 1998, reversible addition–fragmentation chain transfer (RAFT) polymerization has gained widespread recognition as a valuable one-pot technique for synthesizing complex macromolecules [17]. Its controlled polymerization characteristics make RAFT polymerization indispensable for developing sophisticated polymer topologies and multifunctional copolymers [18].

PDMAEMA has been successfully paired with various alkyl methacrylates, resulting in linear diblock [19,20] and triblock copolymers forming stimuli-responsive nanostructures in solutions. Our group has previously focused on the synthesis of the triblock copolymer PDEGMA-b-PDMAEMA-b-PLMA, which self-assembles in aqueous conditions to create spherical micellar structures. These structures consist of a stimuli-responsive PDMAEMA and PDEGMA corona surrounding a hydrophobic PLMA core [21]. In a similar fashion, PDMAEMA-b-PLMA-b-POEGMA amphiphilic triblock terpolymers have successfully encapsulated curcumin within spherical micelles [22]. Additionally, Jung et al. developed a triblock copolymer composed of hydrophilic PDMAEMA and PEG, which is end-capped by thermoresponsive tri(ethylene glycol) methyl ether methacrylate (TEGMA) units and hydrophobic n-butyl methacrylate (BuMA) moieties, resulting in multiresponsive micelles [23]. Similarly, Constantinou et al. synthesized terpolymers consisting of TEGMA, BuMA, and DMAEMA units resulting in spherical micelles [24]. Fan et al. synthesized ternary random copolymers that incorporate hydrophobic hexyl methacrylate moieties along with ionizable units of DMAEMA and MAA resulting in pH-sensitive micelles [25]. Non-linear copolymers are less researched. Skandalis et al. synthesized, utilizing the "arm first" method, amphiphilic mikto-arm $PDMAEMA_xPLMA_y$ stars [26]. In order to fully explore the potential of amphiphilic materials, targeted research should focus on simplified systems that effectively demonstrate their advantageous properties. Additionally, it is crucial to develop straightforward and efficient methods for synthesizing these novel polymer materials. By doing so, we can facilitate their scalability for a multitude of applications.

This work reports the synthesis of four amphiphilic random/statistical copolymers composed of DMAEMA and LMA, including two with a linear structure and two featuring a hyperbranched topology, utilizing RAFT polymerization. Research was conducted to assess the influence of pH variations and ionic strength on the behavior of the stimuli-sensitive polymeric nanostructures. This study is particularly significant as it explores, beyond the synthesis, the impact of macromolecular composition and non-linear, non-blocky architecture on the self-assembly process and the formation of amphiphilic copolymer nanostructures in aqueous media under different conditions.

2. Materials and Methods

2.1. Materials

DMAEMA, LMA, hydroquinone monomethyl ether (MEHQ) inhibitor remover, 2,2 azobisisobutyronitrile (AIBN), 4-cyano-4-(phenyl-carbonothioylthio)-pentanoic acid (CPAD), pyrene, and all solvents, including 1,4-dioxane (99.8% pure) and tetrahydrofuran (THF), were supplied by Sigma Aldrich (St. Louis, MO, USA). Ethylene glycol dimethacrylate (EGDMA) monomer, utilized as the branching agent, was purchased from Merck (Darmstadt, Germany). Water for injection (WFI) was purchased from DEMO ABEE (Athens, Greece). The monomers were purified using a column packed with MEHQ inhibitor remover. Methanol was used to recrystallize AIBN. All the solvents used were of analytical grade.

2.2. Synthesis of Amphiphilic Copolymers

Linear P-(LMA-co-DMAEMA) was synthesized via RAFT polymerization (Figure 1). Briefly, AIBN (radical initiator), CPAD (chain transfer agent), 1,4-dioxane (reaction solvent), DMAEMA (hydrophilic monomer), and LMA (hydrophobic monomer) were added into a round-bottomed flask in appropriate quantities (see Table 1). The contents of the flask were then stirred vigorously utilizing a magnetic stirrer to ensure thorough mixing. Subsequently, the solution was de-aerated by bubbling with N_2 for 20 min. Then the flask was set in an oil bath at 70 °C under magnetic stirring and the polymerization was allowed to proceed for 24 h. The mixture was then cooled to freezing temperature (~−20 °C) to quench the reaction. Finally, the frozen product was exposed to the ambient atmosphere to complete the polymerization process. The product was precipitated in a tenfold volume of cold hexane to remove unreacted monomers and other impurities. The precipitated polymer was separated by decantation, collected using THF, and dried in a vacuum oven for 72 h at an ambient temperature. The hyperbranched analogs were synthesized in the same manner (see Figure 1), with the exception of the addition of EGDMA as the branching agent in a 1.2:1 molar ratio to CPAD.

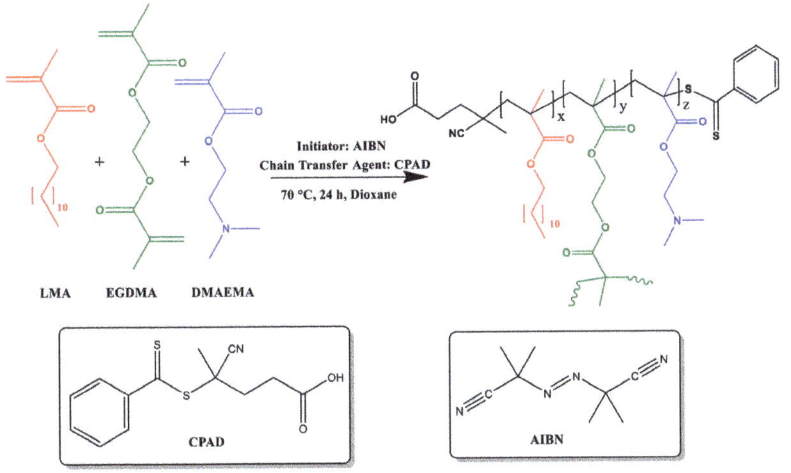

Figure 1. Synthesis route of hyperbranched H-P(LMA-co-DMAEMA) copolymer.

Table 1. Polymer synthesis and characterization data.

Polymer	Code	mmol Ratios [a]	M_w (g/mol) ($\times 10^4$) [b]	M_w/M_n [b]	%wt DMAEMA [c]
P-(LMA-co-DMAEMA)	P1	3.9:6.4:0.2:0.1	9.7	1.17	56
P-(LMA-co-DMAEMA)	P2	2.4:8.9:0.2:0.1	7.6	1.2	70
H-P(LMA-co-DMAEMA)	H1	3.9:6.4:0.24:0.1:0.2	14.9	1.4	54
H-P(LMA-co-DMAEMA)	H2	2.4:8.9:0.24:0.1:0.2	20.1	1.9	70

[a] LMA:DMAEMA:CPAD:AIBN:EGDMA; [b] determined by SEC; [c] determined by ^{1}H-NMR. P denotes linear copolymer and H hyperbranched copolymer.

2.3. Colloidal Dispersion Preparation

Nanoprecipitation was selected as the aqueous polymer solution preparation technique because of its many benefits. The technique calls for two miscible solvents, one of which is typically water in excess. This technique's fundamental concept involves the transition phase that occurs upon dispersion of the polymer solution in organic solvent into the aqueous medium. The polymer is dissolved in THF and afterwards, this solution is agitated and "violently" mixed/dispersed in an aqueous medium. The nanoparticles develop rapidly in an effort to evade the water molecules, thus forming nanodroplets in the aqueous phase. The organic solvent eventually evaporates with heat, creating an aqueous colloidal solution with newly formed nanoparticles [27]. Each copolymer was dissolved in THF ($C_{polymer} = 5 \times 10^{-3}$ g/mL). The solution was then injected into the appropriate volume of distilled water while being vigorously stirred at 55 °C. To ensure proper THF evaporation, heat was applied for a minimum of two hours before being left overnight at room temperature. Water was added as needed. The final copolymer concentration in every sample was 10^{-3} g/mL in 15 mL total volume of water.

2.4. *Methods*

2.4.1. Size Exclusion Chromatography

Each copolymer's molecular weight and molecular weight distribution were determined by employing a Waters SEC system. Three μ-Styragel mixed bed columns (with pore sizes varying from 10^2 to 10^6 Å), a Waters 1515 isocratic pump, and a Waters 2414 refractive index detector (equilibrated at 40 °C) constitute the system. The eluent, THF, was set at a flow rate of 1.0 mL/min and contained 5% triethylamine as a mobile phase additive. The average molecular weights of the linear monodisperse polystyrene standards used to calibrate the column set ranged from 1200 g mol^{-1} to 152,000 g mol^{-1}. The data were collected and analyzed using the Waters Breeze software (Breeze v2.0, Waters Corporation, Milford, MA, USA).

2.4.2. Proton Nuclear Magnetic Resonance Spectroscopy

Vnmrj software (VNMRJ 2.2C, Varian, Palo Alto, CA, USA) was utilized to acquire spectra using a Varian 300 (300 MHz) spectrometer. Both linear and hyperbranched polymers were dissolved in $CDCl_3$ ($C_{polymer}$ = 1–4 mg/mL). Parts per million (ppm) are used to express chemical shifts, with TMS acting as an internal reference. MestReNova software (MestReNova 14.0.0, Mestrelab Solutions, Bajo, Spain) was used to analyze the acquired spectra.

2.4.3. Fourier Transform Infrared Spectroscopy

A Fourier transform instrument (Bruker Equinox 55, Bruker Optics GmbH, Ettlingen, Germany) equipped with a single-bounce attenuated total reflectance (ATR) diamond accessory (Dura-Samp1IR II by SensIR Technologies, Chapel Hill, NC, USA) was used to obtain mid-infrared spectra in the 500–4000 cm^{-1} region. On average, 100 scans were obtained with a resolution of 2 cm^{-1}. A press was used to measure the solid polymers.

2.4.4. Dynamic Light Scattering

An ALV/CGS-3 Compact Goniometer System (ALV GmbH, Hessen, Germany) was used to conduct DLS studies. The JDS Uniphase 22 mW He–Ne laser in this system operates at a wavelength of 632.8 nm. The goniometer was fixed at a 90° measurement angle. The system was connected to a digital ALV-5000/EPP multitau correlator with 288 channels. The cumulant method and the CONTIN algorithm were used to analyze the autocorrelation functions, averaging five measurements in 30 s duration each. Hydrophilic syringe PVDF filters (0.45 μm pore size) were used to filter all samples before conducting measurements.

2.4.5. Electrophoretic Light Scattering

The results were obtained using a Nano ZetaSizer system (Malvern, Worcestershire, UK) equipped with a 4 mW He–Ne laser that operated at a wavelength of 633 nm and a scattering angle of 173°. The data established herein represent an average of 10 consecutive scans at ambient temperature. The Smoluchowski equation was used to analyze the collected data.

2.4.6. UV–Vis Spectroscopy

A Perkin Elmer Lambda 19 UV–Vis–NIR spectrometer (Perkin Elmer, Waltham, MA, USA) was used to record the UV–Vis spectra in the 200–600 nm range. The scan speed was set to 240 nm/min for each measurement. Given the utilization of a double-beam spectrometer, a reference cuvette containing the dispersion medium served as the reference for all measurements.

2.4.7. Fluorescence Spectroscopy

A Spectrofluorometer Fluorolog-3 Jobin Yvon-Spex (model GL3-21, Horiba, Kyoto, Japan) was used to record fluorescence spectra at ambient temperature throughout a wavelength range of 350–700 nm. The excitation wavelength was set at 335 nm, and the emission and excitation slits were both set at 2 nm in order to acquire pyrene spectra. The intensity of the first peak (I_1) in the pyrene emission spectrum was divided by the intensity of the third peak (I_3) to determine the I_1/I_3 ratio. A series of 11 copolymer concentrations were prepared with consecutive dilutions, with polymer concentrations ranging from 10^{-3} g mL^{-1} to 10^{-8} g mL^{-1}. Each sample contained 0.1% v/v of a 1 mM pyrene solution in acetone (c_{pyrene} = 1 μM).

2.4.8. Transmission Electron Microscopy

Transmission electron microscopy (TEM) images were obtained using a Tecnai F20 X TWIN microscope (FEI Company, Hillsboro, OR, USA) equipped with a field emission gun, operating at an acceleration voltage of 200 kV. Images were recorded on the Gatan Rio 16 CMOS 4 k camera (Gatan Inc., Pleasanton, CA, USA) and processed with Gatan Microscopy Suite (GMS) software version 3.5 (Gatan Inc., Pleasanton, CA, USA). For each measurement, 6 μL of solution was placed on a copper grid covered with carbon film and air dried at room temperature before conducting measurements.

3. Results and Discussion

3.1. Size Exclusion Chromatography Analysis

Figure 2 illustrates the synthesized copolymers' SEC traces, validating each synthetic route's success (see Table 1). Specifically, the linear copolymers produced a single sharp peak with a narrow distribution indicating unimodal polymer molecular weight distributions. On the contrary, the hyperbranched analogs generated monomodal molecular weight distributions with significantly larger polydispersity indexes. The SEC results show significantly higher weight-average molecular weights than stoichiometric calculations. This is due to the hyperbranched copolymer's globular structure, which causes a significant difference in hydrodynamic volume compared to the linear standards used for calibration [28]. The dispersity values of the hyperbranched copolymers were higher than those of the linear analogs due to the branching process during polymerization.

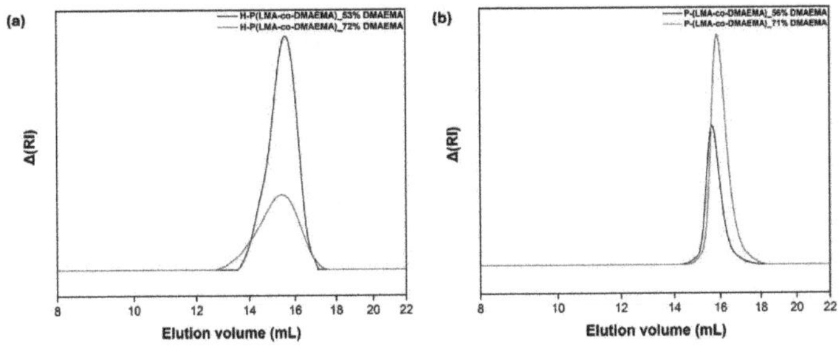

Figure 2. SEC curves of the hyperbranched (**a**) and linear (**b**) LMA/DMAEMA copolymers.

3.2. Proton Nuclear Magnetic Resonance Spectroscopy Analysis

The chemical composition and structural identity of the aforementioned copolymers were determined qualitatively and quantitatively via proton NMR spectroscopy as shown in Figure 3. More specifically, stoichiometric compositions were calculated based on the major peaks on the NMR spectra, namely the -CH_3 of the dimethylamino group of

DMAEMA found at 2.28 ppm and the -CH_2- of the LMA chain found at 1.27 ppm. The experimental compositions were estimated to be nearly identical to the targeted values. Yet, due to the similar chemical environment of the different hydrogen atoms present, quantitative analysis was notably challenging, a fact that justifies the deviation from the theoretical values.

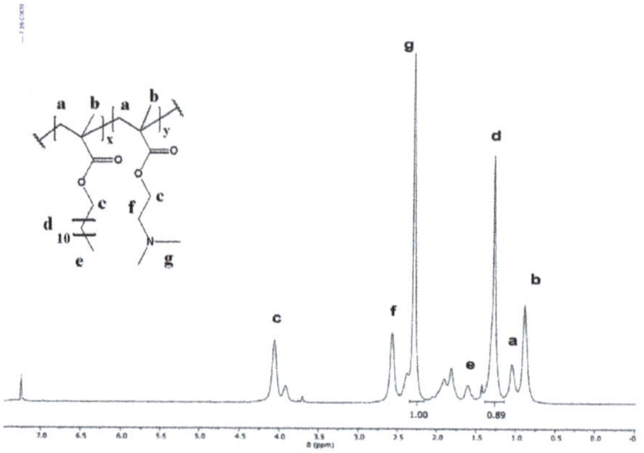

Figure 3. ^{1}H-NMR spectrum of P2 in $CDCl_3$. Letters above the spectra peaks correspond to the relevant H nuclei shown in the chemical structure of the copolymer in the inset.

3.3. *Fourier Transform Infrared Spectroscopy Analysis*

FT-IR in conjunction with ^{1}H-NMR spectra further confirm the chemical composition of the resulting copolymers. Beginning from the right, the primary peaks seen in Figure 4 were as follows: The C-O-C methacrylate stretch and the stretching vibration of the C-N bond in the DMAEMA unit correspond to the split in the 1145–1180 cm^{-1} range. The double intermediate peak (1240–1260 cm^{-1}) can be attributed to C-H wagging or rocking vibration. C-H scissoring accounts for the 1460 cm^{-1} medium peak. The stretching vibration of C-O was linked to the sharp distinctive peak at 1765 cm^{-1}. The characteristic bands of P(DMAEMA) are associated with (C-H(-N(CH_3)$_2$) stretching and are situated between 2820 and 2710 cm^{-1}. The strong peak at 2925 cm^{-1} indicates asymmetric C-H stretching, while the neighboring peak represents symmetric stretching in the same group. Furthermore, the FT-IR spectra's lack of a distinct C=C stretch band indicates that the monomers—including EGDMA vinyl units—are entirely consumed during polymerization [29,30].

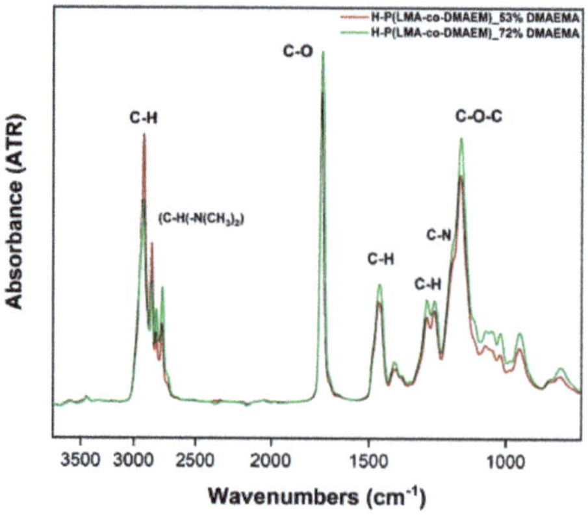

Figure 4. ATR-FTIR spectra of the hyperbranched copolymers (in the solid state).

3.4. Zeta Potential Analysis

In colloidal suspensions, zeta potential is a key parameter that influences biodistribution and the nanosuspensions' stability. High zeta potential levels can enhance electrostatic repulsion forces between particles that help to maintain the stability of nanoformulations [31,32]. The existence of tertiary amine groups in the DMAEMA segments is responsible for the high positive values (see Table 2) observed throughout all samples. These suspensions are perfect for applications in nanomedicine since formulations with zeta potential values in this range (larger than a 30 mV absolute value) are regarded as pharmaceutically stable [33]. A high zeta potential value contributes to enhanced stability by reducing the likelihood of aggregation, thereby ensuring the effective delivery of therapeutic agents at the nanoscale. This stability is crucial for maintaining the efficacy and safety of nanomedicine formulations in various biomedical applications.

Table 2. ζ-potential values of copolymer solutions at pH = 7, $C_{polymer} = 10^{-3}$ g/mL.

Sample	ζ-Potential (mV)
P1	+49
P2	+44
H1	+50
H2	+38

Based on the results, branching does not seem to affect zeta potential significantly. One should bear in mind that branched copolymers have a large number of -COOH groups originating from CTA fragments in the structure, while linear ones have only one -COOH group at the end of the chain. The protonation–deprotonation equilibria of carboxylic and amine chains and the degree of aggregation determine the overall charge of the structures in aqueous media. It seems that the higher number of amine groups originating from the DMAEMA segments dominate and determine in both cases the zeta potential values. In terms of the composition, the increase of DMAEMA from 54% to 70% in H1 and H2, respectively, induces a 24% decrease in zeta potential, which is unexpected. However, there is no hard evidence that could aid in ascribing this change to a specific phenomenon. Polar groups are expected to reside on the surface of the aggregates in contact with water, but the overall conformation of the associated chains within the aggregates determines the surface charge of the structures formed. Apparently, amphiphilic macromolecules arrange differently within the aggregates in the linear and branched copolymer cases.

3.5. Response to Solution pH Changes

PDMAEMA is a weak cationic polyelectrolyte ($pK_a = 7.5$) which is largely water soluble at both neutral and acidic pH. The amine groups are partially protonated under these conditions, while the inverse is true at higher pH values [34]. This unique property makes PDMAEMA particularly suitable for drug delivery applications, as it leverages the lower acidity found in tumor tissues to facilitate sustained drug release [35]. To assess whether these characteristics were preserved in the copolymers, light scattering techniques were utilized at three distinct pH values: 3, 7, and 10 (refer to Figures 5 and 6 and Table 3). These results allow for several conclusions. In most cases, two populations are observed, indicating the formation of distinct in size nanostructures through self-assembly, probably due to the random sequence of hydrophilic/hydrophobic segments in the copolymers. Aggregates with small and large numbers of polymer chains coexist. The linear copolymers are quite similar in terms of R_h in neutral conditions. In the case of H1, uniformly sized nanostructures with a rather large size dispersity and sufficient mass are formed at pH 7 while smaller-sized structures dominate at pH 3. Furthermore, there is a noticeable mass difference, as indicated by the scattered light intensities, between the 30–70 grouping (copolymers P2 and H2) and the 50–50 grouping (copolymers P1 and H1). This is explained by the fact that a higher percentage of hydrophobic LMA segments promotes more compact nanostructures with a higher mass, which raises the observed scattered light intensity values. Interestingly, hyperbranched and linear nanostructures differ in mass due to the former's significantly higher number of -COOH polar end groups, which enhance the hydrophilicity of these systems and promote greater interactions with aqueous media. These structures noticeably rearrange when the pH is varied, as evidenced by the considerable variations in intensity, size, and PDI observed in each solution. Smaller populations are observed, particularly in acidic conditions (R_h = 1–9 nm), which in some cases can be attributed to non-aggregated copolymer chains present in the suspension. Additionally, a decrease in mass and an increase in size can be observed, which is associated with the PDMAEMA pendant groups' higher charge density, which results in intense repulsion between the hydrophilic chains [36]. As more and more solvent accumulates in the nanostructures, this action causes swelling. Interestingly, hyperbranched and linear nanostructures differ in mass due to the former's significantly higher number of -COOH polar end groups, which enhance the hydrophilicity of these systems at pH 10 and promote greater interactions with aqueous media. As the pH of the solution increases, a greater proportion of the carboxylic acid groups within the polymer become ionized. This ionization leads to enhanced electrostatic repulsion between the negatively charged carboxylate groups, which subsequently contributes to the expansion of the hydrophobic domains in the copolymer aggregates. At these pH levels, PDMAEMA's hydrophobicity increases.

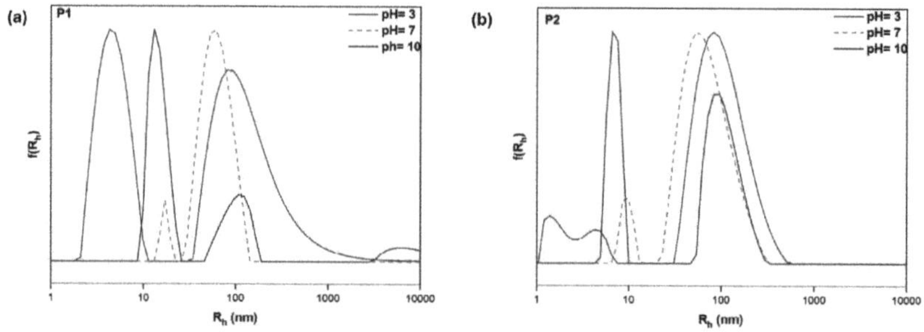

Figure 5. DLS size distributions for P1 (**a**) and P2 (**b**) linear copolymer aqueous solutions, $C_{polymer} = 10^{-3}$ g/mL.

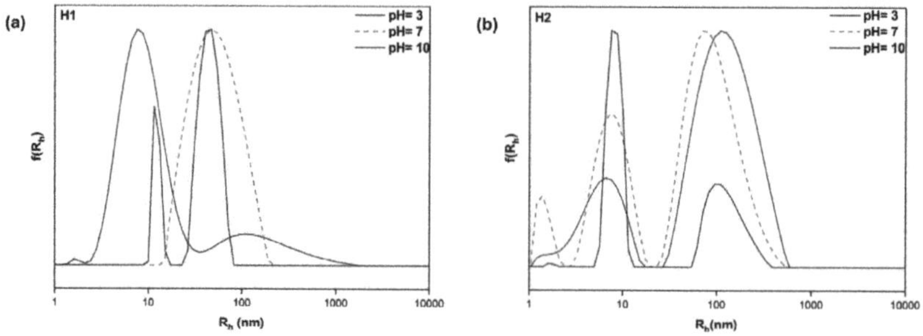

Figure 6. DLS size distributions for H1 (**a**) and H2 (**b**) hyperbranched copolymer aqueous solutions.

Table 3. DLS data emphasizing pH variations in the formed copolymer nanostructures, $C_{polymer} = 10^{-3}$ g/mL.

Sample	pH	Intensity (kcps)	PDI	R_h (nm)
P1	3	282	0.472	4 133
	7	86,800	0.257	17 62
	10	3064	0.426	14 100
P2	3	119	0.558	1 4 98
	7	2091	0.391	9 68
	10	816	0.519	7 106
H1	3	195	0.401	9 108
	7	71,400	0.249	50
	10	1409	0.354	12 45
H2	3	198	0.547	6 122
	7	212	0.515	1 7 89
	10	607	0.473	8 122

3.6. Behavior in Salt Solutions

Amphiphilic polyelectrolyte self-assembly is affected by ionic strength, as the addition of salt alters the strength of the electrostatic interactions between charged moieties. Specifically, higher ionic strength typically screens these electrostatic interactions, making them weaker and allowing for greater rearrangement or aggregation of the polyelectrolyte chains. Therefore, amphiphilic ionic copolymers should exhibit variable structural characteristics in response to fluctuations in the solution's ionic strength [37]. Topology seems to produce small effects, as copolymer compositional analogs produced similar behavioral motifs. The intensity values in the 50–50 composition grouping (copolymers P2 and H2 see Figures 7 and 8) increase and then quickly decline with salt concentration. This could be explained as the structures "tightening up" due to an overabundance of ions, which causes swelling. These structures appear to stabilize following the salt concentration increase. Regarding the other group, a mass increase is noted, which may be explained by the fact that hydrophobic interactions are the primary driving force behind self-assembly, which results in more compact structures.

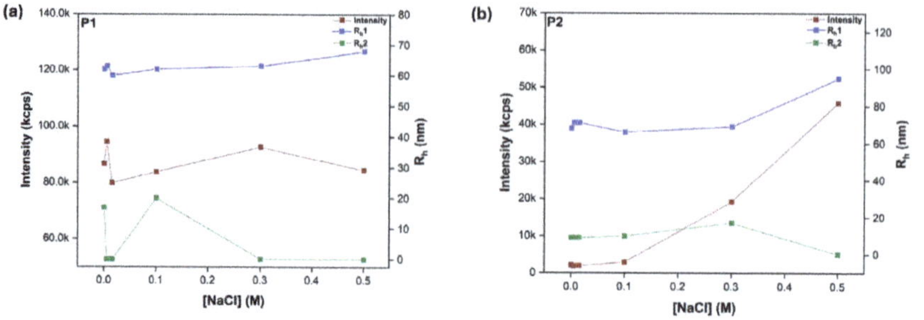

Figure 7. Collective DLS data as a function of salinity. (**a**) P1 and (**b**) P2 linear copolymers, $C_{polymer} = 10^{-3}$ g/mL, pH = 7.

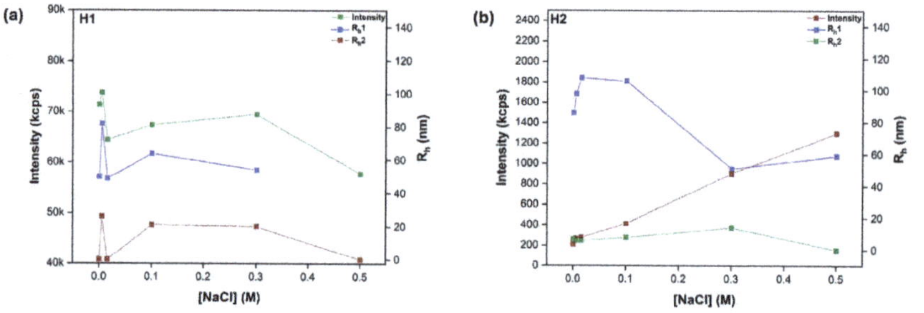

Figure 8. Collective DLS data as a function of salinity. (**a**) H1 and (**b**) H2 hyperbranched copolymers, $C_{polymer} = 10^{-3}$ g/mL, pH = 7.

3.7. Microenvironment Polarity Through Pyrene Fluorescence

Pyrene has been used as a fluorescent probe because it can provide a thorough understanding of amphiphilic copolymer self-assembled nanostructures in aqueous conditions [38]. In addition to serving as a sufficient measure of hydrophobicity, the I_1/I_3 value is often utilized to determine the critical aggregation concentration (CAC) values. This involves plotting the I_1/I_3 value against the logarithm of the polymer concentration. Specifically, the CAC is defined as the point where two tangent lines intersect over the steep decline observed in the I_1/I_3 value as the concentration of the copolymer increases [39]. As expected, H1 and P1 present lower CAC values, as the higher LMA content adds to the hydrophobicity of the nanostructures (see Figures 9 and 10). This improved stability can be linked to the stronger interactions among the hydrophobic regions in the nanostructures, resulting in more robust structural integrity and allowing them to maintain their form upon dilution in high volumes. As anticipated, the comparison of I_1/I_3 across varying pH levels reveals a distinct trend: higher values are observed in acidic conditions, while lower values are found in basic conditions (see Table 4). This indicates that the hydrophobicity of the nanostructures changes significantly depending on the acidity or alkalinity of the environment, just as observed in the DLS measurements.

Table 4. Pyrene fluorescence data in copolymer aqueous solutions, $C_{polymer} = 10^{-3}$ g/mL, C_{pyrene} =1 μM.

Sample	pH	I_1/I_3
	3	1.29
P1	7	0.98
	10	1.11

Table 4. *Cont.*

Sample	pH	I_1/I_3
P2	3	1.43
	7	0.96
	10	1.07
H1	3	1.26
	7	1.11
	10	1.10
H2	3	1.44
	7	1.11
	10	1.17

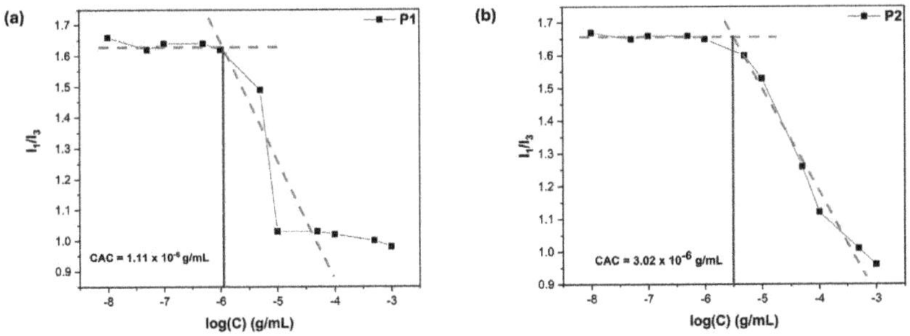

Figure 9. I_1/I_3 vs. polymer concentration plots for CAC determination regarding (**a**) P1 and (**b**) P2 copolymer aqueous solutions, $C_{pyrene} = 1\ \mu M$. (Green lines refer to the two tangent lines described in the text).

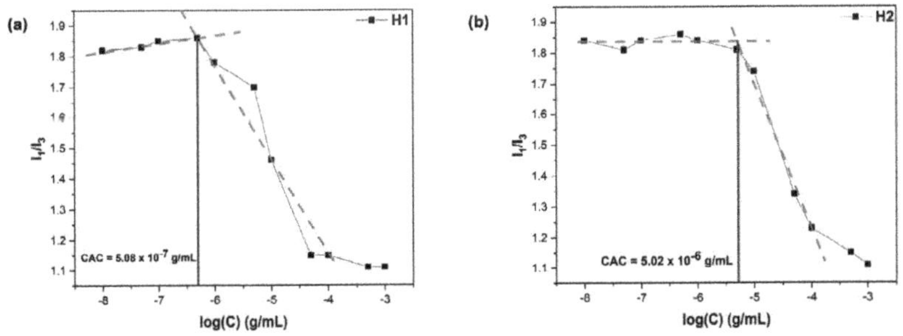

Figure 10. I_1/I_3 vs. polymer concentration plots for CAC determination regarding (**a**) H1 and (**b**) H2 copolymer aqueous solutions. (Green lines refer to the two tangent lines described in the text).

3.8. *Transmission Electron Microscopy Imaging*

Transmission electron microscopy (TEM) is recognized as a powerful imaging tool for studying the structural features of copolymer nanostructures. This technique allows for precise observation of the morphology of polymer nanostructures and the interfaces between different inner domains [40]. In the P1 solutions, particles of both spherical and irregular shape were identified, with noticeable high-contrast rod-like structures present throughout (see Figure 11). These rod-like structures are likely LMA domains, which exhibit a higher contrast due to their increased compactness/hydrophobicity compared to the hydrophilic, water swollen, outer DMAEMA domains in the self-assembled structures. The particle sizes range from 10 to 220 nm, with an average size of 50 nm based on the image analysis of 200 objects.

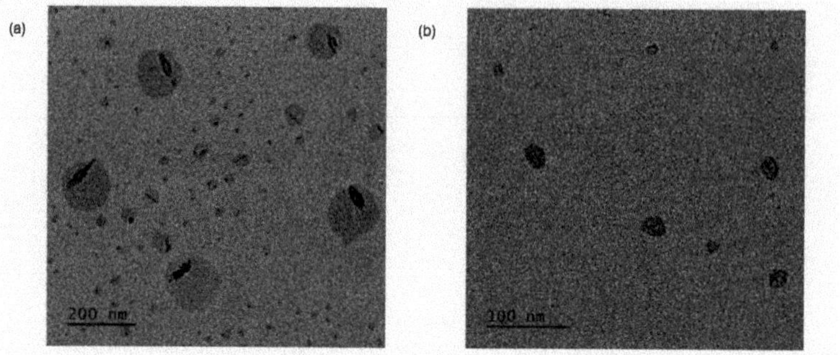

Figure 11. TEM images from (**a**) P1 and (**b**) P2 copolymer solutions.

Particles in the P2 solutions exhibit both spherical and irregular shapes, with sizes ranging from 50 to 80 nm. Some particles are larger, measuring between 150 and 200 nm, while smaller particles, approximately 5 to 15 nm, can be found on their surfaces. Smaller spherical aggregates seem to cluster together to form larger structures, while the lower contrast gaps observed in larger particles can be assigned to the DMAEMA nanodomains swollen by water molecules.

Regarding H1, particles are mainly characterized by a spherical shape. Moreover, small spherical particles resembling those in the P2 sample, with diameters ranging from 5 to 15 nm, have also been identified. Just as in the case of P2, DMAEMA nanodomain gaps are again observed.

Particles exhibiting both spherical and irregular shapes were identified in the H2 copolymer solutions, with a lower portion of their surfaces displaying enhanced contrast. Such high contrast domains are randomly distributed within the particle area and may be rich in LMA segments. The average size of these particles, determined from the measurement of 200 objects, was approximately 27 nm. The correlation between this topology and the nanostructures is emphasized by the formation of seemingly vesicular structures in the H2 hyperbranched copolymers (e.g., Figure 12b upper part). Moreover, similar to the findings in P2, small spherical high-contrast domains were observed on the surfaces of larger particles, measuring between 5 and 15 nm in size. Overall, the large variety of nanostructures observed in the copolymer solutions should be attributed to the statistical arrangement of hydrophobic segments within the macromolecules, as well as to the existing compositional heterogeneity of the amphiphilic copolymers.

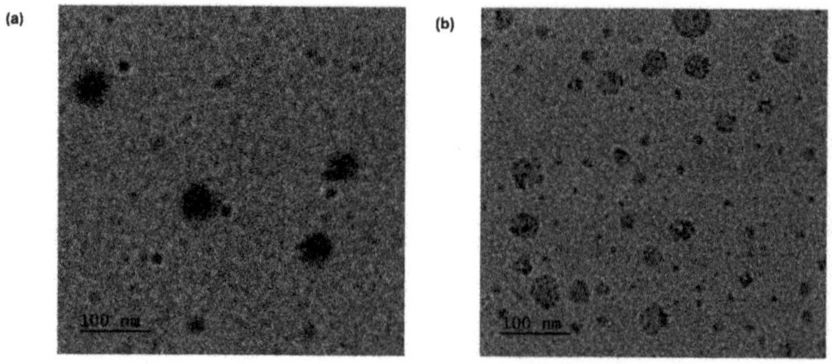

Figure 12. TEM images from (**a**) H1 and (**b**) H2 copolymer solutions.

4. Conclusions

To summarize, the findings of this work contribute to the general knowledge regarding the synthesis of linear and hyperbranched amphiphilic polyelectrolyte-type copolymers and their self-assembly behavior through the formation of nanostructures in solutions. The pH and salt-responsive properties of the PDMAEMA component together with variations in copolymer composition allow for fine-tuning of the structure of the formed nanoparticles. The nanoparticulate systems formulated and characterized exhibited a high surface charge, which renders them colloidally stable and allows them to interact with anionic species (including nucleic acids and proteins). Their amphiphilicity allowed for the encapsulation of pyrene, which gave further information about the micropolarity of the microenvironment within the nanoparticles and its variation with the solution pH, indicating their medicinal application as nanocarriers of hydrophobic compounds. Such systems could be engineered for the transfer and targeted release of their functional payload in acidic cancerous cell environments. Finally, TEM imaging provided insight into the footprint of such nanostructure collections and emphasized the impact of comonomer statistical distribution within the macromolecular chains as well as their global topology.

Supplementary Materials: The following supporting information can be downloaded at: https://www.mdpi.com/article/10.3390/polym17050701/s1. Figure S1. ATR-FT-IR spectra of the neat linear copolymers; Figure S2. ^{1}H-NMR spectrum of P1 in $CDCl_3$; Figure S3. ^{1}H-NMR spectrum of H1 in $CDCl_3$; Figure S4. ^{1}H-NMR spectrum of H2 in $CDCl_3$.

Author Contributions: Conceptualization, S.P.; methodology, A.M.G. and S.P.; formal analysis, A.M.G., A.F., and S.P.; investigation, A.M.G. and A.F.; resources, B.T. and S.P.; data curation, A.M.G.; writing—original draft preparation, A.M.G.; writing—review and editing, A.M.G., A.F., B.T., and S.P.; supervision, S.P. All authors have read and agreed to the published version of the manuscript.

Funding: This research received no external funding.

Institutional Review Board Statement: Not applicable.

Data Availability Statement: Data produced in this study are included in the manuscript and the Supplementary Materials.

Conflicts of Interest: The authors declare no conflicts of interest.

References

1. Li, J.C.; Reddy, V.S.; Jayathilaka, W.A.D.M.; Chinnappan, A.; Ramakrishna, S.; Ghosh, R. Intelligent Polymers, Fibers and Applications. *Polymers* **2021**, *13*, 1427. [CrossRef] [PubMed]
2. Mu, M.; Ebara, M. Smart Polymers. In *Polymer Science and Nanotechnology: Fundamentals and Applications*; Elsevier: Amsterdam, The Netherlands, 2020; pp. 257–279. [CrossRef]
3. Mane, S.R.; Sathyan, A.; Shunmugam, R. Biomedical Applications of PH-Responsive Amphiphilic Polymer Nanoassemblies. *ACS Appl. Nano Mater.* **2020**, *3*, 2104–21117. [CrossRef]
4. Perin, F.; Motta, A.; Maniglio, D. Amphiphilic Copolymers in Biomedical Applications: Synthesis Routes and Property Control. *Mater. Sci. Eng. C* **2021**, *123*, 111952. [CrossRef]
5. Zhou, M.; Wen, L.; Wang, C.; Lei, Q.; Li, Y.; Yi, X. Recent Advances in Stimuli-Sensitive Amphiphilic Polymer-Paclitaxel Pro-drugs. *Front. Bioeng. Biotechnol.* **2022**, *10*, 875034. [CrossRef]
6. Kashapov, R.; Gaynanova, G.; Gabdrakhmanov, D.; Kuznetsov, D.; Pavlov, R.; Petrov, K.; Zakharova, L.; Sinyashin, O. Self-Assembly of Amphiphilic Compounds as a Versatile Tool for Construction of Nanoscale Drug Carriers. *Int. J. Mol. Sci.* **2020**, *21*, 6961. [CrossRef] [PubMed]
7. Chen, Y.; Liu, W.; Zeng, G.; Liu, Y.J. Microporous PDMAEMA-Based Stimuli-Responsive Hydrogel and Its Application in Drug Release. *J. Appl. Polym. Sci.* **2017**, *134*, 45326. [CrossRef]

8. Saraiva, G.K.V.; de Souza, V.V.; de Oliveira, L.C.; Noronha, M.L.C.; Masini, J.C.; Chaimovich, H.; Salinas, R.K.; Florenzano, F.H.; Cuccovia, I.M. Characterization of PMMA-b-PDMAEMA Aggregates in Aqueous Solutions. *Colloid. Polym. Sci.* **2019**, *297*, 557–569. [CrossRef]
9. Huang, Y.; Yong, P.; Chen, Y.; Gao, Y.; Xu, W.; Lv, Y.; Yang, L.; Reis, R.L.; Pirraco, R.P.; Chen, J. Micellization and Gelatinization in Aqueous Media of PH- and Thermo-Responsive Amphiphilic ABC ($PMMA_{82}$-*b*-$PDMAEMA_{150}$-*b*-$PNIPAM_{65}$) Triblock Copolymer Synthesized by Consecutive RAFT Polymerization. *RSC Adv.* **2017**, *7*, 28711–28722. [CrossRef]
10. Dong, Z.; Wei, H.; Mao, J.; Wang, D.; Yang, M.; Bo, S.; Ji, X. Synthesis and Responsive Behavior of Poly(N,N-Dimethylaminoethyl Methacrylate) Brushes Grafted on Silica Nanoparticles and Their Quaternized Derivatives. *Polymer* **2012**, *53*, 2074–2084. [CrossRef]
11. Kumar, K.; Mogha, N.K.; Yadav, R.; Venkatesu, P. Insulin-Induced Conformational Transition of Fluorescent Copolymers: A Perspective of Self-Assembly between Protein and Micellar Solutions of Smart Copolymers. *Phys. Chem. Chem. Phys.* **2020**, *22*, 9573–9586. [CrossRef]
12. Cai, Y.; Ding, P.; Ni, J.; Zhou, L.; Ahmad, A.; Guo, X.; Cohen Stuart, M.A.; Wang, J. Regulated Polyelectrolyte Nanogels for Enzyme Encapsulation and Activation. *Biomacromolecules* **2021**, *22*, 4748–4757. [CrossRef] [PubMed]
13. Agarwal, S.; Zhang, Y.; Maji, S.; Greiner, A. PDMAEMA Based Gene Delivery Materials. *Mater. Today* **2012**, *15*, 388–393. [CrossRef]
14. Qiu, X.; Lu, L.; Wang, J.; Tang, G.; Song, G. Fabrication, Thermal Properties and Thermal Stabilities of Microencapsulated n-Alkane with Poly(Lauryl Methacrylate) as Shell. *Thermochim. Acta* **2015**, *620*, 10–17. [CrossRef]
15. Duc, N.T.; Ha, P.T.T.; Van Khoi, N.; Huy, N.Q.; Mien, N.T.; Hoan, D.C.; Chau, C.N.; Tung, N.T. Oil Absorbent Based on Lauryl Methacrylate Grafting onto Cellulose of Water Hyacinth (*Eichhornia crassipes*). *Vietnam. J. Chem.* **2024**, *62*, 169–178. [CrossRef]
16. Suraj Belgaonkar, M.; Kandasubramanian, B. Hyperbranched Polymer-Based Nanocomposites: Synthesis, Progress, and Applications. *Eur. Polym. J.* **2021**, *147*, 110301. [CrossRef]
17. Perrier, S. 50th Anniversary Perspective: RAFT Polymerization—A User Guide. *Macromolecules* **2017**, *50*, 7433–7447. [CrossRef]
18. Tian, X.; Ding, J.; Zhang, B.; Qiu, F.; Zhuang, X.; Chen, Y. Recent Advances in RAFT Polymerization: Novel Initiation Mecha-nisms and Optoelectronic Applications. *Polymers* **2018**, *10*, 318. [CrossRef]
19. Chrysostomou, V.; Pispas, S. Stimuli-responsive amphiphilic PDMAEMA-b-PLMA copolymers and their cationic and zwitterionic analogs. *J. Polym. Sci. Part A Polym. Chem.* **2018**, *56*, 598–610. [CrossRef]
20. Wang, J.; Ge, Y.; Wang, Y.; Sun, R.; Yang, X.; Xue, H.; Ma, X.; Liu, J.; Hu, K. Pd nanoparticles supported in PDMAEMA-b-PLMA micelles: A superb catalytic platform for Suzuki-Miyaura cross-coupling in water. *Eur. Polym. J.* **2024**, *202*, 112650. [CrossRef]
21. Giaouzi, D.; Pispas, S. PDEGMA-*b*-PDMAEMA-*b*-PLMA triblock terpolymers and their cationic analogues: Synthesis, stimuli responsive self-assembly and micelleplex formation. *Polym. Chem.* **2024**, *15*, 1536–1551. [CrossRef]
22. Skandalis, A.; Selianitis, D.; Sory, D.R.; Rankin, S.M.; Jones, J.R.; Pispas, S. Poly(2-(dimethylamino) ethyl methacrylate)-*b*-poly(lauryl methacrylate)-*b*-poly(oligo ethylene glycol methacrylate) triblock terpolymer micelles as drug delivery carriers for curcumin. *J. Appl. Polym. Sci.* **2022**, *139*, e52899. [CrossRef]
23. Jung, F.A.; Panteli, P.A.; Ko, C.H.; Kang, J.-J.; Barnsley, L.C.; Tsitsilianis, C.; Patrickios, C.S.; Papadakis, C.M. Structural Properties of Micelles Formed by Telechelic Pentablock Quaterpolymers with pH-Responsive Midblocks and Thermoresponsive End Blocks in Aqueous Solution. *Macromolecules* **2019**, *52*, 9746–9758. [CrossRef]
24. Constantinou, A.P.; Zhao, H.; McGilvery, C.M.; Porter, A.E.; Georgiou, T.K. A Comprehensive Systematic Study on Thermoresponsive Gels: Beyond the Common Architectures of Linear Terpolymers. *Polymers* **2017**, *9*, 31. [CrossRef]
25. Fan, F.; Piao, J.-G.; Zhao, Y.; Jin, L.; Li, M.; Wang, Y.; Yang, L. Bioinspired Membrane-Disruptive Macromolecules as Drug-Free Therapeutics. *ACS Appl. Bio Mater.* **2020**, *3*, 1267–1275. [CrossRef]
26. Skandalis, A.; Pispas, S. Synthesis of (AB) n-, AnBn-, and AxBy-type amphiphilic and double-hydrophilic star copolymers by RAFT polymerization. *J. Polym. Sci. Part A Polym. Chem.* **2019**, *57*, 1771–1783. [CrossRef]
27. Zielinska, A.; Carreiró, F.; Oliveira, A.M.; Neves, A.; Pires, B.; Nagasamy Venkatesh, D.; Durazzo, A.; Lucarini, M.; Eder, P.; Silva, A.M.; et al. Polymeric Nanoparticles: Production, Characterization, Toxicology and Ecotoxicology. *Molecules* **2020**, *25*, 3731. [CrossRef]
28. Hartlieb, M.; Floyd, T.; Cook, A.B.; Sanchez-Cano, C.; Catrouillet, S.; Burns, J.A.; Perrier, S. Well-Defined Hyperstar Copolymers Based on a Thiol–Yne Hyperbranched Core and a Poly(2-Oxazoline) Shell for Biomedical Applications. *Polym. Chem.* **2017**, *8*, 2041–2054. [CrossRef]
29. Salerno, A.; Diéguez, S.; Diaz-Gomez, L.; Fajrin, A.; Marliana, S.D.; Handayani, D.S. Synthesis of Eugenol–Lauryl Methacrylate Copolymers via Cationic Polymerization. *IOP Conf. Ser. Mater. Sci. Eng.* **2018**, *349*, 012003. [CrossRef]
30. Balafouti, A.; Pispas, S. Hyperbranched Copolymers of Methacrylic Acid and Lauryl Methacrylate H-P(MAA-Co-LMA): Syn-thetic Aspects and Interactions with Biorelevant Compounds. *Pharmaceutics* **2023**, *15*, 1198. [CrossRef] [PubMed]

31. Onugwu, A.L.; Nwagwu, C.S.; Onugwu, O.S.; Echezona, A.C.; Agbo, C.P.; Ihim, S.A.; Emeh, P.; Nnamani, P.O.; Attama, A.A.; Khutoryanskiy, V.V. Nanotechnology Based Drug Delivery Systems for the Treatment of Anterior Segment Eye Diseases. *J. Control. Release* **2023**, *354*, 465–488. [CrossRef]
32. Bhattacharjee, S. DLS Zeta Potential—What They Are What They Are Not? *J. Control. Release* **2016**, *235*, 337–351. [CrossRef] [PubMed]
33. Sharma, S.; Shukla, P.; Misra, A.; Mishra, P.R. Interfacial and Colloidal Properties of Emulsified Systems: Pharmaceutical and Biological Perspective. In *Colloid and Interface Science in Pharmaceutical Research and Development*; Elsevier: Amsterdam, The Netherlands, 2014; pp. 149–172. [CrossRef]
34. Van De Wetering, P.; Moret, E.E.; Schuurmans-Nieuwenbroek, N.M.E.; Van Steenbergen, M.J.; Hennink, W.E. Structure-Activ-ity Relationships of Water-Soluble Cationic Methacrylate/Methacrylamide Polymers for Nonviral Gene Delivery. *Bioconjugate Chem.* **1999**, *10*, 589–597. [CrossRef] [PubMed]
35. Huang, J.; Xiao, Z.; Chen, G.; Li, T.; Peng, Y.; Shuai, X. A PH-Sensitive Nanomedicine Incorporating Catalase Gene and Photo-sensitizer Augments Photodynamic Therapy and Activates Antitumor Immunity. *Nano Today* **2022**, *43*, 101390. [CrossRef]
36. Car, A.; Baumann, P.; Duskey, J.T.; Chami, M.; Bruns, N.; Meier, W. PH-Responsive PDMS-b-PDMAEMA Micelles for Intracel-lular Anticancer Drug Delivery. *Biomacromolecules* **2014**, *15*, 3235–3245. [CrossRef]
37. Borisova, O.; Billon, L.; Zaremski, M.; Grassl, B.; Bakaeva, Z.; Lapp, A.; Stepanek, P.; Borisov, O. Synthesis and PH- and Salinity-Controlled Self-Assembly of Novel Amphiphilic Block-Gradient Copolymers of Styrene and Acrylic Acid. *Soft Matter* **2012**, *8*, 7649–7659. [CrossRef]
38. Yuan, J.; Xu, J. Synthesis of Amphiphilic Block Copolymer and Its Application in Pigment-Based Ink. *Materials* **2024**, *17*, 330. [CrossRef]
39. Yan, M.; Li, B.; Zhao, X. Determination of Critical Aggregation Concentration and Aggregation Number of Acid-Soluble Colla-gen from Walleye Pollock (*Theragra chalcogramma*) Skin Using the Fluorescence Probe Pyrene. *Food Chem.* **2010**, *122*, 1333–1337. [CrossRef]
40. Bürger, J.; Kunnathully, V.S.; Kool, D.; Lindner, J.K.N.; Brassat, K. Characterisation of the PS-PMMA Interfaces in Microphase Separated Block Copolymer Thin Films by Analytical (S)TEM. *Nanomaterials* **2020**, *10*, 141. [CrossRef]

polymers

MDPI

Article

Optimizing the Synthesis of CO_2-Responsive Polymers: A Kinetic Model Approach for Scaling Up

Emil Pashayev and Prokopios Georgopanos *

Helmholtz-Zentrum Geesthacht, Institute of Membrane Research, Max-Planck-Straße 1, 21502 Geesthacht, Germany; emil.pashayev@hereon.de
* Correspondence: prokopios.georgopanos@hereon.de

Abstract: The kinetic model is a crucial tool for optimizing polymer synthesis protocols and facilitating the scaled-up production processes of the CO_2-responsive polymer poly(N-[3-(dimethylamino)propyl]-acrylamide)-b-poly(methyl methacrylate)(PDMAPAm-*b*-PMMA), which is supposed to be implemented in direct air capture (DAC) technology. This study presents a simulation of the kinetic model developed for the Reversible Addition–Fragmentation Chain-Transfer (RAFT) polymerization of N-[3-(dimethylamino)propyl]-acrylamide (DMAPAm), alongside an investigation into the kinetics of this polymerization using the simulation as an analytical tool, as well as the application of the simulation for the upscaling of RAFT polymerization. Ultimately, the kinetic model was validated through two kinetic experiments, confirming its reliability. It was subsequently employed to optimize the synthesis recipe and to predict the properties of PDMAPAm homopolymers, thereby supporting the upscaling of PDMAPAm-*b*-PMMA diblock copolymer synthesis. In the end, the preliminary results of the CO_2-responsiveness of the diblock copolymer were determined with a simple experiment.

Keywords: CO_2-responsive polymers; kinetic modeling of RAFT polymerization; upscaling

1. Introduction

Carbon capture is a key area of research that is increasingly considered for application as a technology to mitigate CO_2 emissions [1–4]. In a recent approach, carbon dioxide present in the atmosphere, even at relatively low concentrations, can be captured using a novel approach known as direct air capture (DAC) [2,3,5]. In DAC systems, it is essential to establish a stronger interaction between CO_2 molecules and the solid adsorbent [3,5], and diblock copolymers can be considered adsorbents for DAC because of their capacity to form films with high porosity [6–8] and can also be scalable [9–12]. Diblock copolymers that incorporate amine functional groups are particularly effective at capturing CO_2 due to the strong attraction that certain chemical groups to for carbon dioxide. A range of acrylate polymers, including poly(2-(dimethylamino)ethyl methacrylate) (PDMAEMA) [13,14], poly(N,N-dimethylallylamine) (PDMAAm) [15,16], poly(2-aminoethyl methacrylate) (PAEM) [17,18], and poly(N-[3-(dimethylamino)propyl]-acrylamide) (PDMAPAm) [19], which possess amine groups (responsive to CO_2), can act as the foundational block for subsequent chain growth into diblock or triblock copolymers.

Controlled radical polymerization (CRP) is one of the most effective techniques for synthesizing well-defined block copolymers, because this polymerization technique enables the precise targeting of molecular weights and block ratios [20–23]. To date, three primary techniques of CRP have been explored and documented in the literature: atom transfer radical polymerization (ATRP), nitroxide-mediated polymerization (NMP), and Reversible Addition–Fragmentation Chain-Transfer Polymerization (RAFT) [24–26]. These established polymerization techniques allow for effective control over the reaction rate, dispersity, and molecular weight distribution (MWD) of the resulting polymer [27–30].

Compared to ATRP and NMP, RAFT-mediated polymerization offers the versatility to work with a wide variety of monomers, including butyl acrylate, (meth)acrylates, (meth)acrylamides, acrylonitrile, styrene [30–32] and its derivatives, butadiene, vinyl acetate, N-vinylpyrrolidone [27], and vinyl pyridines [20–22,33], and at the same time, the kinetics of it can be simulated, as shown in several publications [9–11,34,35]. Due to these flexibilities, RAFT polymerization was selected as an optimal polymerization technique for the synthesis of PDMAPAm homopolymer in a previous publication [19].

The quest to comprehend the kinetics of the polymerization process and its impact on the characteristics of the final products has been integral to the evolution of the RAFT polymerization method since its invention [36,37]. Reaction kinetic models are a vital component in this area of research, as evidenced by numerous studies in the literature. So far, several studies have been conducted regarding the kinetic modeling of bulk and solution RAFT polymerization of styrene [35,38–42], methyl methacrylate (MMA) [42–45], dodecyl methacrylate [46], 4-vinyl pyridine [9,10,12], NIPAM [47,48], and DMAEMA [49–51]. In some of them, the model has been used as a tool for upscaling the polymerization process [9,11].

Therefore, this research demonstrates the development of a kinetic model, the application of the validated kinetic model in upscaling of the synthesis of the poly(N-[3-(dimethylamino)propyl]-acrylamide)-*b*-poly(methyl methacrylate) (PDMAPAm-*b*-PMMA) diblock copolymer, and also preliminary results regarding the CO_2-responsiveness of the diblock copolymer. The model investigates the polymerization kinetics and predicts the key properties of the homopolymer (PDMAPAm), such as molar mass, dispersity, and livingness. These properties are crucial for chain extension with MMA to synthesize PDMAPAm-*b*-PMMA. Additionally, the kinetic model is applied to scale up the synthesis of CO_2-responsive PDMAPAm-*b*-PMMA from a 10 mL lab flask to a 100 mL glass reactor, which can facilitate the fabrication of CO_2 membrane adsorbers on a larger scale for CO_2 capture.

2. Materials and Methods

2.1. Materials

N-[3-(Dimethylamino)propyl]acrylamide (stabilized with MEHQ) (DMAPAm, >98.0%, TSI, Zwijndrecht, Belgium) was first dissolved in 1,4 dioxane (≥99.8%, Merck, Darmstadt, Germany) and percolated through a column of basic aluminum oxide (>98%, Sigma-Aldrich) before use to remove the inhibitor. Then, 2,2′-Azobis(2-methylpropionitrile) (AIBN, 98%, Sigma-Aldrich, Taufkirchen, Germany, stored at 4 °C) and 4-cyano-4-[(dodecylsulfanylthiocarbonyl)-sufanyl]pentanoic acid (CDTPA, 97%, Sigma-Aldrich, Taufkirchen, Germany, stored at 4 °C) were added into the monomer and solvent mixture. Methyl methacrylate (99%, stabilized with ≤30 ppm MEHQ, Sigma-Aldrich, Steinheim, Germany) was also freshly percolated with the same procedure for the same reason. Nitrogen was purchased by Linde (Hamburg, Germany, 99.999% purity). The copolymer was precipitated in n-hexane (99%, Sigma-Aldrich, Taufkirchen, Germany). The proton nuclear magnetic resonance (^1H NMR) measurements were carried out in deuterated chloroform-d1 ($CDCl_3$, 99.8% containing 0.03% (v/v) TMS, Sigma-Aldrich, Taufkirchen, Germany, stored at 4 °C). The gel permeation chromatography (GPC) measurements were carried out with the use of N,N-dimethyl acetamide (DMAc, ≥99.9%, Sigma-Aldrich, Taufkirchen, Germany) as the solvent and eluent of the system.

2.1.1. Synthesis of PDMAPAm via RAFT Polymerization

For the kinetic experiments of PDMAPAm synthesis via RAFT polymerization, CDTPA (36.3 mg, 0.09 mmol, 1 eq.), AIBN (14.93 mg, 0.09 mmol, 1 eq.), and percolated DMAPAm (2800 mg, 18.00 mmol, 200 eq.) were dissolved in 2.4 mL of 1,4-dioxane (20 wt% monomer content) and separately added into small 10 mL flasks. The mixture in the flasks was degassed with nitrogen in a cold-water bath at 15 °C for 20 min (see the detailed synthesis protocol in the Supplementary Materials, Table S1). Following this, polymerization was conducted in a 10 mL glass flask using a thermo-shaker set at 70 °C and 250 rpm for 4 h, and every half an hour, the reaction in one of the flasks was stopped. The reaction was terminated by ice-cooling (0 °C) and exposure to air. Samples for ^{1}H NMR and GPC were taken to determine the monomer conversion and molecular weight of the synthesized polymer, respectively. The final product was precipitated in n-hexane, followed by filtration and drying in a vacuum oven at 35 °C and a reduced pressure of 1 mbar overnight.

For upscaling, the DMAPAm was polymerized in a 100 mL glass reactor by applying the same recipe as mentioned above. A total of 10 g of homopolymer was synthesized (see more in the Supplementary Materials, Table S2), which was used for the synthesis of the diblock copolymer PDMAPAm-*b*-PMMA.

2.1.2. Synthesis of PDMAPAm-b-PMMA via RAFT Polymerization

To synthesize the PDMAPAm-*b*-PMMA diblock copolymer, dried PDMAPAm (2600 mg, 0.126 mmol, 1 eq.), which served as the macro-RAFT/macro-stabilizer; MMA (1000 mg, 10 mmol, 100 eq.); and AIBN (17 mg, 0.126 mmol, 1 eq.) were dissolved in 15 mL of 1,4-dioxane (refer to the detailed synthesis protocol in the Supplementary Materials, Table S3). The mixture was degassed with nitrogen in a cold-water bath at 15 °C for 20 min. Polymerization was then conducted in a 30 mL glass flask using a thermo-shaker at 70 °C and 250 rpm for 24 h. The reaction was terminated by ice-cooling (0 °C) and exposure to air. Samples for ^{1}H NMR and GPC were taken to determine the MMA conversion and the molecular weight of the synthesized diblock copolymer, respectively. The polymer was precipitated in n-hexane and dried in a vacuum oven at 35 °C and 1 mbar for 24 h.

2.2. *Analytics*

2.2.1. Proton Nuclear Magnetic Resonance Spectroscopy (^{1}H NMR)

^{1}H NMR spectroscopy experiments were performed using a Bruker AV500 spectrometer (Bruker, Rheinstetten, Germany) and a Spinsolve Carbon 80 benchtop spectrometer (80 MHz, Magritek GmbH, Aachen, Germany). The spectra obtained with the Bruker AV500 were recorded by applying a 10 ms 90° pulse at a sample temperature of 298 K. Sixteen scans were recorded with a relaxation delay of 3 s. In a similar way, the experiments with the Spinsolve Carbon 80 were accomplished using the proton+ protocol with 4 scans, an acquisition time of 6.4 s, a repetition time of 1 min, and a pulse angle of 90°. The sample concentrations were 20 g L^{-1} and $CDCl_3$. The ^{1}H NMR spectra were analyzed with the software MestReNova 10.0 (Mestrelab Research, Santiago de Compostela, Spain). 1,3,6-trioxane was used as an internal standard. DMAPAm conversion from macro-RAFT synthesis was estimated by comparing the integration of double-bond proton areas in the sample before and after the reaction, which was around 90%. Using the same method, the conversion of methyl methacrylate in the second block synthesis, also measured in $CDCl_3$, was calculated from the decrease in the integral of the monomer peaks, which was also approximately 90%.

2.2.2. Gel Permeation Chromatography (GPC)

The apparent molecular weight distributions of the PDMAPAm homopolymers (macro-RAFT agents) and PDMAPAm-*b*-PMMA diblock copolymers were measured via GPC at 50 °C with DMAc with the addition of lithium chloride (0.1 M) as an eluent. A Waters 717 plus instrument (Waters, Milford, MA, USA) equipped with PSS GRAM columns (PSS GmbH, Mainz, Germany) [GRAM pre-column (dimensions 8–50 nm) and two GRAM columns of different porosity (3000 and 1000)] with dimensions of 8 × 300 mm and a particle size of 10 µm was used. The samples were measured at a flow rate of 1 mL min^{-1} using a VWR-Hitachi 2130 pump (VWR Hitachi, Darmstadt, Germany) and a VWR-Hitachi L2490 RI (refractive index) detector (VWR Hitachi, Darmstadt, Germany). The GPC equipment was calibrated with narrow PMMA standards, and the data were analyzed using PSS WinGPC UniChrom software, version 1.0, Agilent Technologies, Inc. 2023 (PSS GmbH, Mainz, Germany).

3. Results

3.1. Preliminary Remarks on Kinetic Experiments and Polymer Synthesis

As described in previous work [19], the synthesis of PDMAPAm-*b*-PMMA involved two-step polymerization of the two monomers, as shown in Figure 1. First, poly(N-[3-(dimethylamino)propyl]acrylamide) (PDMAPAm) was synthesized at 70 °C by RAFT solution polymerization of N-[3-(dimethylamino)propyl]acrylamide in 1,4-dioxane. Subsequently, PDMAPAm served as a macroRAFT agent for the synthesis of a PDMAPAm-*b*-PMMA diblock copolymer by extending the chain via RAFT solution polymerization.

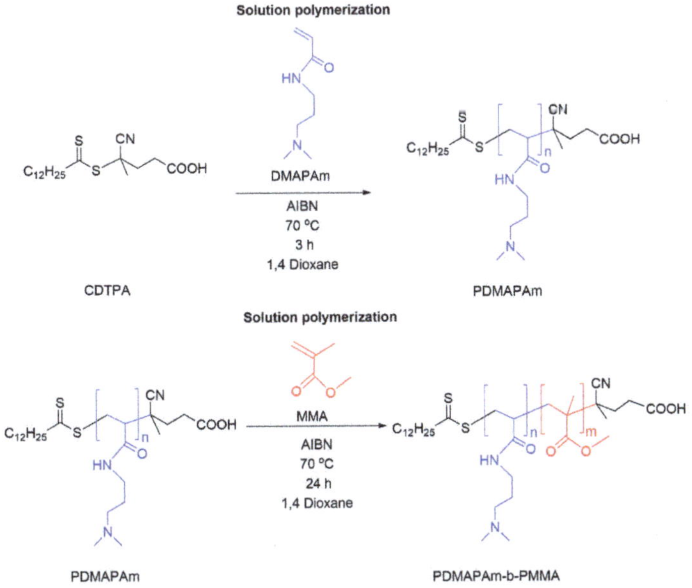

Figure 1. Synthesis of PDMAPAm-*b*-PMMA using PDMAPAm as macro-RAFT agent.

The aim was to synthesize a PDMAPAm-*b*-PMMA diblock copolymer with $\overline{M}_{w,app}$ = 40kDa and a 70:30 = PDMAPAm/PMMA ratio in order to have a large amount of

PDMAPAm, which is a functional material/block that shows CO_2 responsivity. To achieve this goal, it was very important to develop a kinetic model of polymerization that allows the study of the kinetics of polymerization and optimization of the synthesis protocol in order to obtain a polymer with the desired polymer properties, obtaining a great advantage of digitalized synthesis and reducing possible experimental attempts.

3.2. *Mark–Houwink–Sakurada Parameters*

The GPC evaluation of PDMAPAm and PDMAPAm-*b*-PMMA is performed with the PMMA standard, which is not a 100% fit to the molecular structure of the PDMAPAm-based copolymer. Therefore, the Mark–Houwink parameters of PDMAPAm are required to determine the actual apparent molar mass of the synthesized polymer. Using a similar procedure to that described in our previous publication [9], the Mark–Houwink parameters of PDMAPAm are determined by measuring the intrinsic viscosity (see Supplementary Information, Figures S1 and S2) of PDMAPAm with different molar masses dissolved in THF (Table 1).

Table 1. Determined Mark–Houwink–Sakurada parameters of PDMAPAm in THF.

$K_{PDMAPAm}(\frac{dL}{g})$	$a_{PDMAPAm}$
1.29×10^{-4}	6.25×10^{-1}

The Mark–Houwink parameters a and K vary depending on the specific polymer–solvent system. For solvents, an a-value of 0.5 indicates a theta solvent, while an a-value of 0.8 is characteristic of good solvents [52,53]. It is evident from the results that the Mark–Houwink parameters of PDMAPAm are also in this range and agree well with the Mark–Houwink parameters of other N-containing polymers [54–56]. For PDEAEMA, PDMAEMA, PDMAPMAE [54], PNEAEMA, and PMOMA [56], the value of a changes between 6.00×10^{-1} and 7.5×10^{-1}, and the a-value of PDMAPAm also falls within this range.

As for the K value of the other N-containing polymers, it ranges from 2.10×10^{-4} to 0.5 $\times 10^{-4}$ $\frac{dL}{g}$, whereas the figure is equal to 1.29×10^{-4} for our homopolymer. By knowing the Mark–Houwink parameters of PDMAPAm and PMMA [57,58] determined by viscosimetry measurement, the real $\overline{M}_w$ of PDMAPAm is calculated using Equation (1) [59].

$$\overline{M}_{w,PDMAPAm} = \left[\frac{K_{PMMA}}{K_{PDMAPAm}}\right]^{\left(\frac{1}{a_{PDMAPAm}+1}\right)} \cdot \left[\overline{M}_{w,PMMA}\right]^{\left(\frac{a_{PMMA}+1}{a_{PDMAPAm}+1}\right)} \quad (1)$$

In Figure 2, the weight-average molecular weights of PDMAPAm determined using the Mark–Houwink parameters of PMMA, with those determined using the estimated Mark–Houwink parameters of PDMAPAm, are presented. It can be seen from the diagrams that the evaluations with PMMA and PDMAPAm standards generally provide similar results for molecular weights below 20 kDa. The molar masses evaluated against the estimated PMMA parameters overlap with the molar masses evaluated against the estimated PDMAPAm parameters at higher RAFT-to-initiator ratios, but a slight discrepancy is observed as the ratio decreases. This occurs because at higher molecular masses than 20 kDa, the difference between the weight-average molecular masses determined using PMMA and PDMAPAm standards increases due to the different Mark–Houwink parameters of PDMAPAm and PMMA in THF.

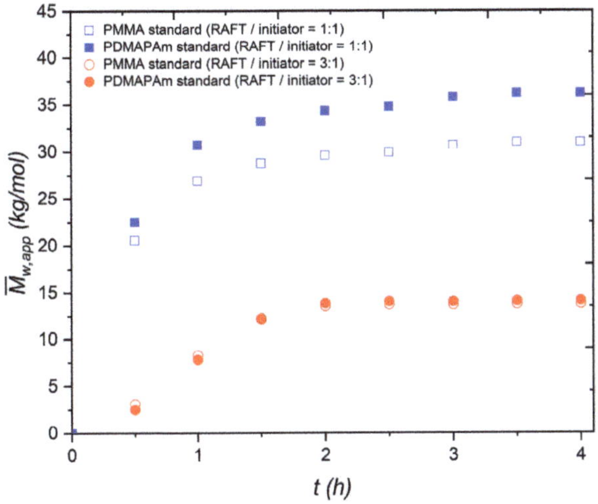

Figure 2. Correction of GPC evaluation. Evolution of weight-average molar masses of PDMAPAm from RAFT/initiator = 1:1 and RAFT/initiator = 3:1 experiments evaluated with PMMA and PDMAPAm standards.

3.3. Model Description

3.3.1. Structure of Model

The model developed by De Rybel et al. [41] and F. Kandelhard et al. [9] for the RAFT polymerization of styrene and 4-vinyl pyridine was selected as the base for this study due to its comprehensive consideration of all critical reaction steps, which are initiation, re-initiation, propagation, termination by recombination, cross-termination, pre-equilibrium, and RAFT equilibrium reactions (Scheme 1). Since the monomer concentration in solvent is 20% w/w, the gel effect for the termination reaction was not considered. There were two approaches to cross-termination in the literature. The first one is proposed by Barner-Kowollik et al. [23], who posit that the intermediate radical generated during the RAFT polymerization process is relatively stable and has a long lifespan. This model is referred to as the slow fragmentation model, which suggests negligible cross-termination of the intermediate radical and a high equilibrium constant (approximately k_{add}/k_{frag} = 10^4 to 10^8) for the reaction between a growing radical and a dormant/RAFT-intermediated radical. The second one, which is proposed by Monteiro et al. [60], suggests that there is considerable cross-termination of the intermediate radical with other free radicals in the solution. This model is known as the intermediate radical termination model, predicting equilibrium constants that are several orders of magnitude lower than those of the slow fragmentation model (around a maximum of 10^2). In this study, the model of Monteiro et al. [60] is considered because it is assumed that k_{add}/k_{frag} will not be higher than 10^2 and cross-termination of radical RAFT species with two chain lengths ($R_nX^*R_m$) is possible with an oligomeric growing radical (R_i) ($i_c = 2$ & $\alpha_c = 0.33$) [61], as shown in Equations (2) and (3). Additionally, $R_nX^*R_0$ can also cross-terminate with primary radicals and growing radicals, as assumed in the previous publications [9,41].

$$k_{t_{cross,i}} = k_{t_{cross,1}} (i+1)^{-\alpha_c} \quad if\ i \leq i_c \tag{2}$$

$$k_{t_{cross,i}} = 0 \quad if\ i > i_c \tag{3}$$

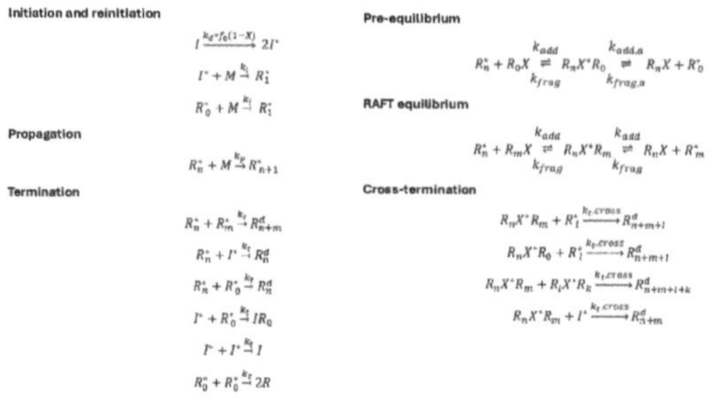

Scheme 1. Reaction steps of kinetic model. I, M, R_n^d, R_0X, and R_nX are initiator, monomer, terminated polymer, RAFT agent, and RAFT intermediate with one chain length, respectively. I^*, R_0^*, R_i^*, $R_nX^*R_0$, and $R_nX^*R_m$ are initiator radical, small leaving group left from RAFT agent, growing radical, RAFT intermediate radical with one chain length, and RAFT intermediate radical with two chain lengths.

3.3.2. Reaction Rates

The rate constants for the reactions incorporated within the model are represented in Table 2. Since the initiator, AIBN, is used in this experiment, the k_d and f values are taken from our previous publication and $k_i = 10{\cdot}k_p$ is assumed again [9]. In addition, the rate coefficients for the pre-equilibrium reaction steps are also added from a previous publication [9]. The kinetics of DMAPAm have not yet been studied, and therefore, there are no data on the rate coefficients of propagation and termination during the recombination of this monomer. However, DMAPAm can be considered a member of the family of N-containing methacrylates, similar to other monomers described in one of the recent publications of K. B. Kockler et al. [54]. This means that the propagation rate coefficient of DMAPAm should also be in the range described in this publication, which is assumed to be 2.50×10^3 [L mol^{-1} s^{-1}] for now, and this assumption will be checked at the end by comparing the rate coefficient of termination, estimated by applying parameter estimation, with the termination rate coefficients of other N-containing methacrylates belonging to the same family [54].

Table 2. Rate coefficients used for reaction kinetic modeling.

Rate Coefficient	A [Variable Unit]	E_A [K]	k (at 70 °C) [Variable Unit]	Reference
k_d	1.58×10^{15} [s^{-1}]	1.55×10^4	3.64×10^{-5} [s^{-1}]	[9]
f_0	-	-	0.6	[9]
k_p	2.44×10^6 [L mol^{-1} s^{-1}]	2.36×10^3	2.5×10^3 [L mol^{-1} s^{-1}]	[54]
k_i		$k_i = 10{\cdot}k_p$		[9]
$k_{t,cross}$				
k_t		Estimated in this work		
k_{add}				
k_{frag}				
$k_{add,a}$	-	-	1.00×10^4 [s^{-1}]	[9]
$k_{frag,a}$	-	-	2.00×10^4 [L mol^{-1} s^{-1}]	[9]

3.3.3. Model Validation

The kinetic model was developed using Predici, version 11. Laplace. 1.5, 2023, a software tool designed for the numerical simulation of polymerization reactions, describing the critical reaction steps mentioned in the previous section.

Altogether, two reaction kinetic experiments were performed with two different RAFT/initiator ratios (CDTPA/AIBN = 1:1 and 3:1). The reaction rate coefficients of cross-termination, termination by recombination, and addition–fragmentation reactions were estimated by fitting the kinetic model to the experimental data of monomer conversion ($X_{exp.}$ %), the theoretical number-average molar mass ($\overline{M}_{n,th}$, which was estimated using the equation existing in the literature [7,20,21] based on monomer conversion determined with ^{1}H-NMR measurement), and the apparent weight-average molar mass ($\overline{M}_{w,app}$, measured with GPC). For this fitting, a parameter estimation mode in the software Predici was used. In the parameter estimation, the data from the two experiments were inserted into the model, and the data sets of the experiments were fitted together, as shown in Figure 3.

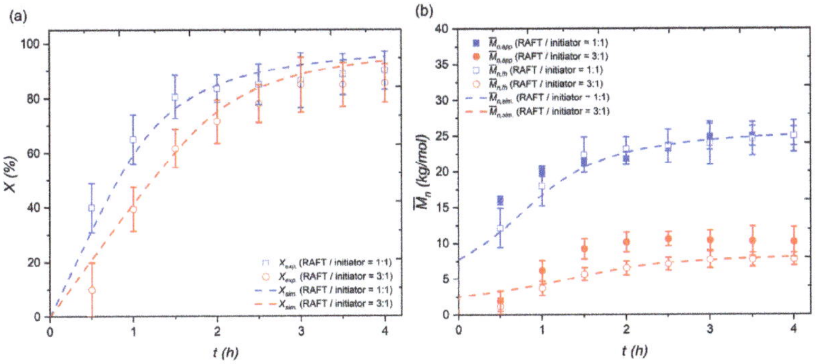

Figure 3. (**a**) Evolution of simulated ($X_{sim.}$) and experimental ($X_{exp.}$) DMAPAm conversion for two different RAFT/initiator ratios of 1:1 and 3:1. (**b**) Evolution of simulated ($\overline{M}_{n,sim.}$), apparent ($\overline{M}_{n,app}$), and theoretical ($\overline{M}_{n,th}$) number-average molar masses within period of 4 h for two different RAFT/initiator ratios of 1:1 and 3:1.

As depicted in Figure 3, the model accurately predicts the evolution of conversion and average molar mass. In particular, Figure 3a represents a complete overlap between the experimental ($X_{exp.}$) and simulated monomer conversion ($X_{sim.}$) data, whereas Figure 3b demonstrates that the experimental number-average molar mass ($\overline{M}_{n,exp.}$) data align closely with the simulated ($\overline{M}_{n,sim.}$) data throughout the entire reaction time. There is a minimal deviation of the apparent number-average molecular weight not only from the simulated $\overline{M}_n$ but also from the theoretical $\overline{M}_n$, which could be due to the fact that the used PMMA standard in this study is only valid at $\overline{M}_{w,app}$ < 20 kDa, as described in the literature [57], and this PMMA standard was chosen for comparison in our study because these Mark–Houwink parameters of PMMA were also determined by viscosimetry [57]. Therefore, better overlapping of $\overline{M}_{n,app}$ with $\overline{M}_{n,th}$ can be seen at a higher molar mass than this, especially in the case of RAFT/initiator = 1:1. In addition to this, at the beginning of the reaction, based on the simulation result, the $\overline{M}_n$ value in both cases increases very fast, which may be related to the fast decomposition of AIBN at 70 °C, which leads to the formation of a high amount of initiator radical at the start, as described in previous publications [11,62,63]. In addition, after 3 h of reaction time, the monomer conversion and the number-average molecular weight increase at a negligibly slow rate, which means the reaction can be terminated earlier than after 24 h [19].

From the fitting mentioned before, the values of rate coefficients of cross-termination, termination by recombination, and addition—fragmentation reactions were estimated (Table 3).

Table 3. Estimated values of rate coefficients of cross-termination, termination by recombination, and addition–fragmentation reactions from parameter estimation (PE).

$k_{t,cross}$	k_t	k_{add}
8.00×10^9 [L mol^{-1} s^{-1}]	4.00×10^6 [L mol^{-1} s^{-1}]	1.00×10^4 [L mol^{-1} s^{-1}]

The estimated rate coefficients agree quite well with the values available in the literature [9,35,41,60]. The assumption made regarding the possible cross-termination of the intermediate RAFT radical (which possesses two chain lengths) with the oligomeric growing radical was proven to be correct, as the equilibrium constant for the RAFT equilibrium ($K = \frac{k_{add}}{k_{frag}}$) was calculated to be around 5, which is lower than 10^2. This means that this model follows the intermediate radical termination model of Monteiro et al. [60], predicting equilibrium constants that are several orders of magnitude lower than those of the slow fragmentation model of Barner-Kowollik et al. [23]. In other words, some of the radical RAFT species are consumed for a cross-termination reaction with oligomeric growing radicals [61], leading to an equilibrium constant several orders of magnitude lower than those of the slow fragmentation model [23].

Another assumption was concerning the value of the propagation rate coefficient of DMAPAm, which was assumed to be 2.5×10^3 [L mol^{-1} s^{-1}], which is in the range of propagation rate coefficients of N-containing methacrylates. The termination rate coefficient of DMAPAm was estimated to be 4.00×10^6 [L mol^{-1} s^{-1}], which is the same as the termination rate coefficient of DMAEMA [51], which also belongs to N-containing methacrylates. This means that the assumption regarding the kp value was correct, since the final estimated value of k_t is also within the values of the termination rate coefficients of other N-containing methacrylates.

To investigate the kinetics of the living reaction, a change in the number-average molar mass ($\overline{M}_n$) of the polymer as a function of the monomer conversion (X%) is observed. Figure 4a indicates this change based on the experimental number-average molar mass ($\overline{M}_{n,app}$) (from GPC) and the monomer conversion (X_{exp}.) (from ^{1}H NMR), whereas Figure 4b demonstrates this change using the simulated number-average molar mass ($\overline{M}_{n,sim.}$) and monomer conversion (X_{sim}.). It can be clearly seen that both the experimental and simulated data show that $\overline{M}_n$ changes linearly with the monomer conversion, which means that the system exhibits pseudo-living reaction kinetics [11,64,65]. This linear increase is due to the controlled nature of RAFT polymerization, where the chain transfer agents regulate the growth of polymer chains, ensuring that all chains grow simultaneously and at a similar or uniform rate [66–68], which, in the end, makes the molecular weight distribution remain narrow [64,65].

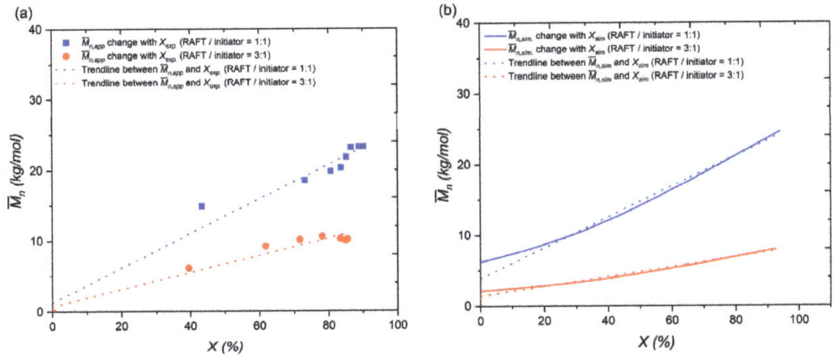

Figure 4. Change in number-average molar mass with monomer conversion. (**a**) Apparent number-average molar mass ($\overline{M}_{n,app}$) (from GPC) change with monomer conversion ($X_{exp.}$) (from proton NMR) for both RAFT/initiator ratios of 1:1 and 3:1 and their corresponding linear trendlines. (**b**) Simulated number-average molar mass ($\overline{M}_{n,sim.}$) change with simulated monomer conversion ($X_{sim.}$) for both RAFT/initiator ratios of 1:1 and 3:1 and their corresponding linear trendlines.

3.4. Model Prediction and Application for Scaling Up

Using the validated model, the apparent weight-average molecular weight and the livingness of the PDMAPAm homopolymer were predicted and compared with the experimentally determined $\overline{M}_{n,app}$ values and the estimated homopolymer livingness from the experimental results. The simulation of the homopolymer livingness in the model is based on the equation existing in the literature [69]. In contrast, the experimental livingness of the PDMAPAm homopolymer could be determined either by comparing the theoretical $\overline{M}_n$ of the diblock copolymer with the apparent $\overline{M}_n$ or by analyzing the GPC raw data of the diblock copolymer synthesized from the macroRAFT of both kinetic experiments, as described in a previous publication [4]. Since the GPC data of diblock copolymers did not show any second peaks corresponding to any potential macroRAFT rest, the experimental livingness was estimated only by comparing the theoretical $\overline{M}_n$ of the diblock copolymer with the apparent $\overline{M}_n$ (see more in Supplementary Materials, Equation (S1)). The results of the comparison are shown in Figure 5.

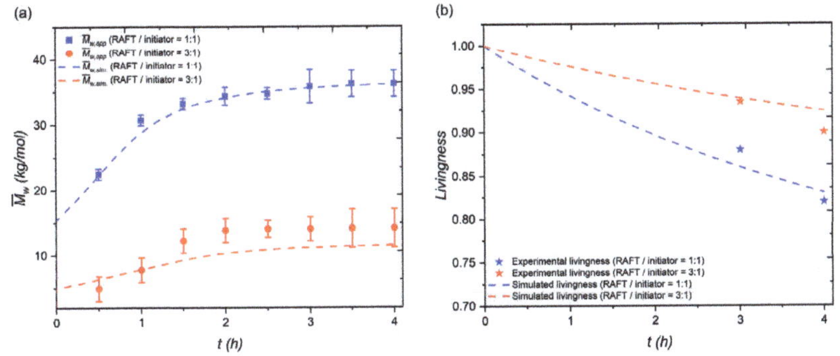

Figure 5. (**a**) Evolution of simulated ($\overline{M}_{w,sim}$) and experimental ($\overline{M}_{w,app}$) weight-average molar masses for two different RAFT/initiator ratios of 1:1 and 3:1. (**b**) Evolution of simulated livingness and experimentally determined livingness (based on ratio GPC peaks) for two different RAFT/initiator ratios of 1:1 and 3:1.

It can be seen from Figure 5 that, in general, the apparent weight-average molecular weight (Figure 5a) and the livingness of the homopolymer (Figure 5b) within the reactions were predicted correctly for both kinetic experiments.

On closer inspection, there is a slight deviation in the simulated $\overline{M}_w$ from the experimental values in the kinetic experiment with a higher RAFT agent concentration, where the maximum $\overline{M}_w$ is close to 15 kDa, which is still outside the PMMA calibration range of GPC. The deviation is almost the same as in the $\overline{M}_n$ comparison, which means that this could be related to the chosen PMMA standard (valid above 20 kDa) for the evaluation.

Furthermore, the model predicts the livingness, in other words, the chain-end fidelity of PDMAPAm, very well. It was determined that after 4 h of reaction, for a RAFT/initiator ratio of 3:1, the experimental livingness is around 90%, whereas the value reaches at 83% for the lower-RAFT-concentration case, as illustrated in Figure 5; these values are equivalent to 4–5 DMAPAm monomeric units, which could be difficult to identify with GPC. These values of experimentally determined livingness agree quite well with the predicted livingness of the PDMAPAm from both kinetic experiments. There is a slight deviation in the predicted livingness for the lower-RAFT-concentration experiment, but the deviation is so small that it can be considered negligible.

Moreover, based on the model prediction, the livingness of the PDMAPAm homopolymer decreases with reaction time for both RAFT/initiator ratios, as shown in Figure 5b, because more radicals undergo termination during the polymerization, as stated in previous publications [9,10]. As for comparing the reaction kinetic experiments with each other, the simulation shows that a higher RAFT concentration leads to higher livingness of the homopolymer in the end. This trend is also followed by the experimental livingness of the synthesized PDMAPAm. Additionally, after 3 h of reaction time, no substantial change is observed in the value of the average molar masses. However, the livingness of the homopolymer still decreases with the same gradient. At the same time, experimentally, it can be seen that macroRAFT livingness is higher at the end of 3 h than at the end of 4 h of reaction time. Therefore, the same average molar mass is obtained by stopping the reaction earlier with even higher livingness, as experimentally proven in a previous publication [19].

In Figure 6, the simulated molecular weight distributions (MWDs) are correlated with the experimental MWDs of the polymerization product measured with GPC for both experiments after 3 h of reaction. It can be clearly seen from the data that the model-predicted dispersity for both reactions is well estimated, because the breadth of the simulated and experimental MWDs for both reactions is the same as that calculated by the model. However, generally, the experimental MWDs and MWDs simulated by the model do not overlap for the experimental RAFT/initiator ratio of 3:1, and the peak point, M_p (representing the most probable molecular weight of the polymer chains in the sample), of the experimental and simulated MWDs are not close to each other. As mentioned before, this discrepancy was observed between simulated $\overline{M}_n$ and experimental $\overline{M}_n$, and this was justified with the evaluation standards of the GPC results.

Considering the predicted $\overline{M}_w$, livingness, and optimal reaction time for both the kinetic experiment protocols and comparing the prediction results of the kinetic model with our aim stated at the start of this study, the synthesis protocol with a RAFT/initiator ratio of 1:1 is considered to be optimal, since it leads to more-or-less acceptable livingness, which leads to the desired weight-average molar mass for the homopolymer (25 kDa) at the end of the 3 h of reaction time and, at the same time, a PDMAPAm/PMMA block ratio of 70:30.

Using the optimal recipe (RAFT/initiator ratio of 1:1), the synthesis of the homopolymer PDMAPAm was increased from 10 mL lab-scale synthesis to 100 mL glass reactor synthesis. Afterwards, PDMAPAm was used as a macroRAFT to synthesize a PDMAPAm-*b*-PMMA diblock copolymer with the desired molar mass and PDMAPAm/PMMA block ratio. Based on the ^1H NMR result, the conversion of MMA for the synthesis of the second block was calculated to be over 90%.

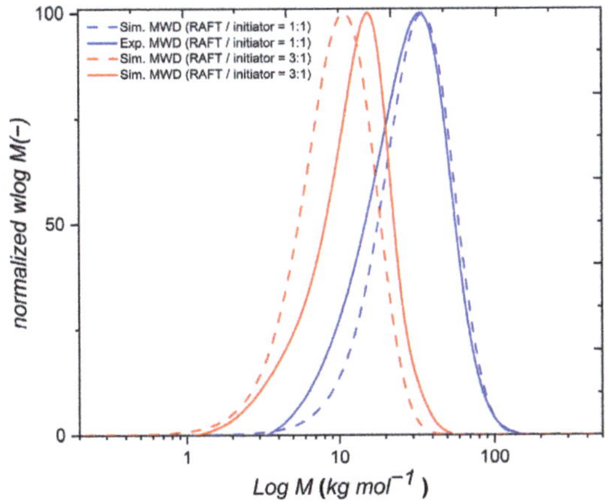

Figure 6. Molecular weight distributions (MWDs) of PDMAPAm. Simulated (Sim.) MWD represented by dashed lines and experimental (Exp.) MWD shown as continuous lines for both RAFT/initiator ratios of 1:1 and 3:1 at end of 3 h reaction time.

In Figure 7, the MWDs of the synthesized PDMAPAm and PDMAPAm-*b*-PMMA in the glass reactor are presented as measured with GPC. Comparing the MWDs of both the homopolymer and diblock copolymer, it is obvious from the graph that the MWD of PDMAPAm shifts slightly to higher molar masses as its chain is extended with the second monomer MMA. This shift clearly shows the growth of the second block [7,9,10,19].

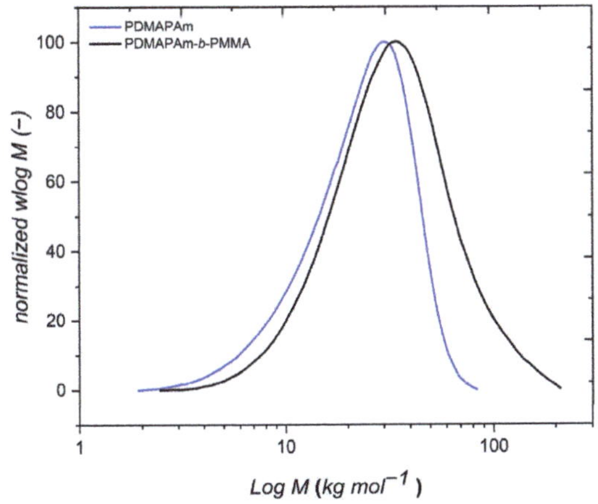

Figure 7. Molar mass distribution curve of PDMAPAm and PDMAPAm-*b*-PMMA measured with GPC.

Moreover, Table 4 shows that the weight-average molar mass of the synthesized PDMAPAm-*b*-PMMA is around 40 kDa, and the ratio of the PDMAPAm block to the

PMMA block is around 70:30, in line with the aim stated at the beginning. This means that the targeted molecular characteristics for the diblock polymer were achieved.

Table 4. The apparent number and weight-average molecular weight along with the dispersity of the synthesized PDMAPAm-*b*-PMMA.

Polymer	$\overline{M}_{n,app}$ (kDa)	$\overline{M}_{w,app}$ (kDa)	Đ
$PDMAPAm_{165}$	19	26	1.4
$PDMAPAm_{163}$-*b*-$PMMA_{130}$	26	39	1.5

3.5. Preliminary Results of CO_2-Responsiveness of Polymer

To demonstrate the preliminary results of the CO_2-responsive behavior of the PDMAPAm-*b*-PMMA diblock copolymer, a small experiment was performed.

First, the pH of commercial water by "Viva con Agua" (Hamburg, Germany) was measured to be around 8.25, and then, by degassing, it was saturated with CO_2, which reduced the pH to 5.29 due to the formation of carbonic acid ($CO_2 + H_2O \rightleftarrows H_2CO_3$), which dissociates into bicarbonate (HCO_3^-) and hydrogen ions (H^+) [70], lowering the pH.

The diblock copolymer was incrementally introduced into the CO_2-saturated water, and the resultant pH changes in the solution were systematically measured at room temperature (23 °C) (Figure 8).

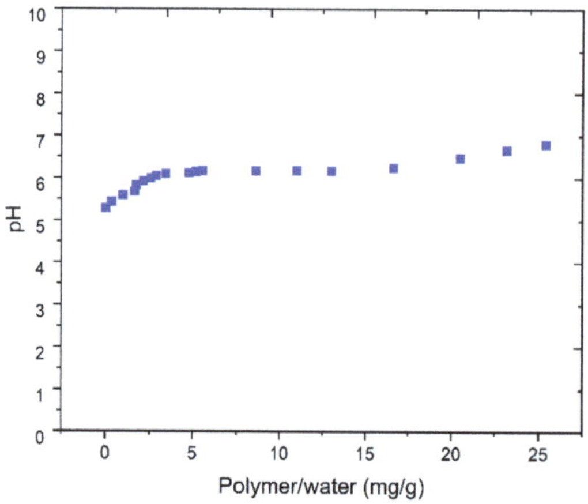

Figure 8. Resultant pH changes in CO_2-saturated water with dissolved diblock copolymer.

It can be clearly seen from Figure 8 that as the PDMAPAm-*b*-PMMA diblock copolymer is added into the CO_2-saturated water, its pH increases from 5.29 to 7.10. This might be related to the fact that as soon as the diblock copolymer is added into the CO_2-saturated water, the tertiary amine group on the PDMAPAm block can theoretically react with CO_2 and water ($R_3N + CO_2 + H_2O \rightleftarrows R_3NH^+ + HCO_3^-$) [71–73] and form bicarbonate ions. As a result, this can reduce the proton concentration in the water and results in an increase in the pH value. This interaction of the tertiary amine group with water and CO_2 and the formation of bicarbonate ions show that the diblock copolymer is CO_2-reponsive, as mentioned in several publications [71,72], and could be a potential material for CO_2 capture [71,72].

4. Conclusions

A reaction kinetic model for the RAFT polymerization of DMAPAm was developed and validated with two different reaction kinetic experiments. Good agreement was obtained between simulated and experimental monomer conversion and the number-average molar masses. To validate the developed model along with more concrete parameter estimation, the Mark–Houwink parameters of PDMAPAm were estimated in order to adapt the molar masses of the polymer. Within the validation, it was determined that propagation and termination by the recombination rate coefficients of DMAPAm follow the values of the corresponding rate coefficients of other N-containing monomers. At the same time, it was identified that the Mark–Houwink parameters of PDMAPAm fall within the range of the parameters of N-containing methacrylate polymers. This means that DMAPAm also belongs to the family of N-containing monomers, and the developed model is applicable to the RAFT polymerization of other N-containing monomers. In addition, both simulated and experimental $\overline{M}_n$ as a function of the monomer conversion relation showed that the system has pseudo-living reaction kinetics, referring to the existence of the RAFT kinetics of the system. After the validation, the model was used to predict livingness, in other words, the chain-end group fidelity; weight-average molar mass; and MWD of the homopolymer. Using the kinetic model, the synthesis protocol of the homopolymer PDMAPAm was optimized to achieve the desired $\overline{M}_w$ and livingness with the highest monomer conversion in an optimal reaction time. In the end, the optimal recipe was implemented to synthesize the CO_2-responsive diblock copolymer of PDMAPAm-b-PMMA with the desired molar mass (40 kDa) and molar mass ratio of the PDMAPAm block to the PMMA block (70:30) on a larger scale, which can be implemented in direct air capture technology.

In the future, to make the prediction of the simulation more precise and the examination of the reaction kinetics more thorough, a calorimeter can be utilized to determine the polymerization rate experimentally, in other words, to determine the exact propagation rate coefficient of the polymerization, and to investigate the energy balance of the RAFT polymerization of DMAPAm, which could also open a door to optimizing the reaction in terms of energy consumption and to bring the process to industry.

Supplementary Materials: The following supporting information can be downloaded at: https://www.mdpi.com/article/10.3390/polym17081115/s1, Table S1. Recipe of kinetic experiment for PDMAPAm synthesis; Table S2. Recipe for scale-scaling up of PDMAPAm synthesis; Table S3. Recipe of PDMAPAm-b-PMMA synthesis; Figure S1. Reduced viscosity change with polymer concentration in THF for PDMAPAm homopolymers with different molar masses (15 kDa, 25 kDa, 42 kDa, 94 kDa); Figure S2. Plot of intrinsic viscosity versus molar mass of both PMMA and PDMAPAm homopolymers [57,58].

Author Contributions: Conceptualization, P.G.; Data Curation, E.P.; Formal Analysis, E.P.; Funding Acquisition, P.G.; Investigation, E.P.; Methodology, E.P.; Project Administration, P.G.; Resources, P.G.; Software, E.P.; Supervision, P.G.; Validation, E.P. and P.G.; Visualization, E.P.; Writing—Original Draft, E.P.; Writing—Review and Editing, P.G. All authors have read and agreed to the published version of the manuscript.

Funding: This work was supported by the Initiative and Networking Fund of the Helmholtz Association, project DACStorE (KA2-HSC-12).

Institutional Review Board Statement: Not applicable.

Data Availability Statement: The characterization data are available upon request from the authors.

Acknowledgments: The authors thank Silvio Neumann and Maren Brinkmann for performing the NMR and GPC measurements, as well as Felix Kandelhard for the discussion on the polymerization kinetics. Scientific discussions with Volker Abetz are gratefully acknowledged.

Conflicts of Interest: The authors declare no conflict of interest.

Abbreviations

The following abbreviations are used in this manuscript:

DMAPAm	N-[3-(dimethylamino)propyl]-acrylamide
MMA	Methyl methacrylate
PDMAPAm-*b*-PMMA	Poly(N-[3-(dimethylamino)propyl]-acrylamide)-b-poly(methyl methacrylate)
RAFT	Reversible Addition−Fragmentation Chain-Transfer
CDTPA	4-cyano-4-[(dodecylsulfanylthiocarbonyl)-sufanyl]pentanoic acid
AIBN	2,2′-Azobis(2-methylpropionitrile)
$CDCl_3$	Deuterated chloroform
THF	Tetrahydrofuran
4VP	4-Vinylpyridine
^{1}H NMR	Proton Nuclear Magnetic Resonance Spectroscopy
GPC	Gel permeation chromatography

References

1. Bui, M.; Adjiman, C.S.; Bardow, A.; Anthony, E.J.; Boston, A.; Brown, S.; Fennell, P.S.; Fuss, S.; Galindo, A.; Hackett, L.A. Carbon capture and storage (CCS): The way forward. *Energy Environ. Sci.* **2018**, *11*, 1062–1176. [CrossRef]
2. Ozkan, M.; Nayak, S.P.; Ruiz, A.D.; Jiang, W. Current status and pillars of direct air capture technologies. *Iscience* **2022**, *25*, 103990. [CrossRef] [PubMed]
3. Erans, M.; Sanz-Pérez, E.S.; Hanak, D.P.; Clulow, Z.; Reiner, D.M.; Mutch, G.A. Direct air capture: Process technology, techno-economic and socio-political challenges. *Energy Environ. Sci.* **2022**, *15*, 1360–1405. [CrossRef]
4. Breyer, C.; Fasihi, M.; Bajamundi, C.; Creutzig, F. Direct air capture of CO_2: A key technology for ambitious climate change mitigation. *Joule* **2019**, *3*, 2053–2057. [CrossRef]
5. Murdock, C.R.; Didas, S.; Jones, C.W. Direct capture of CO_2 from ambient air. *Chem. Rev.* **2016**, *116*, 11840–11876.
6. Zhou, Z.; Liu, T.; Khan, A.U.; Liu, G. Block copolymer–based porous carbon fibers. *Sci. Adv.* **2019**, *5*, eaau6852. [CrossRef] [PubMed]
7. Nieswandt, K.; Georgopanos, P.; Abetz, V. Well-defined polyvinylpyridine-block-polystyrene diblock copolymers via RAFT aqueous-alcoholic dispersion polymerization: Synthesis and isoporous thin film morphology. *Polym. Chem.* **2021**, *12*, 2210–2221. [CrossRef]
8. Shevate, R.; Kumar, M.; Karunakaran, M.; Canlas, C.; Peinemann, K.-V. Surprising transformation of a block copolymer into a high performance polystyrene ultrafiltration membrane with a hierarchically organized pore structure. *J. Mater. Chem. A* **2018**, *6*, 4337–4345. [CrossRef]
9. Kandelhard, F.; Pashayev, E.; Schymura, J.; Georgopanos, P. Kinetic Modeling of the Synthesis of Poly (4-vinylpyridine) Macro-Reversible Addition-Fragmentation Chain Transfer Agents for the Preparation of Block Copolymers. *Ind. Eng. Chem. Res.* **2023**, *62*, 8696–8708. [CrossRef]
10. Pashayev, E.; Kandelhard, F.; Georgopanos, P. Experimental & Modelling Digital Twin Approach for Polymer Synthesis via Re-initiated Oxygen inhibited RAFT Polymerization. *Macromol. React. Eng.* **2023**, *17*, 2200068.
11. Kandelhard, F.; Schuldt, K.; Schymura, J.; Georgopanos, P.; Abetz, V. Model-Assisted Optimization of RAFT Polymerization in Micro-Scale Reactors—A Fast Screening Approach. *Macromol. React. Eng.* **2021**, *15*, 2000058. [CrossRef]
12. Pashayev, E.; Kandelhard, F.; Georgopanos, P. Sauerstoffempfindlichkeit der RAFT-Polymerisation–Eine Modellierungsstudie. *Chem.-Ing.-Tech.* **2024**, *96*, 805–812. [CrossRef]
13. Shieh, Y.-T.; Lin, Y.-T.; Cheng, C.-C. CO_2-switchable behavior of chitosan-g-poly [(2-dimethylamino) ethyl methacrylate] as an emulsifier. *Carbohydr. Polym.* **2017**, *170*, 281–288. [CrossRef] [PubMed]
14. Zhang, C.; Zhou, J.; Ye, X.; Li, Z.; Wang, Y. CO_2-responsive membranes prepared by selective swelling of block copolymers and their behaviors in protein ultrafiltration. *J. Membr. Sci.* **2022**, *641*, 119928. [CrossRef]
15. Cunningham, M.F.; Jessop, P.G. An introduction to the principles and fundamentals of CO_2-switchable polymers and polymer colloids. *Eur. Polym. J.* **2016**, *76*, 208–215. [CrossRef]

16. Riabtseva, A.; Ellis, S.N.; Champagne, P.; Jessop, P.G.; Cunningham, M.F. CO_2-responsive branched polymers for forward osmosis applications: The effect of branching on draw solute properties. *Ind. Eng. Chem. Res.* **2021**, *60*, 9807–9816. [CrossRef]
17. Demirci, S.; Sutekin, S.D.; Guven, O.; Sahiner, N. Poly (2-aminoethyl methacrylate) based microgels catalyst system to be used in hydrolysis and methanolysis of $NaBH_4$ for H_2 generation. *Int. J. Hydrog. Energy* **2023**, *48*, 23002–23012. [CrossRef]
18. Ji, W.; Panus, D.; Palumbo, R.N.; Tang, R.; Wang, C. Poly (2-aminoethyl methacrylate) with well-defined chain length for DNA vaccine delivery to dendritic cells. *Biomacromolecules* **2011**, *12*, 4373–4385. [CrossRef]
19. Pashayev, E.; Georgopanos, P. CO_2-Responsive Copolymers for Membrane Applications, Synthesis, and Performance Evaluation. *Macromol. Mater. Eng.* **2025**, *310*, 2400290. [CrossRef]
20. Nieswandt, K.; Georgopanos, P.; Abetz, C.; Filiz, V.; Abetz, V. Synthesis of poly (3-vinylpyridine)-block-polystyrene diblock copolymers via surfactant-free RAFT emulsion polymerization. *Materials* **2019**, *12*, 3145. [CrossRef]
21. Nieswandt, K.; Georgopanos, P.; Held, M.; Sperling, E.; Abetz, V. RAFT Emulsion Polymerization of Styrene Using a Poly ((N, N-dimethyl acrylamide)-co-(N-isopropyl acrylamide)) mCTA: Synthesis and Thermosensitivity. *Polymers* **2021**, *14*, 62. [CrossRef]
22. Abetz, V. Isoporous block copolymer membranes. *Macromol. Rapid Commun.* **2015**, *36*, 10–22. [CrossRef] [PubMed]
23. Barner-Kowollik, C.; Quinn, J.F.; Morsley, D.R.; Davis, T.P. Modeling the reversible addition–fragmentation chain transfer process in cumyl dithiobenzoate-mediated styrene homopolymerizations: Assessing rate coefficients for the addition–fragmentation equilibrium. *J. Polym. Sci. Part A Polym. Chem.* **2001**, *39*, 1353–1365. [CrossRef]
24. Ribelli, T.G.; Konkolewicz, D.; Bernhard, S.; Matyjaszewski, K. How are radicals (re) generated in photochemical ATRP? *J. Am. Chem. Soc.* **2014**, *136*, 13303–13312. [CrossRef] [PubMed]
25. Liu, C.; Hong, C.-Y.; Pan, C.-Y. Polymerization techniques in polymerization-induced self-assembly (PISA). *Polym. Chem.* **2020**, *11*, 3673–3689. [CrossRef]
26. Delplace, V.; Guégain, E.; Harrisson, S.; Gigmes, D.; Guillaneuf, Y.; Nicolas, J. A ring to rule them all: A cyclic ketene acetal comonomer controls the nitroxide-mediated polymerization of methacrylates and confers tunable degradability. *Chem. Commun.* **2015**, *51*, 12847–12850. [CrossRef]
27. Konkolewicz, D.; Hawkett, B.S.; Gray-Weale, A.; Perrier, S. RAFT polymerization kinetics: Combination of apparently conflicting models. *Macromolecules* **2008**, *41*, 6400–6412. [CrossRef]
28. Antonopoulou, M.-N.; Whitfield, R.; Truong, N.P.; Wyers, D.; Harrisson, S.; Junkers, T.; Anastasaki, A. Concurrent control over sequence and dispersity in multiblock copolymers. *Nat. Chem.* **2022**, *14*, 304–312. [CrossRef]
29. Shimizu, T.; Whitfield, R.; Jones, G.R.; Raji, I.O.; Konkolewicz, D.; Truong, N.P.; Anastasaki, A. Controlling primary chain dispersity in network polymers: Elucidating the effect of dispersity on degradation. *Chem. Sci.* **2023**, *14*, 13419–13428. [CrossRef]
30. Mishima, E.; Yamago, S. Controlled random and alternating copolymerization of (meth) acrylates, acrylonitrile, and (meth) acrylamides with vinyl ethers by organotellurium-, organostibine-, and organobismuthine-mediated living radical polymerization reactions. *J. Polym. Sci. Part A Polym. Chem.* **2012**, *50*, 2254–2264. [CrossRef]
31. Breulmann, M. *Synthesis, Properties and Applications of Acrylonitrile ± Styrene ± Acrylate Polymers*; John Wiley & Sons: Hoboken, NJ, USA, 2003; p. 341.
32. Saunders, K.; Saunders, K. *Acrylic Polymers*; Springer Science & Business Media: Berlin, Germany, 1988; pp. 125–148.
33. Peinemann, K.-V.; Abetz, V.; Simon, P.F. Asymmetric superstructure formed in a block copolymer via phase separation. *Nat. Mater.* **2007**, *6*, 992–996. [CrossRef] [PubMed]
34. Desmet, G.B.; De Rybel, N.; Van Steenberge, P.H.; D'hooge, D.R.; Reyniers, M.F.; Marin, G.B. Ab-Initio-Based Kinetic Modeling to Understand RAFT Exchange: The Case of 2-Cyano-2-Propyl Dodecyl Trithiocarbonate and Styrene. *Macromol. Rapid communications* **2018**, *39*, 1700403. [CrossRef] [PubMed]
35. De Rybel, N. Deterministic Modeling of the Interplay Between Molecular Diffusion and Chemical Kinetics of Bulk RAFT Polymerization. Doctoral Dissertation, Ghent University, Ghent, Belgium, 2019.
36. Chiefari, J.; Chong, Y.; Ercole, F.; Krstina, J.; Jeffery, J.; Le, T.P.; Mayadunne, R.T.; Meijs, G.F.; Moad, C.L.; Moad, G. Living free-radical polymerization by reversible addition-fragmentation chain transfer: The RAFT process. *Macromolecules* **1998**, *31*, 5559. [CrossRef]
37. Moad, G.; Rizzardo, E.; Thang, S.H. Living radical polymerization by the RAFT process. *Aust. J. Chem.* **2005**, *58*, 379–410. [CrossRef]
38. Hernández-Ortiz, J.C.; Jaramillo-Soto, G.; Palacios-Alquisira, J.; Vivaldo-Lima, E. Modeling of Polymerization Kinetics and Molecular Weight Development in the Microwave-Activated RAFT Polymerization of Styrene. *Macromol. React. Eng.* **2010**, *4*, 210–221. [CrossRef]
39. Smulders, W.; Gilbert, R.G.; Monteiro, M.J. A Kinetic Investigation of Seeded Emulsion Polymerization of Styrene Using Reversible Addition– Fragmentation Chain Transfer (RAFT) Agents with a Low Transfer Constant. *Macromolecules* **2003**, *36*, 4309–4318. [CrossRef]

40. Goto, A.; Sato, K.; Tsujii, Y.; Fukuda, T.; Moad, G.; Rizzardo, E.; Thang, S.H. Mechanism and kinetics of RAFT-based living radical polymerizations of styrene and methyl methacrylate. *Macromolecules* **2001**, *34*, 402–408. [CrossRef]
41. De Rybel, N.; Van Steenberge, P.H.; Reyniers, M.F.; Barner-Kowollik, C.; D'hooge, D.R.; Marin, G.B. An update on the pivotal role of kinetic modeling for the mechanistic understanding and design of bulk and solution raft polymerization. *Macromol. Theory Simul.* **2017**, *26*, 1600048. [CrossRef]
42. De Rybel, N.; Van Steenberge, P.H.; Reyniers, M.-F.; D'hooge, D.R.; Marin, G.B. How chain length dependencies interfere with the bulk RAFT polymerization rate and microstructural control. *Chem. Eng. Sci.* **2018**, *177*, 163–179. [CrossRef]
43. López-Domínguez, P.; Rivera-Peláez, J.E.; Jaramillo-Soto, G.; Barragán-Aroche, J.F.; Vivaldo-Lima, E. Modeling of RAFT polymerization of MMA in supercritical carbon dioxide using the PC-SAFT equation of state. *React. Chem. Eng.* **2020**, *5*, 547–560. [CrossRef]
44. López-Domínguez, P.; Zapata-González, I.; Saldívar-Guerra, E.; Vivaldo-Lima, E. Mathematical modelling of RAFT polymerization. *RAFT Polym. Methods Synth. Appl.* **2021**, *1*, 187–221.
45. Devlaminck, D.J.; Van Steenberge, P.H.; Reyniers, M.-F.; D'hooge, D.R. Deterministic modeling of degenerative RAFT miniemulsion polymerization rate and average polymer characteristics: Invalidity of zero–one nature at higher monomer conversions. *Macromolecules* **2018**, *51*, 9442–9461. [CrossRef]
46. Tou, J.C.; Whiting, L.F. The thermokinetic performance of an accelerating rate calorimeter. *Thermochim. Acta* **1981**, *48*, 21–42. [CrossRef]
47. Luan, B.; Li, C.e.; Moad, G.; Muir, B.W.; Zhu, J.; Patel, J.; Lim, S.; Hao, X. Kinetic modelling of the reversible addition–fragmentation chain transfer polymerisation of N-isopropylacrylamide. *Eur. Polym. J.* **2019**, *120*, 109193. [CrossRef]
48. Liu, Q.; Zhang, P.; Qing, A.; Lan, Y.; Lu, M. Poly (N-isopropylacrylamide) hydrogels with improved shrinking kinetics by RAFT polymerization. *Polymer* **2006**, *47*, 2330–2336. [CrossRef]
49. Quiñonez-Angulo, P.; Ruiz-Villegas, J.; Licea-Claverie, Á.; Ramirez-Jiménez, A.; Miranda-Soto, V.; Zapata-González, I. A kinetic study, thermal analysis and kinetic modeling on homo and copolymerization of 2-(N, N-diethylamino) ethyl methacrylate and PEGMA. *Eur. Polym. J.* **2018**, *109*, 347–359. [CrossRef]
50. Krasia, T.C.; Patrickios, C.S. Amphiphilic polymethacrylate model co-networks: Synthesis by RAFT radical polymerization and characterization of the swelling behavior. *Macromolecules* **2006**, *39*, 2467–2473. [CrossRef]
51. Ajogbeje, O.J.; Lacík, I.; Hutchinson, R.A. The influence of monomer ionization and hydrolysis on the radical polymerization kinetics of 2-(dimethylamino) ethyl methacrylate in aqueous solution. *Polym. Chem.* **2023**, *14*, 2624–2639. [CrossRef]
52. Masuelli, M.A. Mark-Houwink parameters for aqueous-soluble polymers and biopolymers at various temperatures. *J. Polym. Biopolym. Phys. Chem.* **2014**, *2*, 37–43.
53. Masuelli, M.A. Dextrans in aqueous solution. Experimental review on intrinsic viscosity measurements and temperature effect. *J. Polym. Biopolym. Phys. Chem.* **2014**, *1*, 13–21.
54. Kockler, K.B.; Fleischhaker, F.; Barner-Kowollik, C. Free radical propagation rate coefficients of N-containing methacrylates: Are we family? *Macromolecules* **2016**, *49*, 8572–8580. [CrossRef]
55. Sakellariou, G.; Siakali-Kioulafa, A.; Hadjichristidis, N. Synthesis, chain flexibility, and glass-transition temperature of poly (2, 2-diphenylethyl methacrylate). *Int. J. Polym. Anal. Charact.* **2003**, *8*, 269–277. [CrossRef]
56. Kockler, K.B.; Fleischhaker, F.; Barner-Kowollik, C. Investigating the propagation kinetics of a novel class of nitrogen-containing methacrylates via PLP-SEC. *Polym. Chem.* **2016**, *7*, 4342–4351. [CrossRef]
57. Wagner, H.L. The Mark–Houwink–Sakurada relation for poly (methyl methacrylate). *J. Phys. Chem. Ref. Data* **1987**, *16*, 165–173. [CrossRef]
58. Janča, J.; Vlček, P.; Trekoval, J.; Kolínský, M. Universal calibration in gel permeation chromatography for poly (methyl methacrylate) and poly (2-methoxyethyl methacrylate). *J. Polym. Sci. Polym. Chem. Ed.* **1975**, *13*, 1471–1473. [CrossRef]
59. Gruendling, T.; Junkers, T.; Guilhaus, M.; Barner-Kowollik, C. Mark–Houwink Parameters for the Universal Calibration of Acrylate, Methacrylate and Vinyl Acetate Polymers Determined by Online Size-Exclusion Chromatography—Mass Spectrometry. *Macromol. Chem. Phys.* **2010**, *211*, 520–528. [CrossRef]
60. Monteiro, M.J.; de Brouwer, H. Intermediate radical termination as the mechanism for retardation in reversible addition–fragmentation chain transfer polymerization. *Macromolecules* **2001**, *34*, 349–352. [CrossRef]
61. Konkolewicz, D.; Hawkett, B.S.; Gray-Weale, A.; Perrier, S. RAFT polymerization kinetics: How long are the cross-terminating oligomers? *J. Polym. Sci. Part A Polym. Chem.* **2009**, *47*, 3455–3466. [CrossRef]
62. Magri, G.; Barter, M.; Fletcher-Charles, J.; Choi, H.; Slocombe, D.; Richards, E.; Folli, A.; Porch, A.; Murphy, D.M. An in situ study of the thermal decomposition of 2, 2′-azobis (2-methylpropionitrile) radical chemistry using a dual-mode EPR resonator. *Res. Chem. Intermed.* **2023**, *49*, 289–305. [CrossRef]

63. Magri, G.; Barter, M.; Fletcher-Charles, J.; Choi, H.; Slocombe, D.; Richards, E.; Folli, A.; Porch, A.; Murphy, D.M. An in-situ study of the thermal decomposition of AIBN radical chemistry using a dual mode EPR resonator. *Preprint* **2022**. [CrossRef]
64. Chernikova, E.; Zaitsev, S.; Plutalova, A.; Mineeva, K.; Zotova, O.; Vishnevetsky, D. Control over the relative reactivities of monomers in RAFT copolymerization of styrene and acrylic acid. *RSC Adv.* **2018**, *8*, 14300–14310. [CrossRef]
65. Barner-Kowollik, C. *Handbook of RAFT Polymerization*; John Wiley & Sons: Hoboken, NJ, USA, 2008.
66. Moad, G. RAFT polymerization to form stimuli-responsive polymers. *Polym. Chem.* **2017**, *8*, 177–219. [CrossRef]
67. Moad, G.; Rizzardo, E.; Thang, S.H. End-functional polymers, thiocarbonylthio group removal/transformation and reversible addition–fragmentation–chain transfer (RAFT) polymerization. *Polym. Int.* **2011**, *60*, 9–25. [CrossRef]
68. Moad, G. RAFT polymerization–then and now. In *Controlled Radical Polymerization: Mechanisms*; ACS Publications: Washington, DC, USA, 2015; pp. 211–246.
69. Perrier, S. 50th Anniversary Perspective: RAFT Polymerization—A User Guide. *Macromolecules* **2017**, *50*, 7433–7447. [CrossRef]
70. Crooks, J.E.; Donnellan, J.P. Kinetics of the reaction between carbon dioxide and tertiary amines. *J. Org. Chem.* **1990**, *55*, 1372–1374. [CrossRef]
71. Chowdhury, F.A.; Yamada, H.; Higashii, T.; Goto, K.; Onoda, M. CO_2 Capture by Tertiary Amine Absorbents: A Performance Comparison Study. *Ind. Eng. Chem. Res.* **2013**, *52*, 8323–8331. [CrossRef]
72. Lai, N.; Zhu, Q.; Qiao, D.; Chen, K.; Tang, L.; Wang, D.; He, W.; Chen, Y.; Yu, T. CO_2 capture with absorbents of tertiary amine functionalized nano–SiO_2. *Front. Chem.* **2020**, *8*, 146. [CrossRef]
73. Chen, M.; Luo, Q.; Lin, H.; Sun, Q.; Gao, H.; Liu, S.; Li, Y.; Liang, Z. A study on reaction mechanism and kinetics of CO_2 and MEA/DEA-tertiary amines in non-aqueous and water-lean solutions. *Chem. Eng. Sci.* **2023**, *269*, 118431. [CrossRef]

polymers

MDPI

Review

Hybrid Materials Based on Self-Assembled Block Copolymers and Magnetic Nanoparticles—A Review

Galder Kortaberria

Group 'Materials + Technologies' (GMT), Chemical and Environmental Engineering Department, Faculty of Engineering of Gipuzkoa, University of the Basque Country (UPV/EHU), Plaza Europa 1, 20018 Donostia-San Sebastian, Spain; galder.cortaberria@ehu.eus

Abstract: In this review work, the different routes and methods for preparing hybrid materials based on nanostructured block copolymers (BCPs) and magnetic nanoparticles (MNPs) are analyzed, as they can be potentially employed in different sectors like biomedicine, electronic or optoelectronic devices, data storing devices, etc. The first procedure for their preparation consists of the nanostructuring of BCPs in the presence of previously synthesized NPs by modifying their surface for increasing compatibility with the matrix or employing magnetic fields for NP orientation, which can also promote the orientation of nanodomains. Surface modification with surfactants led to the selective confinement of NPs depending on the interaction (mainly hydrogen bonding) degree and their intensity. Surface modification with brushes can be performed by three methods, including grafting from, grafting to, or grafting through. Those methods are compared in terms of success for the positioning and confinement of NPs in the desired domains, showing the crucial importance of brush length and grafting density, as well as of NP amount and modification degree in the self-assembled morphology. Regarding the use of external magnetic fields, the importance of relative amounts of MNPs and BCPs employed and that of the magnetic field intensity for the orientation of the NPs and the nearby BCP domains is shown. The second procedure, consisting of the in situ synthesis of NPs inside the nanodomains by a reduction in the respective metallic ions or employing metal-containing BCPs for the generation of MNP patterns or arrays, is also shown. In all cases, the transference of magnetic properties to the nanocomposite was successful. Finally, a brief summary of some aspects about the use of BCPs for the synthesis, encapsulation, and release of MNPs is shown, as they present potential biomedical applications such as cancer treatment, among others.

Keywords: magnetic nanoparticles; block copolymers; nanocomposites; templates; patterns

1. Introduction

Organic/inorganic nanocomposites are hybrid materials in which an inorganic filler or nanoparticles (NPs) are dispersed or selectively placed into the organic part, usually a polymeric matrix. Those inorganic fillers or NPs transfer their specific optical, electric, or magnetic properties to the hybrid material [1–3], with potential applications in different sectors such as solar cells, sensors, or magnetic storage devices, among others [4–6]. Some of the most important hybrid materials are those in which block copolymers (BCPs) constitute the organic part.

BCPs can self-assemble, generating nanostructures such as lamellar, cylindrical, or spherical ones, among others. For this reason, they can be employed to lead to the self-assembly of NPs inside those nanostructures, resulting in materials designed at the nanoscale that can present improved properties [7]. BCPs can be employed either for the synthesis of NPs or for their placement in high-ordered nanostructures, resulting in nanomaterials with specific optical, magnetic, or electronic properties. The concept of synergistic assembly should be mentioned at this point, as BCPs can rearrange their nanostructures due to their interactions with the inorganic fillers.

Many works can be found in the literature related to hybrid materials based on several BCP matrices and different inorganic NPs, such as silver [4,8], gold [9,10], TiO_2 [1,11], or quantum dots like CdSe [3,12–14], among others, but there is still work ongoing and to be conducted in this field, and is the object of many investigations.

On the other hand, magnetic nanoparticles (MNPs) have been deeply investigated by many researchers in many fields, such as magnetic fluids [15], catalysis [16], biotechnology [17], magnetic resonance imaging [18], and data storage [19]. Different types of MNPs have been synthesized with different compositions and phases, including iron oxides [20]; pure metals such as Fe and Co [21,22]; spinel-type ferromagnets such as $MgFe_2O_4$, $MnFe_2O_4$, and $CoFe_2O_4$ [23,24]; and alloys such as $CoPt_3$ and FePt [25,26]. During the last few years, several investigations have been conducted on the synthesis of iron oxides such as Fe_3O_4 (magnetite), α-Fe_2O_3 (hematite), γ-Fe_2O_3 (maghemite), FeO (wüstite), ε-Fe_2O_3, and β-Fe_2O_3 [27], with magnetite and maghemite being the most employed ones.

Regarding the procedures employed for the fabrication of hybrid BCP–MNP nanocomposites, they can be divided into two main types. The first one consists of first synthesizing the NPs with a controlled size (or using commercial ones), and then nanostructuring the BCP in their presence, taking into account that the presence of NPs can modify matrix morphologies due to interactions or by selective positioning. The second procedure consists of the in situ synthesis of MNPs inside the nanostructured BCP during the nanostructure formation in controlled conditions.

For the first procedure, several strategies can be employed. One of them is the modification of the MNP surface with surfactants or with polymeric chains (brushes) grown on their surface. The coating of the NP surface with surfactants leads to interactions (hydrogen bonding, hydrophilic or hydrophobic attraction) with a specific block of the BCP in order to be placed at those domains, in addition to avoiding the formation of aggregates, which is a trend with most of NPs [28–31]. This is the easiest procedure, as there are many surfactants available and the chemistry is relatively simple. Another strategy is the growing of polymeric chains (brushes) that present affinity (or even are the same) with one of the blocks for the selective placement of MNPs at the desired domains [3,12,13,32–35]. There are three different methods for brush growing, known as grafting to, grafting from, and grafting through. All those procedures will be explained and compared in their respective sections of the manuscript, including a table for a better understanding.

The last of the strategies for the first main procedure is the employment of magnetic fields for the alignment or selective placement of MNPs in nanostructured BCPs [29,31,36,37]. In this case, a magnetic field whose intensity is controlled is applied during nanocomposite formation. Several authors showed that, by controlling the intensity (usually with the distance between magnets) of the magnetic field and the amount of MNPs employed, alignment can be achieved, in some cases orienting the BCP domains surrounding the MNPs.

The second main pathway employed for hybrid nanocomposites consists of synthesizing MNPs in situ, inside the nanodomains corresponding to one of the blocks, by a reduction in the corresponding metallic ions. In this vein, during the last few years, there has been ongoing research on the employment of BCPs as templates for the synthesis of MNP arrays with specific locations, or even for MNP encapsulation and posterior release, with potential applications in biomedicine or drug delivery [38–45]. The scheme for both main pathways can be seen in Figure 1.

Examples of the proper dispersion and placement of MNPs without any modification or magnetic field can also be found [46], as well as of the control of particle/solvent and polymer/solvent interactions in order to promote the selective placement of MNPs at the desired domains of the block copolymer [47].

In the present work, the studies performed by many authors on the synthesis and characterization of BCP/MNP hybrid systems by different strategies or methods are reviewed, showing the main findings and potential applications of the generated materials. The review will be divided into sections regarding the different strategies employed by authors: surface modification with surfactants, the growing of polymer brushes, the use of the magnetic field, and the role of BCPs as templates for MNP synthesis, encapsulation, or release. This last part will only be introduced and shown briefly, as there are many ongoing studies, and the aim of the present review is mostly to show the rest of the procedures for preparing thin-film composites based on hybrid BCP/MNP materials.

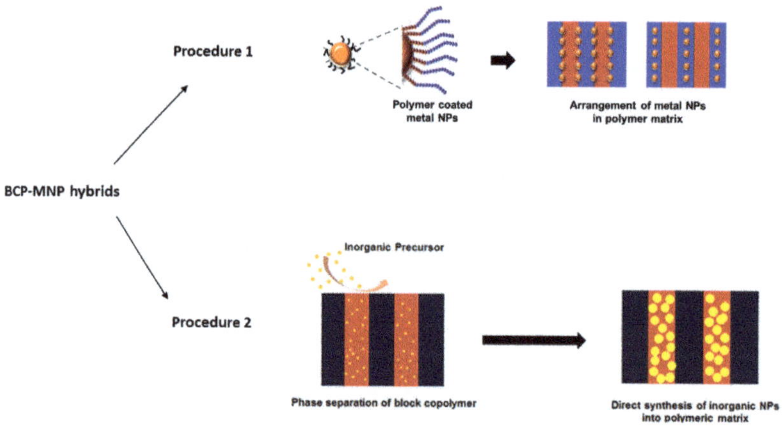

Figure 1. Schematic representation of the two main procedures for the preparation of BCP–MNP hybrid materials. (Adapted with permission from [48], Frontiers 2015).

2. Surface Modification with Surfactants

This method, considered the simplest one, is based on the use of previously synthesized MNPs, with the surface modified with surfactants, which are added to the BCP that is afterwards assembled into different nanostructures [28–31]. Surfactant molecules that coated the surface of the MNPs present higher affinity towards one of the blocks, leading to the confinement of the MNPs into the nanodomains corresponding to that block. However, several authors have pointed out that, especially for ferromagnetic nanofillers such as iron oxides, cobalt, and nickel, the strong magnetic attractions among them led to the aggregation of MNPs.

Several studies can be found on the use of surfactants for dispersing MNPs in nanostructured BCPs. Char et al. [28] employed poly(styrene-b-isoprene) (PS-b-PI) assembled into PI cylinders and poly(styrene-b-4-vinylpiridine) (PS-b-P4VP) assembled into PS cylinders as templates for the placement of oleic acid-coated Fe_2O_3 MNPs. Their main finding was that, independently of the employed BCP, the placement of MNPs and the formation of aggregates depended on the solvents employed for casting nanocomposite films: When solvents solving both blocks were employed, the aggregation of nanofillers occurred during film formation. On the other hand, with the use of solvents selective for one of the blocks (forming the cylinders), MNPs were selectively placed into the cylinders and the inversion of micelles occurred during film formation. They concluded that depending on the degree of interaction between the MNPs and one of the blocks, nanofillers should be placed either at the interfaces (disrupting the nanostructure when the MNPs were increased) or inside the minor domains.

Park et al. [31] also employed oleic acid-capped Fe_2O_3 nanoparticles for dispersal into a poly(3-hexylthiopene-b-ethylene glycol) (P3HT-b-PEG) copolymer by a magnetic field. For this reason, this study will be analyzed in the section devoted to magnetic field application.

Miles et al. [30] employed the hydrogen bonding interactions among gallic acid-capped FePt nanoparticles and P2VP nanodomains of a PS-b-P2VP diblock copolymer to selectively place them into those nanodomains, obtaining materials with good magneto-optical properties and presenting little trade-off in terms of scattering loss. MNPs improved the Faraday rotation properties. The control of BCP nanostructures and surface functionalization of MNPs allowed for hybrid materials with potential magneto-optical applications to be obtained.

J. Shin et al. [49] analyzed the effect of MNPs on the nanostructure of nanocomposites formed by symmetric poly(styrene-b-1,4-butadiene) (PS-b-PB) BCP with Fe_3O_4 or Au nanoparticles, comparing the effect of NP type as well as their placement at the nanodomains. They first synthesized Au and Fe_3O_4 nanoparticles stabilized with oleyl ligands. The confined co-assembly of symmetric poly(styrene-b-1,4-butadiene) (PS-b-PB) BCP and NPs in evaporative emulsions resulted in particles with various morphologies, including striped ellipsoids, onion-like particles, and their intermediates. The main difference among PS-b-PB/Au and PS-b-PB/Fe3O4 was in the distribution of NPs, which affected the overall particle morphology. In the case of PS-b-PB/Fe_3O_4 ellipsoids, MNPs clustered and segregated to the particle/surrounding interface of the ellipsoids, being selectively placed in the middle of the PB domains in a string-like pattern for PS-b-PB/Fe_3O_4 onion-like particles. The scheme of the employed procedure is shown in Figure 2.

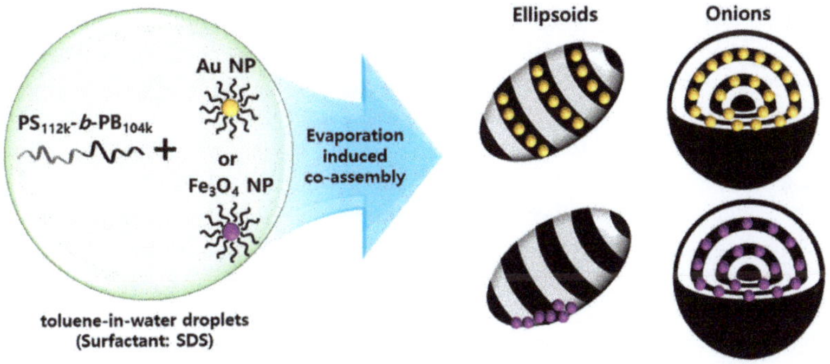

Figure 2. Illustration showing co-assembly of inorganic nanoparticles (Au NP and Fe_3O_4 NPs) with PS-b-PB copolymer by controlled solvent evaporation from a toluene-in-water emulsion. Reproduced with permission from Ref. [49].

3. Surface Modification with Polymeric Brushes

Polymer brushes are polymeric chains anchored or grafted to a surface, with a grafting density high enough for the surface-attached chains to become crowded and expand outward from the surface [50]. Depending on the employed criteria, different classifications of brushes can be found. Depending on the use of linear chains, planar surfaces, or spherical NPs as a substrate for brush growing, brushes can be divided into one-dimensional, two-dimensional, or three-dimensional, respectively. If their composition and architecture are taken into account, brushes can be classified into homopolymeric, mixed homopolymeric, block copolymer, or branched polymer brushes.

Moreover, different methods can be used for grafting polymeric chains to NP surfaces, as shown in Figure 3.

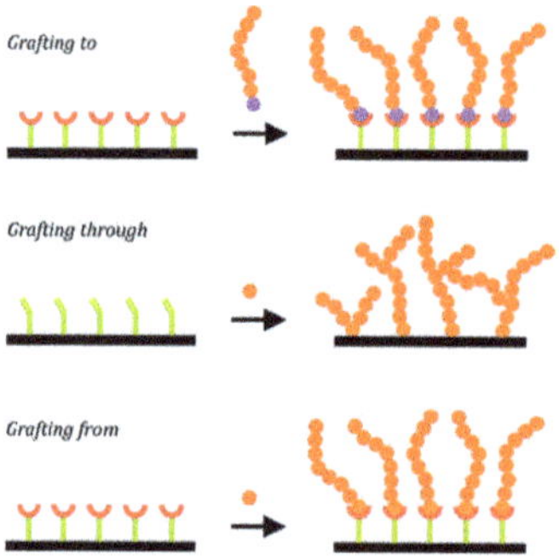

Figure 3. Different techniques for polymer brush grafting.

The first method is known as grafting to: polymers with suitable end-functional groups react with appropriate surface sites on the inorganic nanoparticles [51,52]. The second method, grafting through, consists of linking a monomer by a covalent bond to a surface, from which a surface copolymerization is carried out, incorporating the inorganic part in the polymer chains [13,53]. The growth of chains is initiated in solution, and during propagation, a surface-bound monomer unit can be integrated into the growing chains, with polymer chains being permanently grafted to the surface. The growth of chains can continue with the addition of new units that are either free or attached to the surface. The third method, known as grafting from, consists of grafting initiator molecules (or surfactant molecules) to the surface, from which chains will grow in situ. Several techniques can be used for the polymerization: controlled radical polymerization (CRP), ionic living polymerization, or ring-opening polymerization (ROP) [54,55].

Many works can be found in the literature regarding the use of polymeric brushes for increasing the compatibility with a BCP matrix and placing the MNPs into the desired domains. Those works will be classified in terms of the method employed for the generation of brushes.

3.1. Brushes by the Grafting from Method

Initiator molecules previously anchored to the surface of NPs are responsible for the chain growth, which can occur via different polymerization techniques, obtaining high grafting densities as well as more finely tuned control of the molar mass and polydispersity. Among the variety of radical polymerization methods that can be used (atom transfer radical polymerization (ATRP), reversible addition–fragmentation chain transfer, and nitroxide mediated polymerization), ATRP has been the most widely employed by different authors. In order to obtain uniform growth of chains, the step consisting of an atom transfer is the most important, for which two techniques can be used: the use of a sacrificial free initiator or the employment of a persistent radical for better control of the polymerization.

I. Garcia et al. [56–59] grafted PS or PMMA brushes from the surface of Fe_3O_4 MNPs by the ATRP technique in order to selectively confine them into PS or PMMA nanodomains of different self-assembled BCPs. Firstly, they grafted PS brushes from an NP surface (PS–MN) to add them (1, 2, and 4 wt% NPs) to a commercial poly(styrene-b-butadiene-b-styrene) (SBS) block copolymer [56,57]. A grafting density of 0.9 chains/nm^2 was obtained, in addition to measuring the molar mass of the growth chains after cleavage. After solution mixing, the mixtures were cast onto silicon wafers, nanostructuring them by thermal annealing. By measuring the Tg of both blocks from the BCP and analyzing the effect of MNPs on them (both modified and unmodified ones) and characterizing the obtained morphologies in terms of atomic force microscopy (AFM), they pointed out that PS–MN was located at the PS nanodomains. The improvement in mechanical properties observed for the SBS/PS–MN system compared to that of the SBS/MN one indicated higher interactions and better interfacial adhesion promoted by the PS brushes. Moreover, the increase in the thickness of the PS lamellae, disrupting the morphology, indicated the presence of PS–MN at PS domains. In conclusion, they pointed out that a very good and selective dispersion of PS–MN was obtained in the SBS, maintaining the morphology of a neat copolymer and increasing the mechanical properties.

In a second study, they grafted PMMA brushes (with a grafting density of around 0.1 chains/nm^2) from the surface of magnetic Fe_3O_4 nanoparticles (PMMA–MN) by ATRP to prepare nanocomposites with a poly(2-vinylpyridine-b-methylmethacrylate) (P2VP-b-PMMA) block copolymer by solution mixing and spin coating onto glass wafers [58]. The obtained films were annealed by exposure to CCl_4 vapors (selective for PMMA)—that is, by solvent vapor annealing (SVA). They employed both AFM and magnetic force microscopy (MFM) to analyze the effect of SVA and MNP modification on generated nanostructures or morphologies. By analyzing the effect of exposure time on the morphology of the neat BCP, they found an evolution of the nanostructure: While a hexagonal morphology was obtained after 3 h of exposure, after 9 h, a quasi lamellar morphology normal to the substrate was observed (as CCl_4 is selective for PMMA, it promoted the migration of P2VP from the free surface to the substrate). After 14 h, PMMA migrated to the surface, forming the continuous phase, while P2VP appeared as the discontinuous one. They obtained several conclusions by analyzing the nanocomposites: For 2 wt% PMMA–MN, the hexagonal nanostructure of the neat BCP was maintained, with PMMA–MN being selectively placed at PMMA domains. On the other hand, for the nanocomposite with 4 wt% PMMA–MN, they observed that some agglomerations began to appear. Moreover, the transference of the magnetic properties of the NPs to the nanocomposite thin films was confirmed by MFM.

C. Xu et al. [33] also employed the ATRP polymerization technique for grafting PMMA brushes from the surface of Fe_3O_4 MNPs with the aim of obtaining a good dispersion into a nanostructured PS-b-PMMA block copolymer. Brushes with 2.7, 13.3, and 35.7 kg/mol were obtained, with a grafting density of 0.73 chains/nm^2. Mixtures (with 1 to 16 wt% MNPs) prepared in toluene were spin coated onto silicon wafers to then be thermally annealed for different times at 160 °C. The increase in the molecular weight of the brushes led to a poorer dispersion of MNPs, which tended to form aggregates. They found two types of aggregates: small ones, located at the defects of the lamellae, and larger ones that disrupted the nanostructure, appearing to be encapsulated by onion-ring copolymer structures. The parallel lamellar nanostructure of the neat BCP was maintained for the nanocomposite containing 1 wt% NPs modified with brushes presenting the lowest molar mass, with the nanoparticles appearing to be properly dispersed into the PMMA domains. The higher amount of those NPs avoided the constitution of lamellar morphology, as they tended to aggregate. When NPs grafted with 13.3 kg/mol brushes were employed for nanocomposites, films with 4 wt% NPs presented small aggregates within the PMMA domains, as well as some big agglomerates placed at the boundaries that were unable to be confined into the PMMA domains. For all nanocomposites with NPs presenting brushes of 35.7 kg/mol, aggregates were formed independently of the amount, disrupting the morphology, as their size appeared to be bigger than the lamellar period. The importance of the molar mass of the brushes grafted to the NPs was shown.

S. Douadi-Masrouki et al. [34] also grafted PS brushes from a magnetic γ-Fe_2O_3 nanoparticle surface by ATRP in order to incorporate them into lamellar poly(styrene-b-butyl methacrylate) (PS-b-PBMA) diblock copolymer films. They were synthesized by ATRP symmetrical diblock copolymers in order to obtain lamellar morphologies. Nanocomposites were prepared by solution mixing in toluene, followed by spin coating onto silicon wafers and thermal annealing at 150 °C (well above the T_g of PS and below the order–disorder transition temperature) for 48 h. The neat BCP presented a lamellar nanostructure, with some defects on top of the films. For the NPs, they found that during the synthesis of the brushes, as a consequence of reactions among chains growing from adjacent NPs, they tended to form clusters. However, as the size of those clusters was lower than the lamella thickness, they were flattened between the lamellae, being compatible with self-assembly. They showed nanocomposites presenting the lamellar structure of the neat BCP, with NPs located in the lamellae, despite the NP dispersal being irregular.

Barandiaran et al. [60] also functionalized Fe_2O_3 MNPs with PMMA brushes by ATRP and the grafting from method to prepare nanocomposites with a poly(isoprene-b-methylmethacrylate) (PI-b-PMMA) diblock copolymer. AFM and small-angle X-ray scattering (SAXS) were employed for the characterization of the obtained nanostructures, both for the neat BCP and for the nanocomposites. Morphological characterization of the neat BCP and the nanocomposites was performed in terms of AFM, both for the as-cast films and for those annealed with solvent vapors. The surface perpendicular lamellar nanostructure presented by the as-cast film of the neat BCP was changed to a mixed perpendicular and parallel lamellar morphology for a low NP concentration, with some cylinders appearing for higher NP content. When annealed under solvent vapors, the neat BCP was raised into a cylindrical morphology and maintained for nanocomposites with low NP concentrations. The highest NP concentration, however, led to the disruption of the morphology. By removing the organic part of the nanocomposites by UV irradiation, the authors confirmed the selective placement of the NPs into PMMA domains as a consequence of the presence of PMMA brushes on their surface, as shown in Figure 4. The transference of magnetic properties to the nanocomposites was corroborated in terms of magnetic characterization.

Hartmann et al. [32] analyzed the effect of FePt NPs presenting PMMA brushes grafted from their surface on the nanostructures generated by the self-assembly of a PS-b-PMMa diblock copolymer. They analyzed the placement of NPs within PMMA domains at low NP amounts statistically, concluding that higher NP amounts altered the lamellar nanostructure of the neat BCP. The positioning of NPs into PMMA domains was also analyzed theoretically, and it was found that most of them were placed at lamellar PMMA domains. They concluded that, as the diameter of NPs is smaller than the width of PMMA domains, they did not alter the structure of the BCP morphology, as shown in Figure 5. For NP loadings of higher than 0.5 wt%, they negatively affected the order of the lamellar nanostructure. They demonstrated that the magnetic nature of the NPs inside the nanostructured morphology was maintained with bimodal magnetic force microscopy.

Upadhyaya et al. [35] employed Fe_3O_4 MNPs coated with PMMA brushes (grafting from technique) to prepare mixed-matrix membranes based on a poly(methacrylic acid-b-methyl methacrylate) (PMAA-b-PMMA) block copolymer synthesized by RAFT polymerization. They prepared thin-film membranes by casting or spin coating solutions prepared with different amounts of BCP in tetrahydrofurane and MNPs in water. The pore density was increased by the incorporation of MNPs in all cases, while for the film preparation method, those obtained by casting showed lower flux values compared to those prepared by spin coating, mainly because of the difference in membrane thickness obtained by each procedure. By applying an external magnetic film, the flux values of the membranes prepared by both methods were increased, probably due to the movement of the MNPs inside the organic matrix.

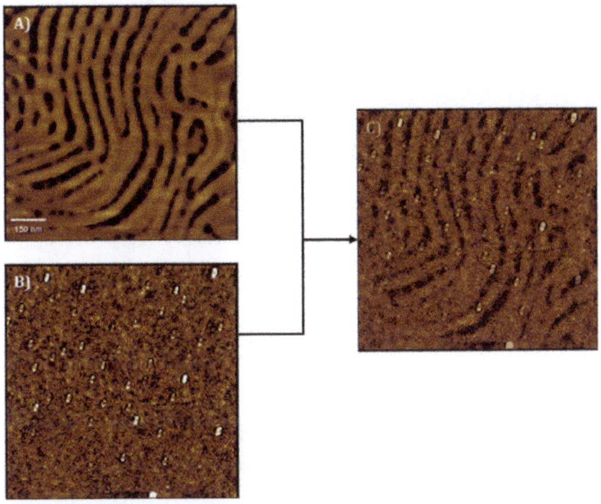

Figure 4. (**A**) AFM-phase image of the as-cast nanocomposite films with 1 wt% nanoparticles, (**B**) AFM-phase image of nanocomposite film with 1 wt% nanoparticles exposed to UV light irradiation for 48 h, and (**C**) superposition of (**A**,**B**). (Reproduced with permission from Ref. [60], copyright Elsevier 2016).

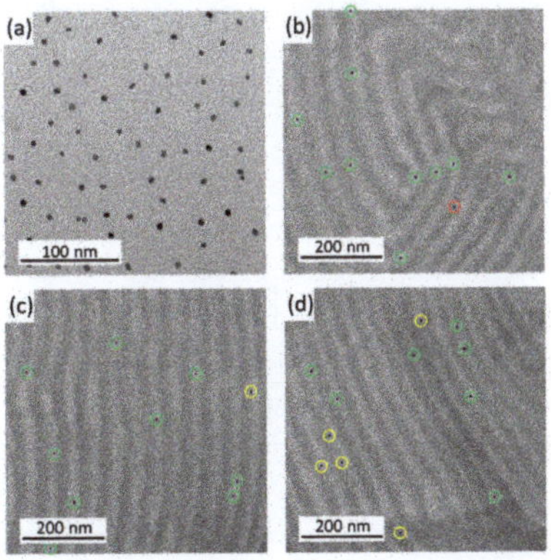

Figure 5. TEM images of the BCP films with 0.1 wt% FePt NP. (**a**) shows the pure FePt NPs, while (**b–d**) are different images of the lamellar PS-b-PMMA morphology with 0.1 wt% FePt NPs. The particles are marked with a green ring in the light PMMA phase, red in the dark PS phase, and yellow on the boundary [32].

3.2. *Brushes by Grafting to Method*

The method consists of the employment of polymers with end-functional groups such as thiol groups, which are grafted to NP surfaces by the functional groups (such as –OH ones, among others) present on them. Though this method is not as extended as the previous one, several authors have employed it, mainly for modifying Au nanoparticles with polymeric brushes for selective placement in BCP nanodomains [61–64], but an interesting study with magnetic nanoparticles can also be found [65].

By this method, I. Barandiaran et al. [65] grafted a poly(methyl methacrylate-b-ε-caprolactone) (PMMA-b-PCl) diblock copolymer to the surface of Fe_2O_3 MNPs to prepare thin-film nanocomposites with poly(styrene-b-ε-caprolactone) (PS-b-PCl). They employed the ATRP polymerization technique to prepare a PMMA-b-PCl copolymer (21,000 g/mol) presenting terminal chlorine groups. MNPs were silanized with 3-aminopropyltryethoxysilane (APTS).

BCPs were attached to the surface of MNPs by the grafting to method. The amine functionalities at the surface of the MNPs underwent an alkylation reaction with the terminal chlorine group, obtaining grafting density values of around 0.04 chains/nm^2. The scheme for both procedures is shown in Figure 6.

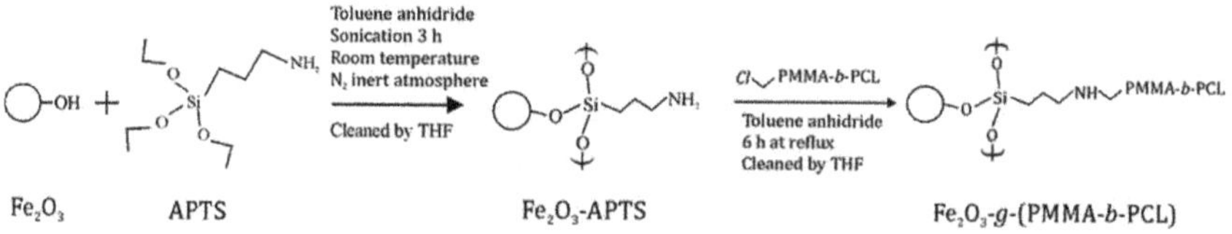

Figure 6. Reaction scheme for the APTS silanization reaction and PMMA-b-PCL copolymer anchoring by the grafting to reaction. (Reproduced with permission from Ref. [65], copyright Elsevier 2014).

Nanocomposite thin films containing 2 and 5 wt% MNPs were prepared by spin coating BCP/MNP solutions in toluene glass substrates. The films were thermally annealed under vacuum at 100 or 120 °C. The neat BCP self-assembled into a worm-like nanostructure for the films annealed at 100 °C, while a lamellar morphology was obtained for those annealed at 120 °C. Nanocomposites resulted in the same morphologies as those of the neat BCP, both at 100 and at 120 °C. By comparing nanocomposites with modified and unmodified MNPs, they found that those with brushes were better dispersed. The use of unmodified MNPs resulted in aggregates, altering the morphology of the BCP. The presence of PCL blocks at the surface seemed to improve the compatibility with the BCP matrix. Regarding the selective placement of MNPs, they claimed that even though some of them appeared at PCL domains, most of them were placed at the interface between the PCL and PS domains, without altering the BCP nanostructure. For samples annealed at 100 °C, MNPs presented average sizes of 25 and 40 nm (for nanocomposites with 2 and 5 wt%, respectively), while for films annealed at 120 °C, average sizes of 22 and 38 nm were measured. As the initial size of the MNPs was around 9 nm, they concluded that the MNPs were not individually dispersed. By TEM characterization they showed that during the modification procedure, the MNPs formed agglomerates surrounded by PMMA-b-PCL, presenting average sizes that were very similar to those found in the nanocomposites. Thus, aggregates were formed during the modification process and not during their dispersion into the BCP. The modification of the surface improved the MNP dispersion without altering the nanostructure of the BCP, as was previously found by other authors.

3.3. Brushes by Grafting Through Method

By this method, molecules presenting functional groups that are able to polymerize (silanes with vinyl groups, for example) are fixed to the surface of MNPs. In this way, NPs present in the polymerization medium are covered by the polymer. Usually, this method results in higher grafting densities compared to the grafting to technique, besides being relatively easier to perform. A kind of network is formed as polymer chains at the surface present bonds among them, due to the multifunctionality of the MNP surface [66].

Following a procedure from Etxeberria et al. [13] employed for grafting brushes to self-synthesized CdSe quantum dots, some authors from the same group used this method to generate PS or PMMA brushes at the surface of Fe_2O_3 MNPs, to then be dispersed into different BCPs [67–69].

In their first work, Barandiaran et al. [67], functionalized the surface of Fe_2O_3 MNPs with PS brushes by this technique (Figure 7) to prepare nanocomposites with PS-b-P4VP.

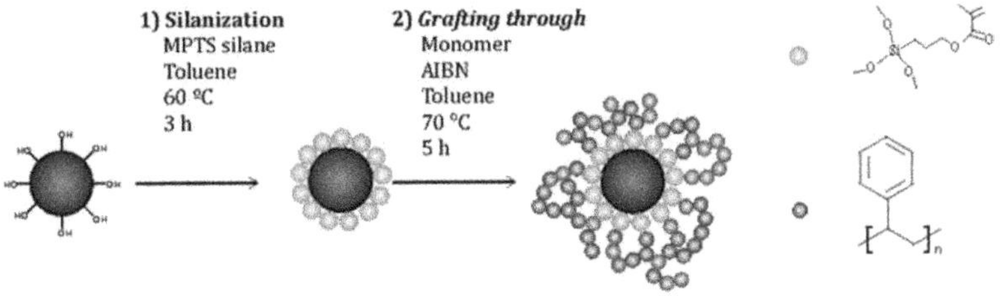

Figure 7. Scheme of the grafting through procedure for NP modification.

The amount of MNPs varied between 1 and 5 wt% at the nanocomposites. Thin films were annealed with saturated dioxane (selective for the PS block) vapors for several exposure times. When samples were annealed for 24 h, the hexagonal morphology of the neat BCP was maintained for the nanocomposite with 1 wt% NPs. However, by increasing the MNP amount, the morphology of the BCP was disrupted. When the films were annealed for 48 h, the neat BCP self-assembled into a stripped morphology, presenting a lamellar structure normal to the substrate. With 1 wt% MNPs, the morphology was maintained, but higher MNP amounts resulted in a nanostructure formed by cylinders perpendicular to the surface, despite some lamellae also were detected. This morphology evolution, in terms of AFM characterization, is shown in Figure 8. MNPs appeared to be well dispersed, without any agglomerations. The reason given by the authors for this evolution of the morphology was the following: as MNPs were selectively placed at PS domains (demonstrated by the elimination of the organic part by UV radiation), there was a reduction in their mobility. SQUID measurements were employed for the magnetic characterization of the nanocomposite films, showing that the magnetic properties were transferred to the nanocomposites.

In a second study [68], they modified the same maghemite MNPs with PMMA brushes via the grafting to technique to prepare nanocomposite thin films with PS-b-PMMA. Nanostructures both for the neat BCP and for the nanocomposites were obtained by SVA, with acetone employed as a selective solvent for PMMA blocks for 16 h, as was previously determined for obtaining a lamellar morphology for the neat BCP. They prepared nanocomposites with both silanized MNPs and those modified with brushes in order to compare the effect of the latter. The use of silanized MNPs resulted in the same lamellar morphology as that of the neat BCP, but with the presence of big agglomerates. Despite those agglomerates tending to gather at PMMA domains, as their size was bigger than that of the PMMA lamellae, they were not selectively placed there. With the purpose of selectively placing them in the PMMA domains, MNPs with PMMA brushes were employed. Even though the lamellar morphology was maintained for nanocomposites with 2 wt%, the morphology was not as regular as that of the neat BCP. No remarkable agglomerates were detected with MNPs at the PMMA domains. For nanocomposites of 5 wt%, the nanostructure started to change to a mixture of lamellae and perpendicular hexagonally packed cylinders. This seemed to indicate that modified MNPs were selectively placed at PMMA domains, altering the volume fraction among blocks and the equilibrium morphology. The authors degraded the organic part by UV radiation, showing that the MNPs were selectively placed. Moreover, SQUID measurements showed that the magnetic properties of nanoparticles were successfully transferred to the nanocomposite thin films.

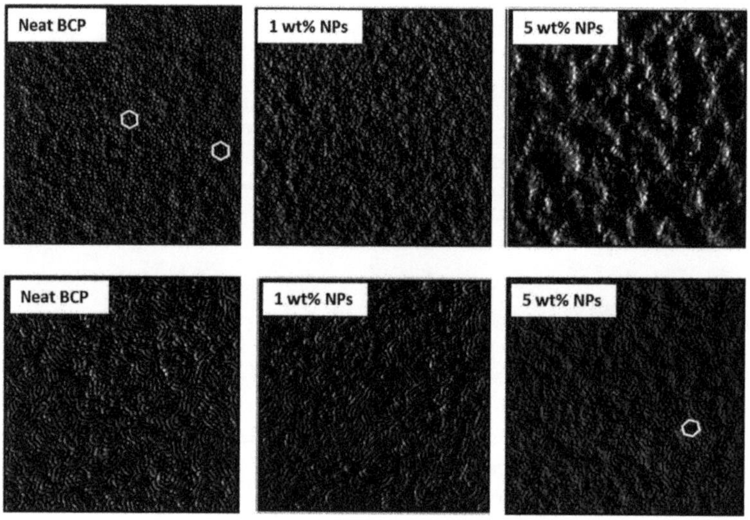

Figure 8. AFM-phase images showing the morphological evolution of the neat BCP and nanocomposites after 24 h (upper images) and 48 h (lower images) of exposure to dioxane vapors.

Finally, in a third study [69], they employed the grafting through technique to modify Fe_2O_3 NPs with PMMA or PS brushes for selective placement at the desired domains of a nanostructured poly(styrene-b-butadiene-b-styrene) (SBS) terpolymer. In order to analyze the effect of functionalization on the dispersion and selective placement of the NPs, nanocomposites were prepared by mixing an SBM copolymer with pristine, PS-, and PMMA-modified nanoparticles. Thin films of the neat BCP and nanocomposites (with 1, 2, and 5 wt% NPs) were prepared by casting SBM solutions in toluene onto silicon wafers. The morphology obtained was analyzed in terms of AFM. The neat BCP was nanostructured into a lamellar morphology, as found for ABA-type BCPs by other authors [4,70]. In this case, the lamellar structure showed an S–B–M–B sequence, with an average interlamellar distance of ~71 nm. For nanocomposites, independently of the modification with PMMA or PS, the lamellar morphology was maintained, with the MNPs appearing to be properly dispersed, without big aggregates. With the aim of analyzing the effect of brush type, nanocomposites with 5 wt% PMMA- and PS-modified MNPs were analyzed in detail. For PS-modified MNPs, domains corresponding to the PS block could be better distinguished in the AFM images, compared to the images obtained for nanocomposites with PMMA-modified MNPs. For nanocomposites with PMMA-modified

MNPs, PS and PMMA domains tended to join together. The comparison of profile images showed that the PS and PMMA domains tended to swell with the addition of MNPs, while for the neat BCP, four domains could be distinguished. This could be a consequence of the placement of MNPs at the PS and PMMA domains, increasing the respective domain volume. The authors related the morphology differences detected between the neat BCP and the nanocomposite thin films to the effect of MNPs on system thermodynamics, altering the interaction among blocks. To show the effect of the surface modification of MNPs, they also prepared nanocomposite thin films under the same conditions as MNPs without any functionalization, concluding that even though the lamellar morphology was maintained, big aggregates were detected, with the dispersion not being as good as for modified MNPs. The transference of magnetic properties to the nanocomposites in all cases was demonstrated by magnetic characterization in terms of SQUID measurements.

4. Alignment with External Magnetic Field

In order to control the location and/or orientation of MNPs in nanocomposites with BCPs, different trials have been carried out, such as the employment of external shear or electric fields, in addition to the previously shown surface modification. In this way, some authors employed an external magnetic field for generating aligned BCP/MNP nanocomposites. Moreover, this magnetic field can also direct their neighboring BCP domains [29,31,36,37].

Hammond et al. [37] presented a method that can be used for the orientation of BCP cylinders, both in thin and in thicker films, and in some cases, even for bulk systems. They showed that rod-shaped NPs could nucleate coaxially oriented BCP cylinders. They analyzed a system based on PS-b-P2VP assembled into P2VP cylinders in a PS matrix and Fe_2O_3 MNPs. Due to the weak ferromagnetic nature of MNPs, they tended to orient towards their major axes perpendicular to the magnetic field. For this reason, nanocomposites were prepared as films with a thickness of 500 nm: MNPs laid in the film plane and perpendicular to the magnetic field, being oriented in the desired direction. In this way, a unique particle orientation was employed. Mixtures containing 0.05 wt% MNPs were cast onto carbon-coated epoxy or mica substrates and placed between two permanent magnets. The solvent was allowed to evaporate slowly.

They concluded that NMPs effectively templated the orientation of nearby BCP cylinders. The application of a magnetic field during film casting led to a partial alignment of the MNPs, resulting in an overall alignment of the nanocomposite. Though the achieved alignment was modest, it can be considered a first step towards a complete alignment, opening a path for the generation of nanocomposites with proper combinations of anisotropic mechanical, magnetic, optical, and electrical properties, among others.

Yao et al. [36] also employed external magnetic fields to prepare nanocomposites based on a PS-b-PMMA copolymer and PS-coated Fe_2O_3 MNPs. By applying the magnetic field during the solvent casting procedure, they obtained nano- and microstructures by magnetic field-guided positioning of maghemite NPs.

For low MNP concentrations, they were selectively placed at PS domains, but by increasing the MNP amount, they tended to aggregate, disrupting the morphology, as they were placed at the surface as wire-shaped stripes. The transference of magnetic properties to the nanocomposites was demonstrated, with the films showing superparamagnetic behavior and shape anisotropy, making them suitable for potential magnetic applications. They were able to generate highly oriented metal oxide wires with different widths and lengths by controlling the MNP amount and the intensity of the magnetic field. The change in the morphology of the BCP from symmetrical to asymmetrical was found to be due to the movement of MNPs within it, resulting in a highly anisotropic structure. The control of all those parameters could allow for the design of hybrid materials with many potential applications.

Park et al. [31] also employed a magnetic field to induce the self-assembly of poly(3-hexylthiopene-b-ethylene glycol) (P3HT-b-PEG) and iron oxide nanoparticles capped with oleic acid at the air–water interface. They claimed to generate, for the first time, thin films based on self-assembled BCP ordered at the nanoscale by the use of magnetic fields. The scheme of the process employed can be seen in Figure 9.

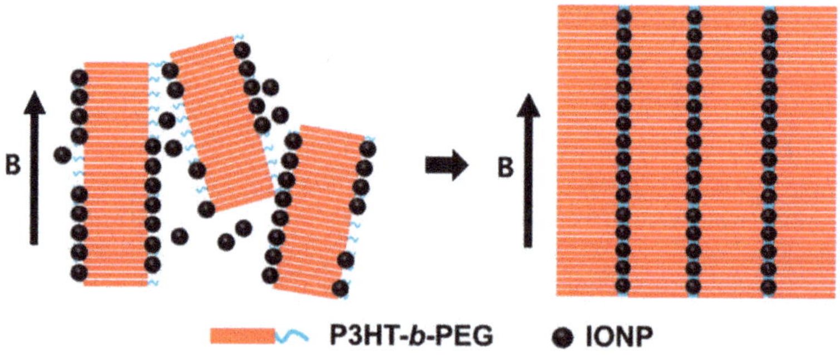

Figure 9. Schematic illustration of the self-assembly process of P3HT-b-PEG and iron oxide NPs in an external magnetic field B. (Reproduced with the permission of Ref. [31], copyright ACS 2022).

The relative amounts of MNPs and BCP employed assured the generation of uniform assembled structures, with the magnetic field intensity controlled by the distance between the magnets. They obtained nanoarrays of polymeric nanowires ordered in the long range and bridged with 1D nanoparticle chains. They pointed out the importance of MNP size. The generation of intermediate magnetic subunits composed by short BCP nanowires, which appeared to be decorated with MNPs, led to the self-assembly. Those subunits, as they claimed, could further assemble form nanoarrays ordered in the long range due to magnetic interactions generated by low-intensity magnetic fields. They could be employed in different potential applications in electronic, optoelectronic, or sensor devices, among others.

Ren et al. [29] induced an order–order transition from hexagonally packed cylinders to long-range-oriented lamellae by the use of magnetic fields in nanocomposites formed by PS homopolymer, poly(styrene-b-ethylene oxide) (PS-b-PEO) copolymer, and a very small amount of Fe_3O_4 MNPs (0.07 vol%) modified on the surface with amino groups. They pointed out that this phase transition was induced by preforming a small-grain local domain comprising a mixture of magnetic field-assisted uniformly arranged NH_2–Fe_3O_4 MNPs and a disordered PS/(PS-b-PEO) BCP melt as a nucleus, in which the interfacial tension at the junction points between PEO and PS blocks was significantly increased due to the enhancement in the block chain immiscibility so as to be able to induce the flat domain interface. This is in contrast with the phase transition kinetics usually found for neat BCPs. In this way, they found that a thermodynamically favored process drove the kinetic pathway, rather than a kinetics-dominated process, due to the alignment of MNPs occurring with higher particle amounts employed.

They also underlined the preferential orientation induced in both nanostructures: lamellar and hexagonally packed. Usually, when non-conductive or magnetically inactive BCPs are employed, a mechanical shear stress is responsible for the orientation, while the orientation can be induced by magnetic fields of low intensity by using very low amounts of MNPs.

Table 1 shows the main advantages and disadvantages related to the procedures described for the development of BCP/MNP hybrid materials.

Table 1. Main advantages and disadvantages of the procedures employed for hybrid material preparation.

Process	Advantages	Disadvantages
Surfactant	Easy to perform Many surfactants available.	NPs tend to form aggregates Less efective than brushes for selective placement.
Grafting to	Acceptable grafting densities. Control of molecular weight of grafted chains. Success in the proper dispersion and selective placement of modified NPs into desired domains.	Limitation on the choice of functional groups available. The thickness of the film, and accordingly the number of functional groups per surface area that can be obtained. Due to the inability of large polymer chains to diffuse to the reactive surface sites, they are sterically hindered by the surrounding bounded chains.
Grafting from	With the development of controlled living radical polymerization methods such as ATRP or RAFT some issues can be improved.	Poor control of molecular weight, polydispersity, end functionality, chain architecture and composition. Stringent conditions and the small number of monomers that can be polymerized in ionic living polymerization method.
Grafting through	Widely applied, even in industrial applications mostly for adhesion promotion. Success in the proper dispersion and selective placement of modified NPs into desired domains.	This approach has not been extensively explored and the details of the mechanism are not well understood. Poor grafting density control.
Magnetic field	Successfully applied even with very small NP amounts. Orientation of both MNPs and surrounding BCP domains. Very promissing procedure for biomedical applications.	Not easy for transference to industrial processes. Higher energy comsuption than other processes. Has to be improved and further investigated.

5. BCPs as Template Patterns for MNP Synthesis, Encapsulation, and Release

Several studies can be found in the literature on the use of BCPs as template patterns or nanoreactors for the generation of MNPs, for their encapsulation and release, for self-assembly inside BCP matrices, or for the generation of magnetic patterns based on metallopolymers or metal-containing BCPs, in some cases assisted by advanced techniques such as nanoimprint lithography (NIL) [38–45].

Regarding the employment of BCPs as a nanoreactor for MNP synthesis, the work of Ahmed et al. [41] must be cited. They employed the BCP nanoreactor route for the self-assembly of magnetic $CoFe_2O_4$ NPs in BCPs at room temperature. They employed diblock copolymers consisting of poly(norbornene) and poly(norbornene-dicarboxcylic acid) blocks (NOR-b-NORCOOH). The use of a template consisting of a self-assembled BCP made it possible to obtain MNPs at room temperature, as demonstrated by wide-angle X-ray diffraction and low-temperature Mossbauer-phase characterization. They processed the nanocomposites by wet chemical methods, obtaining a self-assembly of MNPs within the NORCOOH block at room temperature. The $CoFe_2O_4$ NPs were uniformly dispersed within the polymer matrix. They analyzed the magnetic properties of the obtained nanocomposite films, finding that they were superparamagnetic at room temperature and ferrimagnetic at 5 K.

Different studies can be found on the use of BCPs for the fabrication or encapsulation of MNPs in solution—in some cases, for their posterior release [38–40,42]. In this way, Hickey et al. [40] controlled the self-assembly structure of MNPs and amphiphilic block copolymers in solution by controlling MNP–solvent and MNP–BCP interactions. They employed poly(acrylic acid-b-styrene) (PAA-b-PS) copolymer and oleic acid-capped Fe_2O_3 MNPs.

By controlling solvent–MNP and BCP–MNP interactions, they obtained three different nanostructures: magneto-polymersomes densely packed with MNPs, magneto core–shells or core–shell-type BCP assemblies in which the MNPs appeared to be placed at the interface between the core and the shell, and magneto-micelles or BCP micelles in which the MNPs were homogeneously incorporated.

As MNP addition modified the volume ratio among the blocks, the self-assembly structure was altered, producing magneto-polymersomes instead of the classical magneto-micelles obtained using BCPs that self-assemble into micellar nanostructures. Moreover, polymers forming vesicles usually form magneto-micelles as a consequence of MNP solubilization inside BCP nanostructures. The MNP–BCP interaction also controlled the MNP arrangement. When N,N-dimethylformamide (DMF) was employed, in which PS was not well solvated, MNPs segregated from the PS, forming radial assemblies. On the contrary, by employing a good solvent for the PS and MNPs such as THF, they found a homogeneous distribution of MNPs into the BCP. They concluded that the morphology of the assembled BCP in which the MNPs were encapsulated greatly affected their magnetic relaxation properties.

Lecommandouxa et al. [39] also developed magnetic micelles and vesicles by self-assembly of polypeptide-based BCP and MNPs in water. Iron oxide NPs with magnetic response even to low-intensity magnetic fields were employed as the inorganic part, while the organic part was formed by poly(butadiene-b-glutamic acid) (PB-b-PGA) diblock copolymers able to self-assemble into micelles or vesicles. As the PGA block underwent a helix–coil transition, the formed peptosomes could act as stimuli-responsive capsules, varying their hydrodynamic diameter by as much as 50% in response to a pH change.

They obtained 3D spherical micelles (micelles with a spherical shape) with a hydrophobic ferrofluid inside or hollow vesicles with a (2D, sheet-like structure) magnetic membrane, in which the MNPs appeared to be confined inside the layer of PB blocks. They concluded that the obtained soft magnetic objects could be used for drug delivery by the encapsulation of water-soluble species. The release of this content would be achieved by pH changes or by the use of magnetic fields, which could open the envelope.

Skandalis et al. [38] employed micelle-forming poly(lauryl methacrylate-b-oligo ethylene glycol methacrylate) (PLMA-b-POEGMA) amphiphilic block copolymers for the encapsulation of magnetic iron oxide nanoparticles with water as the solvent. The self-assembly of BCPs in water resulted in micelles and could contain MNPs of up to 10 wt% in their PLMA cores, forming mixed nanostructures, as shown in Figure 10. Comparing the size and distribution of micelles for the BCP and the nanocomposite, they were significantly broader for the latter. By magnetophoresis, the authors demonstrated that the solutions maintained the magnetic properties, as MNPs continued being magnetic after the encapsulation. They also checked that even after the encapsulation of a drug (indomethacin) together with the MNPs, the magnetic properties were maintained.

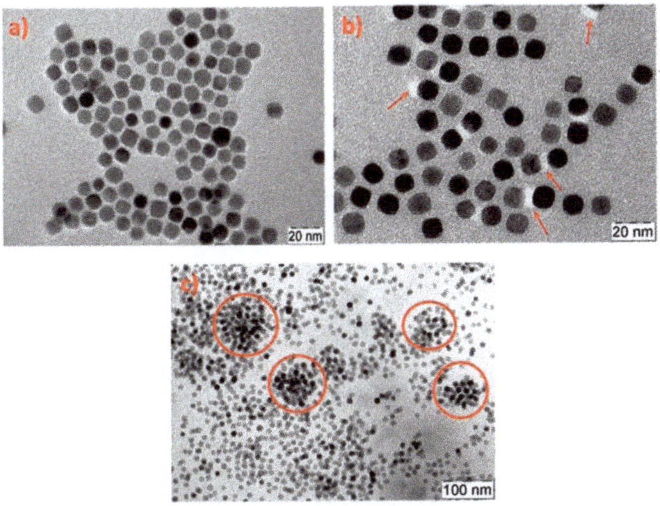

Figure 10. TEM images for (**a**) iron oxide NP-mixed, (**b**) PLMA-b-POEGMA/10% NP-mixed (red arrows indicate areas with a higher concentration of polymeric chains), and (**c**) PLMA-b-POEGMA/10% NP-mixed nanosystems after solvent evaporation (red circles indicate areas with large aggregation). (Reproduced with permission from Ref. [38]).

Morcrette et al. [42] fabricated NPs (with a very narrow size dispersion of around 7%) released in solution with the use of BCP as a patterned template. They employed a combined "top-down" and "bottom-up" approach with the use of a self-assembled PS-PMMA BCP formed on a sacrificial layer. The self-assembly of PS-b-PMMA required an intermediate layer composed of a "random" PS-r-PMMA copolymer with the same composition as the BCP. The reason for employing this sacrificial layer was substrate surface neutralization, which is necessary for obtaining a vertical cylinder-type morphology rather than a horizontal one, which is obtained without the neutralization of the surface. A cross-linking process, induced by a thermal annealing performed after the spin coating, allowed for the grafting of the layer to the sacrificial germanium substrate (previously deposited onto a silicon substrate, which is important for the subsequent lift-off of NPs). Then, the PS-b-PMMA was spin coated and thermally annealed to provide mobility to the polymer chains for self-assembly. PMMA cylinders were removed by employing acetic acid, obtaining a hexagonal template of holes in the PS matrix. After that, they removed the neutral copolymer layer for the deposition of the magnetic material on the cited template by evaporation or DC sputtering. Finally, after removing the PS template by plasma exposure, a packed hexagonal array of NPs was obtained. The method was successfully employed for the fabrication of different NPs such as gold-coated Ni, FeNi, and Fe_2O_3. The use of the germanium oxide sacrificial layer, compatible with BCP mask fabrication, made the release of MNPs in solution possible. In addition to the main potential applications in imaging (MRI, MPI), as anisotropic MNPs with a small size could be obtained, the generated materials could also be employed in other biomedical applications, such as treatments for cancer cell destruction.

Recently, Zhang et al. [44] also employed self-assembled BCPs (in this case, bimetallic or metal-containing BCPs) to obtain magnetic patterns of Fe, Pt NPs, assisted by lithography nanoimprint (NIL). They synthesized four new Fe, Pt-containing BCPs, named P1–P4, selecting a PS-b-P4VP BCP (with the possibility of self-assembly, containing pyridine groups as effective coordination sites for metallic ions) to be employed as a precursor for the generation of monodisperse FePt MNPs without using any surfactant. The post-coordination of the BCP allowed those bimetallic BCPs to be obtained directly, keeping the atomic ratio of Fe to Pt identical. The synthesis procedure is shown in Figure 11.

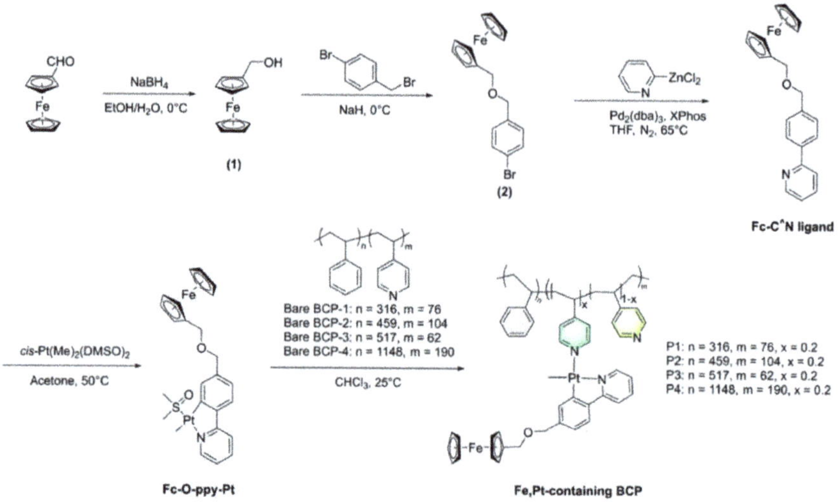

Figure 11. Synthetic route for obtaining Fe, Pt-containing BCPs (reproduced with permission from Ref. [44], copyright Springer Nature 2025).

Among these four bimetallic copolymers, P3 and P4 were the most easily assembled, especially P3, which generated spherical assemblies by solvent vapor annealing with dioxane. For this reason, they chose P3 for the nanoimprint lithography-assisted method, which is shown in detail in the experimental section of their paper [44].

They used the nanoimprint lithography-assisted self-assembly method to obtain large-scale regular patterns, leading to the consequent self-assembly of the BCP into nanodomain arrays without any defect in the previously defined features. After the thermal decomposition (in situ pyrolisis), the patterned metal-containing copolymer was completely transformed into Fe, Pt-based magnetic patterns. They concluded that their materials could present potential applications in ultra-high-density data storage systems.

Metal-containing copolymers were also employed by other authors for the generation of MNPs. Al-Badri et al. [43] synthesized Co-containing novel BCPs by living ring-opening metathesis polymerization. Those novel BCPs could perform microphase separation and, after a heat treatment, presented ferromagnetic properties at room temperature. As the employment of the parent homopolymer generated only paramagnetic materials, the nanostructuring was shown to be a key parameter.

They claimed that the origin of the ferromagnetic properties shown at room temperature (room-temperature ferromagnetic materials (RTFM)) came from the nanostructured confinement of Co within the one-dimensional cylinders of the BCP morphology. The fact that the particles were too small (~5 nm) to present ferromagnetism at room temperature and, moreover, that they were present in both the BCP and the homopolymer, with only the former generating a ferromagnetic material, supported the hypothesis. They further demonstrated the cited hypothesis by synthesizing the BCPs with different morphologies (cylindrical, lamellar, and inverted cylindrical phases). All the nanostructured morphologies led to RFTM, in contrast to the homopolymer, which did not present any nanostructure and did not lead to any RTFM. Moreover, the cylindrical BCP, which was the most strongly confined, presented the best ferromagnetic properties, followed by the inverted cylindrical morphology.

Yiu et al. [45] also employed bimetallic BCP precursors for the generation of FePt MNPs. They incorporated a complex containing Fe and Pt centers into the coordinating block of four different PS-b-P4VP (named P1–P4) copolymers for the synthesis of bimetallic BCPs. They analyzed both thin-film and bulk self-assembly, finding that the spherical and cylindrical morphologies observed corresponded with those of non-metallic BCP, but, of course, in the case of metallic BCPs, the phase-separated cylindrical or spherical nanodomains contained FePt NPs. Once the organic part was pyrolized (at 800 °C in an inert atmosphere), the pyrolisis products of those bimetallic BCPs were analyzed, and they found that the formed FePt NPs presented an fct phase, with average sizes of between 4 and 8 nm. They demonstrated the importance of nanostructuring BCPs for the synthesis of ferromagnetic FePt NPs with a smaller average size with a Fe/Pt stoichiometric ratio close to 1:1 by comparing results obtained with the corresponding P4VP homopolymer. They concluded by pointing out that metallo-BCP lithography was an effective procedure for obtaining FePt NPs from a bimetallic source without the use of any surfactant.

6. Conclusions

This review work shows the implementation of nanostructured BCPs and MNPs, combined in several ways and prepared by several methods, to obtain interesting materials with potential applications in different fields such as biomedicine, optoelectronics, or data storage devices. The main conclusions that can be extracted from the use of different procedures and their success are the following.

The first procedure for their preparation is the nanostructuring of BCPs in the presence of previously synthesized NPs by modifying their surface (with surfactants of polymeric brushes) to increase their compatibility and selectively place them at nanodomains or by employing external magnetic fields for orientation.

By employing surfactants, the degree of interaction between MNPs and one of the blocks determines the dispersion and selective NP placement, as well as the effect on BCP morphology. In any case, the magnetic properties are effectively transferred to the nanocomposite.

The method consisting of the grafting of polymeric brushes has been shown to be more effective than the use of surfactants. Brushes can be grafted to the surface of NPs by three main methods, known as grafting from, grafting through, and grafting to.

By comparing the three methods, the last one has been found to be the least effective for NP confinement. The other two methods are effective, but the dispersion and placement strongly depends on the grafting density and the molar mass of the brushes. The selective placement of NPs at certain domains can affect the morphology of the BCP, which can also be disrupted for higher NP contents if they tend to form aggregates.

In all cases, the magnetic properties are effectively transferred to the nanocomposites, resulting in materials with potential optoelectronic applications.

The use of external magnetic fields for the orientation of NPs and even that of the nearby domains has been employed with success, strongly depending on the amount of NPs and the magnetic field intensity. By properly controlling both parameters, long-range-ordered nanoarrays of polymeric nanowires bridged with NP chains can be obtained.

The second procedure, consisting of the in situ synthesis of NPs inside the nanodomains by a reduction in the respective metallic ions or by employing metal-containing BCPs for the generation of MNPs patterns or arrays, has been successfully applied to obtain selective dispersion and the transference of magnetic properties.

Finally, the synthesis of metallopolymers or metal-containing BCPs to generate magnetic patterns of different MNPs has also been explored by different authors, in some cases assisted by the use of advanced techniques such as nanoimprint lithography. Obtaining MNPs with better size control and properties by using nanostructured BCPs instead of homopolymers has shown the importance of nanostructuring and confinement for a more accurate generation of MNPs with ferromagnetic properties.

Funding: This research was funded by Ministerio de Ciencia, Innovacion y Universidades, grant number PID2021-126417NB-I00, and by the Gobierno Vasco/Eusko Jaurlaritza, grant number IT1690-22.

Institutional Review Board Statement: Not applicable.

Data Availability Statement: The data presented are available on request from the corresponding author.

Conflicts of Interest: The authors declare no conflicts of interest.

References

1. Gutierrez, J.; Tercjak, A.; Garcia, I.; Mondragon, I. The effect of thermal and vapor annealing treatments on the self-assembly of TiO_2/PS-b-PMMA nanocomposites generated via the sol–gel process. *Nanotechnology* **2009**, *20*, 225603. [CrossRef] [PubMed]
2. Xia, X.; Metwalli, E.; Ruderer, M.A.; Körstgens, V.; Busch, P.; Böni, P.; Müller-Buschbaum, P. Nanostructured diblock copolymer films with embedded magnetic nanoparticles. *J. Phys. Condens. Matter* **2011**, *23*, 254203. [CrossRef]
3. Etxeberria, H.; Tercjak, A.; Mondragon, I.; Eceiza, A.; Kortaberria, G. Electrostatic force microscopy measurements of CdSe-PS nanoparticles and CdSe-PS/poly(styrene-b-butadiene-b-styrene) nanocomposites. *Colloid Polym. Sci.* **2014**, *292*, 229–234. [CrossRef]
4. Barandiaran, I.; Gutierrez, J.; Tercjak, A.; Kortaberria, G. Thin film nanocomposites based on SBM triblock copolymer and silver nanoparticles: Morphological and dielectric analysis. *Macromol. Mater. Eng.* **2017**, *302*, 1700169. [CrossRef]
5. Jang, Y.H.; Xin, X.; Byun, M.; Jang, Y.J.; Lin, Z.; Kim, D.H. An Unconventional Route to High-Efficiency Dye-Sensitized Solar Cells via Embedding Graphitic Thin Films into TiO_2 Nanoparticle Photoanode. *Nano Lett.* **2012**, *12*, 479–485. [CrossRef]
6. Park, S.; Lee, D.H.; Xu, J.; Kim, B.; Hong, S.W.; Jeong, U.; Xu, T.; Russell, T.P. Macroscopic 10-Terabit–per– Square-Inch Arrays from Block Copolymers with Lateral Order. *Science* **2009**, *323*, 1030–1033. [CrossRef]
7. Hoheisel, T.N.; Hur, K.; Wiesner, U.B. Block copolymer-nanoparticle hybrid self-assembly. *Prog. Polym. Sci.* **2015**, *40*, 3–32. [CrossRef]
8. Vural, M.; Behrens, A.M.; Ayyub, O.B.; Ayoub, J.J.; Kofinas, P. Sprayable Elastic Conductors Based on Block Copolymer Silver Nanoparticle Composites. *ACS Nano* **2015**, *9*, 336–344. [CrossRef]
9. Song, D.-P.; Lin, Y.; Gai, Y.; Colella, N.S.; Li, C.; Liu, X.-H.; Gido, S.; Watkins, J.J. Controlled Supramolecular Self-Assembly of Large Nanoparticles in Amphiphilic Brush Block Copolymers. *J. Am. Chem. Soc.* **2015**, *137*, 3771–3774. [CrossRef]
10. Escher, A.; Bravetti, G.; Bertucci, S.; Comoretto, D.; Weder, C.; Steiner, U.; Lova, P.; Dodero, A. Crafting Nanostructured Hybrid Block Copolymer–Gold Nanoparticles by Confined Self-Assembly in Evaporative Droplets. *ACS Macro Lett.* **2024**, *13*, 1338–1344. [CrossRef]

11. Gutierrez, J.; Tercjak, A.; Garcia, I.; Peponi, L.; Mondragon, I. Hybrid titanium dioxide/PS-b-PEO block copolymer nanocomposites based on sol–gel synthesis. *Nanotechnology* **2008**, *19*, 155607. [CrossRef] [PubMed]
12. Etxeberria, H.; Zalakain, I.; Fernandez, R.; Kortaberria, G.; Mondragon, I. Controlled placement of polystyrene-grafted CdSe nanoparticles in self-assembled block copolymers. *Colloid Polym. Sci.* **2013**, *291*, 633–640. [CrossRef]
13. Etxeberria, H.; Zalakain, I.; Mondragon, I.; Eceiza, A.; Kortaberria, G. Generation of nanocomposites based on polystyrene-grafted CdSe nanoparticles by grafting through and block copolymer. *Colloid Polym. Sci.* **2013**, *291*, 1881–1886. [CrossRef]
14. Etxeberria, H.; Fernandez, R.; Zalakain, I.; Mondragon, I.; Eceiza, A.; Kortaberria, G. Effect of CdSe nanoparticle addition on nanostructuring of PS-b-P4VP copolymer via solvent vapor exposure. *J. Colloid Interf. Sci.* **2014**, *416*, 25–29. [CrossRef] [PubMed]
15. Munoz-Menendez, C.; Conde-Leboran, I.; Baldomir, D.; Chubykalo-Fesenkob, O.; Serantes, D. The role of size polydispersity in magnetic fluid hyperthermia: Average vs. local infra/over-heating effects. *Phys. Chem. Chem. Phys.* **2015**, *17*, 27812–27820. [CrossRef]
16. Zhao, X.; Liu, X. A novel magnetic $NiFe_2O_4$@graphene–Pd multifunctional nanocomposite for practical catalytic application. *RSC Adv.* **2015**, *5*, 79548–79555. [CrossRef]
17. Aseri, A.; Garg, S.K.; Nayak, A.; Trivedi, S.K.; Ahsan, J. Magnetic Nanoparticles: Magnetic Nano-technology Using Biomedical Applications and Future Prospects. *Int. J. Pharm. Sci. Rev. Res.* **2015**, *31*, 119–131.
18. Sun, C.; Lee, J.S.H.; Zhang, M. Magnetic nanoparticles in MR imaging and drug delivery. *Adv. Drug Deliv. Rev.* **2008**, *60*, 1252–1265. [CrossRef]
19. Galloway, J.M.; Talbot, J.E.; Critchley, K.; Miles, J.J.; Bramble, J.P. Developing Biotemplated Data Storage: Room Temperature Biomineralization of L10 CoPt Magnetic Nanoparticles. *Adv. Funct. Mater.* **2015**, *25*, 4590–4600. [CrossRef]
20. Pang, Y.L.; Lim, S.; Ong, H.C.; Chong, W.T. Research progress on iron oxide-based magnetic materials: Synthesis techniques and photocatalytic applications. *Ceram. Int.* **2016**, *42*, 9–34. [CrossRef]
21. Yoo, H.G.; Byun, M.; Jeong, C.K.; Lee, K.J. Performance Enhancement of Electronic and Energy Devices via Block Copolymer Self-Assembly. *Adv. Mater.* **2015**, *27*, 3982–3998. [CrossRef] [PubMed]
22. Abel, F.M.; Tzitzios, V.; Hadjipanayis, G.C. New approach for direct chemical synthesis of hexagonal Co nanoparticles. *J. Magnet. Magnet. Mater.* **2016**, *400*, 286–289. [CrossRef]
23. Wan, C.; Li, J. Synthesis of well-dispersed magnetic $CoFe_2O_4$ nanoparticles in cellulose aerogels via a facile oxidative co-precipitation method. *Carbohydr. Polym.* **2015**, *134*, 144–150. [CrossRef]
24. Ibrahim, I.; Ali, I.O.; Salamaa, T.M.; Bahgat, A.A.; Mohamed, M.M. Synthesis of magnetically recyclable spinel ferrite (MFe_2O_4, M = Zn, Co, Mn) nanocrystals engineered by sol gel-hydrothermal technology: High catalytic performances for nitroarenes reduction. *Appl. Catal. B Environ.* **2016**, *181*, 389–402. [CrossRef]
25. Bahmanrokh, G.; Hashim, M.; Matori, K.A.; Navasery, M.; Soltani, N.; Vaziri, P.; Kanagesan, S.; Sabbaghizadeh, R.; Shafie, M.S.E. Recording-media-related morphology and magnetic properties of crystalline CoPt3 and CoPt3-Au core-shell nanoparticles synthesized via reverse microemulsion. *J. Appl. Phys.* **2014**, *116*, 093907. [CrossRef]
26. Jha, D.K.; Varadarajan, K.S.; Patel, A.B.; Deb, P. Direct synthesis of water dispersible superparamagnetic TGA capped FePt nanoparticles: One pot, one shot. *Mater. Chem. Phys.* **2015**, *156*, 247–253. [CrossRef]
27. Teixeira, A.P.C.; Tristão, J.C.; Araujo, M.H.; Oliveira, L.C.A.; Moura, F.C.C.; Ardisson, J.D.; Amorimd, C.C.; Lago, R.M. Iron: A Versatile Element to Produce Materials for Environmental Applications. *J. Braz. Chem. Soc.* **2012**, *23*, 1579–1593. [CrossRef]
28. Char, K.; Park, M.J. Selective distribution of interacting magnetic nanoparticles into block copolymer domains based on the facile inversion of micelles. *React. Func. Polym.* **2009**, *69*, 546–551. [CrossRef]
29. Ren, B.H.; Chu, C.Y.; Jhang, B.S. Magnetic Field-Induced Order−Order Transition from Hexagonally Packed Cylinders to Lamellae with Large-Scale Orientations in a Block Copolymer/Magnetic Nanoparticle Nanocomposite Film. *Macromolecules* **2023**, *56*, 3363–3373. [CrossRef]
30. Miles, A.; Gai, Y.; Gangopadhyay, P.; Wang, X.; Norwood, R.A.; Watkins, J.J. Improving Faraday rotation performance with block copolymer and FePt nanoparticle magneto-optical composite. *Opt. Mater. Express* **2017**, *7*, 2126–2140. [CrossRef]
31. Park, M.; Kang, S.; Nam, C.; Narasimha, K.; Lee, W.B.; Park, S.J. Magnetic field-induced self-assembly of conjugated block copolymers and nanoparticles at the air−water interface. *ACS Appl. Mater. Interf.* **2022**, *14*, 8266–8273. [CrossRef]
32. Hartmann, F.; Bitsch, M.; Niebuur, B.J.; Koch, M.; Kraus, T.; Dietz, C.; Stark, R.W.; Everett, C.; Muller-Buschbaum, P.; Janka, O.; et al. Self-Assembly of Polymer-Modified FePt Magnetic Nanoparticles and Block Copolymers. *Materials* **2023**, *16*, 5503. [CrossRef] [PubMed]
33. Xu, C.; Ohno, K.; Ladmiral, V.; Composto, R.J. Dispersion of polymer-grafted magnetic nanoparticles in homopolymers and block copolymers. *Polymer* **2008**, *49*, 3568–3577. [CrossRef]

34. Douadi-Masroukia, S.; Frka-Petesic, B.; Saved, M.; Charleuxd, B.; Cabuila, V.; Sandre, O. Incorporation of magnetic nanoparticles into lamellar polystyrene-b-poly(n-butyl methacrylate) diblock copolymer films: Influence of the chain end-groups on nanostructuration. *Polymer* **2010**, *51*, 4673–4685. [CrossRef]
35. Upadhyaya, L.; Semsarilar, M.; Fernández-Pacheco, R.; Martinez, G.; Mallada, R.; Coelhoso, I.M.; Portugal, A.M.; Crespo, J.G.; Deratania, A.; Quemener, D. Nano-structured magneto-responsive membranes from block copolymers and iron oxide nanoparticles. *Polym. Chem.* **2017**, *8*, 605–614. [CrossRef]
36. Yao, Y.; Metwalli, E.; Niedermeier, M.A.; Opel, M.; Lin, C.; Ning, J.; Perlich, J.; Roth, S.V.; Muller-Buschbaum, P. Nano- and Microstructures of Magnetic Field-Guided Maghemite Nanoparticles in Diblock Copolymer Films. *ACS Appl. Mater. Interfaces* **2014**, *6*, 5244–5254. [CrossRef] [PubMed]
37. Hammond, M.R.; Dietsch, H.; Pravaz, O.; Schurtenberger, P. Mutual alignment of block copolymer-magnetic nanoparticle composites in a magnetic field. *Macromolecules* **2011**, *43*, 8340–8343. [CrossRef]
38. Skandalis, A.; Sergides, A.; Bakandritsos, A.; Pispas, S. PLMA-b-POEGMA amphiphilic block copolymers as nanocarriers for the encapsulation of magnetic nanoparticles and indomethacin. *Polymers* **2018**, *10*, 14. [CrossRef]
39. Lecommandouxa, S.; Sandreb, O.; Che´cota, F.; Rodriguez-Hernandez, J.; Perzynsk, R. Self-assemblies of magnetic nanoparticles and di-block copolymers: Magnetic micelles and vesicles. *J. Magn. Magn. Mater.* **2006**, *300*, 71–74. [CrossRef]
40. Hickey, R.J.; Haynes, A.S.; Kikkawa, J.M.; Park, S.J. Controlling the self-assembly structure of magnetic nanoparticles and amphiphilic block-copolymers: From micelles to vesicles. *J. Am. Chem. Soc.* **2011**, *133*, 1517–1525. [CrossRef]
41. Ahmed, S.R.; Ogale, S.B.; Papaefthymiour, G.C.; Ramesh, R.; Kofinas, P. Magnetic properties of $CoFe_2O_4$ nanoparticles synthesized through a block copolymer nanoreactor route. *Appl. Phys. Lett.* **2002**, *80*, 1616–1618. [CrossRef]
42. Morcrette, M.; Ortiz, G.; Tallegas, S.; Joisten, H.; Tiron, R.; Baron, T.; Hou, Y.; Lequien, S.; Bsiesy, A.; Dieny, B. Fabrication of monodisperse magnetic nanoparticles released in solution using a block copolymer template. *J. Phys. D Appl. Phys.* **2017**, *50*, 295001. [CrossRef]
43. Al-Badri, Z.M.; Maddikeri, R.R.; Zha, Y.; Thaker, H.D.; Dobriya, P.; Shunmugam, R.; Russel, T.P.; Tew, G.N. Room temperature magnetic materials from nanostructured diblock copolymers. *Nat. Commun.* **2011**, *2*, 482. [CrossRef]
44. Zhang, J.; Chen, Y.; Li, G.; Sun, Z.; Meng, Z.; Wong, W.Y. Nanoimprint lithography-assisted block copolymer self-assembly for hyperfine fabrication of magnetic patterns based on L10-FePt nanoparticles. *Sci. China Chem.* **2025**, *68*, 2027–2034. [CrossRef]
45. Yiu, S.C.; Nunns, A.; Ho, C.L.; Ngai, J.H.; Meng, Z.; Li, G.; Gwyther, J.; Whittell, G.R.; Manners, I.; Wong, W.Y. Nanostructured bimetallic block copolymers as precursors to magnetic FePt nanoparticles. *Macromolecules* **2019**, *52*, 3176–3186. [CrossRef]
46. Horechyy, A.; Zafeiropoulos, N.E.; Nandan, B.; Formanek, P.; Simon, F.; Kiriya, A.; Stamm, M. Highly ordered arrays of magnetic nanoparticles prepared via block copolymer assembly. *J. Mater. Chem.* **2010**, *20*, 7734–7741. [CrossRef]
47. Lo, C.T.; Chao, C.J. Synthesis and Characterization of Magnetic Nanoparticle/Block Copolymer Composites. *Langmuir* **2009**, *25*, 12865–12869. [CrossRef]
48. Kim, M.P.; Yi, G.R. Nanostructured colloidal particles by confined self-assembly of block copolymers in evaporative droplets. *Front. Matter.* **2015**, *2*, 45. [CrossRef]
49. Shin, J.J. Morphological Evolution of Hybrid Block Copolymer Particles: Toward Magnetic Responsive Particles. *Polymers* **2023**, *15*, 3689. [CrossRef]
50. Advincula, R.; Ruehe, J.; Brittain, W.; Caster, K. *Polymer Brushes*; VCH Wiley: Weinheim, Germany, 2004.
51. Duwez, A.S.; Guillet, P.; Colard, C.; Gohy, J.-F.; Fustin, C.A. Dithioesters and Trithiocarbonates as Anchoring Groups for the "Grafting-To" Approach. *Macromolecules* **2006**, *39*, 2729–2731. [CrossRef]
52. Sawall, D.D.; Villahermosa, R.M.; Lipeles, R.A.; Hopkins, A.R. Interfacial Polymerization of Polyaniline Nanofibers Grafted to Au Surfaces. *Chem. Mater.* **2004**, *16*, 1606–1608.
53. Henze, M.; Mädge, D.; Prucker, O.; Rühe, J. "Grafting Through": Mechanistic Aspects of Radical Polymerization Reactions with Surface-Attached Monomers. *Macromolecules* **2014**, *47*, 2929–2937. [CrossRef]
54. Ilgacha, D.M.; Meleshkoa, T.K.; Yakimansky, A.V. Methods of Controlled Radical Polymerization for the Synthesis of Polymer Brushes. *Polym. Sci. Ser. C* **2015**, *57*, 3–19. [CrossRef]
55. Li, W.; Bao, C.; Wright, R.A.E.; Zhao, B. Synthesis of mixed poly(ε-caprolactone)/polystyrene brushes from Y-initiatorfunctionalized silica particles by surface-initiated ring-opening polymerization and nitroxidemediated radical polymerization. *RSC Adv.* **2014**, *4*, 18772–18781. [CrossRef]
56. Garcia, I.; Tercjak, A.; Rueda, L.; Mondragon, I. Self-assembled nanomaterials using magnetic nanoparticles modified with polystyrene brushes and poly(styrene-b-butadiene-b-styrene). *Macromolecules* **2008**, *41*, 9295–9298. [CrossRef]
57. Garcia, I.; Tercjak, A.; Zafeiropoulos, N.E.; Stamm, M.; Mondragon, I. Self-assembling nanomaterials using magnetic nanoparticles modified with polystyrene brushes. *Macromol. Rapid Commun.* **2007**, *28*, 2361–2365. [CrossRef]

58. Garcia, I.; Tercjak, A.; Gutierrez, J.; Rueda, L.; Mondragon, I. Nanostructuration via solvent vapor exposure of poly(2-viny lpyridine-b-methyl methacrylate) nanocomposites using modified magnetic nanoparticles. *J. Phys. Chem. C* **2008**, *112*, 14343–14347. [CrossRef]
59. Garcia, I.; Tercjak, A.; Zafeiropoulos, N.E.; Stamm, M.; Mondragon, I. Generation of core-shell iron oxide magnetic nanoparticles with polystyrene brushes via atom transfer radical polymerization. *J. Polym. Sci. Part A Polym. Chem.* **2007**, *45*, 4744–4750. [CrossRef]
60. Barandiaran, I.; Grana, E.; Katsigiannopoulos, D.; Avgeropoulos, A.; Kortaberria, G. Nanocomposites based on nanostructured PI-b-PMMA copolymer and selectively placed PMMA-modified magnetic nanoparticles: Morphological and magnetic characterization. *Eur. Polym. J.* **2016**, *75*, 514–524. [CrossRef]
61. Park, S.C.; Kim, B.J.; Hawker, C.J.; Kramer, E.J.; Bang, J.; Ha, J.S. Controlled ordering of block copolymer thin films by the addition of hydrophilic nanoparticles. *Macromolecules* **2007**, *40*, 8119–8124. [CrossRef]
62. Chiu, J.J.; Kim, B.J.; Bang, J.; Kramer, E.J.; Pine, D.J. Distribution of nanoparticles in lamellar domains of block copolymers. *Macromolecules* **2007**, *40*, 3361–3365. [CrossRef]
63. Chiu, J.J.; Kim, B.J.; Bang, J.; Kramer, E.J.; Pine, D.J. Control of nanoparticle location in block copolymers. *J. Am. Chem. Soc.* **2005**, *127*, 5036–5037. [CrossRef]
64. Kim, B.J.; Bang, J.; Hawker, C.J.; Kramer, E.J. Effect of areal chain density on the location of polymer-modified gold nanoparticles in a block copolymer template. *Macromolecules* **2006**, *39*, 4108–4114. [CrossRef]
65. Barandiaran, I.; Cappelletti, A.; Strumia, M.; Eceiza, A.; Kortaberria, G. Generation of nanocomposites based on (PMMA-b-PCL)-grafted Fe_2O_3 nanoparticles and PS-b-PCL block copolymer. *Eur. Polym. J.* **2014**, *58*, 226–232. [CrossRef]
66. Jordan, R.; Ulman, A. Surface-initiated anionic polymerization of styrene by means of self-assembled monolayers. *J. Am. Chem. Soc.* **1999**, *121*, 1016–1022. [CrossRef]
67. Barandiaran, I.; Kortaberria, G. Synthesis and characterization of nanostructured PS-b-P4VP/Fe_2O_3 thin films with magnetic properties prepared by solvent vapor annealing. *RSC Adv.* **2015**, *5*, 95840–95846. [CrossRef]
68. Barandiaran, I.; Kortaberria, G. Selective placement of magnetic Fe_3O_4 nanoparticles into the lamellar nanostructure of PS-b-PMMA diblock copolymer. *Eur. Polym. J.* **2015**, *68*, 57–67. [CrossRef]
69. Barandiaran, I.; Kortaberria, G. Magnetic nanocomposites based on polystyrene-block-polybutadiene-block-poly(methyl methacrylate) and Fe_2O_3 nanoparticles. *Eur. Polym. J.* **2016**, *78*, 340–351. [CrossRef]
70. Bai, J.; Shi, Z.; Yina, J.; Tian, M. A simple approach to preparation of polyhedral oligomeric silsesquioxane crosslinked poly(styrene-b-butadiene-b-styrene) elastomers with a unique micro-morphology via UV-induced thiol–ene reaction. *Polym. Chem.* **2014**, *5*, 6761–6769. [CrossRef]

polymers

MDPI

Article

Viscoelastic Response of Double Hydrophilic Block Copolymers for Drug Delivery Applications

Achilleas Pipertzis [1,*], Angeliki Chroni [2], Stergios Pispas [2] and Jan Swenson [1]

[1] Department of Physics, Chalmers University of Technology, 41296 Gothenburg, Sweden; jan.swenson@chalmers.se
[2] Theoretical and Physical Chemistry Institute, National Hellenic Research Foundation, 48 Vassileos Constantinou Ave., 11635 Athens, Greece; angelikechrone@gmail.com (A.C.); pispas@eie.gr (S.P.)
* Correspondence: achilleas.pipertzis@chalmers.se

Abstract: This study investigates the mechanical properties of double hydrophilic block copolymers (DHBCs) based on poly[oligo(ethylene glycol) methacrylate] (POEGMA) and poly(vinyl benzyl trimethylammonium chloride) (PVBTMAC) blocks by employing small amplitude oscillatory shear (SAOS) rheological measurements. We report that the mechanical properties of DHBCs are governed by the interfacial glass transition temperature (T_g^{inter}), verifying the disordered state of these copolymers. An increase in zero shear viscosity can be observed by increasing the VBTMAC content, yielding a transition from liquid-like to gel-like and finally to an elastic-like response for the PVBTMAC homopolymer. By changing the block arrangement along the backbone from statistical to sequential, a distinct change in the viscoelastic response is obvious, indicating the presence/absence of bulk-like regions. The tunable viscosity values and shear-thinning behavior achieved through alteration of the copolymer composition and block arrangement along the backbone render the studied DHBCs promising candidates for drug delivery applications. In the second part, the rheological data are analyzed within the framework of the classical free volume theories of glass formation. Specifically, the copolymers exhibit reduced fractional free volume and similar fragility values compared to the PVBTMAC homopolymer. On the contrary, the activation energy increases by increasing the VBTMAC content, reflecting the required higher energy for the relaxation of the glassy VBTMAC segments. Overall, this study provides information about the viscoelastic properties of DHBCs with densely grafted macromolecular architecture and shows how the mechanical and dynamical properties can be tailored for different drug delivery applications by simply altering the ratio between the two homopolymers.

Keywords: double hydrophilic block copolymer; densely grafted architecture; rheology; viscoelastic response

1. Introduction

Copolymers are made from two or more different monomers [1,2]. The advantage of copolymers compared to homopolymers is that it is much easier to tailor their properties. For instance, their balance between hydrophilicity and hydrophobicity, degradation rate, mechanical strength and solubility can easily be optimized for their specific target applications. Particularly promising polymers to use for the fabrication of copolymers are so-called bottlebrush polymers, which consist of a linear backbone with densely grafted side chains extending outward, creating a highly extended and bulky structure [3–10].

The bottlebrush polymers have attracted considerable attention owing to their highly tunable physical and chemical properties, which depend on side chain length, backbone length, and grafting density [3–10]. In comb polymers, side chains act as a solvent, diluting backbone concentration—a phenomenon known as dynamic tube dilution [6,11–13]. This effect results in a low viscosity and a reduced rubbery plateau. These properties make bottlebrush block copolymers promising for applications in drug delivery, energy storage, nanostructures, catalysis, and 3D printing [14–25]. In the case of copolymers for drug delivery applications, it is, for instance, possible to tune their properties to optimize drug release, targeting molecules and biocompatibility [26,27].

Several studies [1,2,28] have indicated that the block arrangement along the polymer backbone significantly influences viscoelastic and mechanical properties, with sequential copolymers generally exhibiting enhanced phase separation and mechanical strength compared to their statistical counterparts. In contrast, statistical or random copolymers lack long-range compositional order, typically exhibiting homogeneous amorphous morphologies. The absence of microphase separation in these materials results in more viscous and less elastic behavior, with lower shear moduli. Therefore, the block arrangement is a critical parameter in tuning the mechanical and structural performance of copolymeric materials [1,2,28].

The viscoelastic response of polymeric formulas plays a critical role in drug delivery applications [29–34]. The ideal viscoelastic properties ensure a balance between elasticity (for mechanical stability) and viscosity (for flow and processability). One key requirement is a shear-thinning behavior (i.e., reduced viscosity at higher frequencies), which enables ease of injection or spreading [29–31]. The viscoelastic properties also affect the diffusion mechanism, which is critical for obtaining a stimuli-responsive release profile. Another vital requirement of the polymeric material is a high degree of biodegradability and biocompatibility, without altering the release characteristics [33,34]. Therefore, tailoring the viscoelastic behavior through the use of copolymers with varied compositions is essential to meet the specific needs for drug delivery applications.

Double hydrophilic block copolymers (DHBCs), composed of two chemically distinct water-soluble blocks, are important in polymer science, pharmacy, biophysics and biochemistry [35–38]. They offer an alternative option to traditional amphiphilic block copolymers and self-assemble in aqueous conditions in response to changes in ionic strength, temperature, pH, or complexation with specific (bio)molecules. DHBCs with charged blocks, known as polyelectrolytes, are promising candidates for biomacromolecule delivery via electrostatic complexation [39]. For enhancing water solubility and stability in aqueous media, a neutral block, such as poly(ethylene glycol) or poly(oligo(ethylene glycol) methacrylate) (POEGMA), is used. A recent review article by Singh et al. provides an overview of the synthesis and biomedical applications of POEGMA-based materials [33]. Diblock and statistical copolymers consisting of POEGMA and poly(vinyl benzyl trimethylammonium chloride) (PVBTMAC) have been successfully synthesized using a reversible addition–fragmentation chain transfer (RAFT) polymerization process [37,38]. Recent research studies have focused on these copolymers' abilities to form electrostatic complexes with hydrophilic magnetic nanoparticles, short DNA and negatively charged proteins (i.e., insulin) [37,38]. Additionally, the relaxation dynamics and the self-assembly of these DHBCs were investigated under dry conditions [40]. It was shown that the weak segregation strength between the two hydrophilic blocks results in homogeneous dynamics that are governed by the interfacial glass transition temperature (T_g^{inter}) [40].

Herein, the previously described block and random DHBCs were studied by means of rheological measurements. We report that the PVBTMAC homopolymer exhibits an elastic response up to high temperatures, reflecting its strong (i.e., rigid) nature. The effect of the VBTMAC block on the zero shear viscosity is presented, and the rheological data are discussed with respect to classical free volume theories of glass formation. The values of fractional free volume, thermal expansion coefficient, fragility, and apparent activation energy at the glass transition temperature (T_g) are reported and compared with those of the PVBTMAC parent homopolymer. The results indicate that a synergistic combination of hydrophilicity, biodegradability and tunable rheological properties (such as shear-thinning behavior and viscosity) can be achieved through optimization of the copolymer composition and block arrangement along the backbone, rendering the studied DHBCs promising candidates for drug delivery applications.

2. Materials and Methods

2.1. Synthesis

The synthetic procedure and molecular characterization of the studied DHBCs are highlighted in previous studies [37,38]. Specifically, the charged DHBCs with densely grafted macromolecular architecture were prepared by RAFT polymerization, an advantageous method for controlling the molar mass (M_w) and achieving polydispersity values close to unity [41,42]. The chemical structure of the investigated DHBCs is depicted in Figure 1.

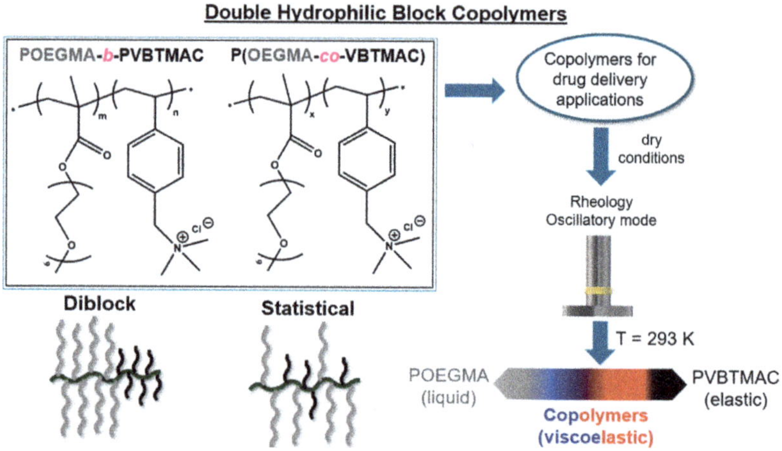

Figure 1. Chemical structure of the dried block and random DHBCs along with a schematic of the block sequence along the backbone and a description of the motivation of this study.

2.2. Rheology

Rheological measurements into the linear viscoelastic regime (LVR) can provide quantitative insight into the viscoelastic response of the studied polymeric materials [43]. An MCR-302 twin mode rheometer by Anton Paar was used for the identification of the viscoelastic properties of grafted copolymers under dry conditions. Measurements were made with the environmental test chamber as a function of temperature by using the heating system (P-PTD200). The samples were prepared on the lower rheometer plate with a diameter of 8 mm. Specifically, the upper plate was brought into contact, and the sample thickness was adjusted accordingly. Typically, the gap between plates ranged between 0.3 mm and 0.6 mm among the investigated DHBCs. At each temperature the linear viscoelastic region of the copolymers was determined by performing dynamic strain sweeps from 0.01 to 50% (the upper limit varies regarding the state of the material) at $\omega = 10$ rad·s^{-1}, as shown in Figure S1. Subsequently, isothermal frequency sweeps in an angular frequency range of $0.1 < \omega < 100$ rad·s^{-1} were carried out in temperature steps of 10 K by employing the extracted strain amplitude of the dynamic strain sweep. Before each isothermal measurement, a thermal stabilization of 20 min was employed to ensure thermal equilibrium. Data collected under the minimum torque of the instrument were excluded. Master curves were constructed by using the principle of time–temperature superposition (*tTs*). For estimating the zero shear viscosity, we used an extrapolation procedure for the data that do not reach the Newtonian limit at low frequencies. Empirical models can be utilized to fit the complex viscosity response and extrapolate to low frequencies to estimate the value of zero shear viscosity, η_0. Specifically, the Carreau–Yasuda (CY) model is given by the following equation [44,45]:

$$\eta^*(\omega) = \eta_0\left[1 + (\lambda\omega)^{\alpha}\right]^{\frac{n-1}{\alpha}}, \quad (1)$$

where λ is the relaxation time, α indicates the width of the transition region between Newtonian and power-law behavior, and n is the power-law index (i.e., a measure of the shear-thinning nature of a polymer melt). All these parameters can be obtained by fitting the measured data. Another empirical model that can be used for fitting the viscosity data is the modified Cross (MC) model, according to [46]

$$\eta^*(\omega) = \frac{\eta_0}{1 + [C_0\omega]^{1-n}}, \quad (2)$$

where C_0 and n can be extracted by fitting the data to Equation (2).

3. Results and Discussion

Quantitative insight into the viscoelastic response of the dried DHBCs can be obtained through small amplitude oscillatory shear (SAOS) rheological measurements. The effect of temperature under isochronal conditions (i.e., at a fixed angular frequency of 10 rad·s^{-1}) on the mechanical properties of the dried copolymers is depicted in Figure S2. The PVBTMAC homopolymer exhibits elastic behavior up to extremely high temperatures, indicating its rigid/strong nature. The temperature dependence of the loss factor verifies the existence of two T_gs, associated with the backbone and side chain vitrification [40]. The glass transition temperatures of the PVBTMAC homopolymer and the statistical copolymers obtained from rheological measurements are higher compared to those determined by calorimetric and dielectric techniques [40]. This discrepancy arises primarily from two factors: (i) differences in the characteristic timescales/frequencies of the methods (i.e., the dielectric T_g is typically defined at a relaxation time of $\tau = 100$ s) and (ii) the rheological T_g, herein determined at the peak of the loss tangent (tanδ), reflects the midpoint of the transition from the glassy modulus (G_g) to the rubbery or entanglement plateau (G_e). Consequently, the rheological T_g corresponds to the end region of the glass transition. Regarding the statistical copolymers, at a fixed temperature, they exhibit reduced shear moduli that are attributable to the influence of the soft OEGMA segments. Additionally, the viscoelastic response is governed by T_g^{inter}, verifying their disordered and well-mixed state, as observed by X-ray diffraction [40].

The dynamic frequency sweeps into the linear viscoelastic response under isothermal conditions (i.e., 293 K) for the statistical copolymers with various compositions are shown in Figures 2 and S3.

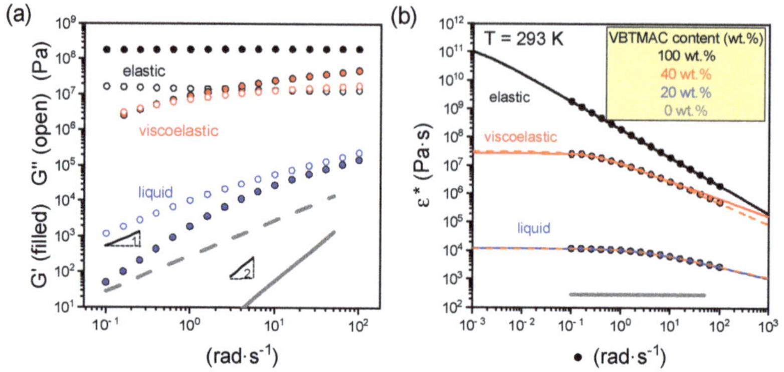

Figure 2. Angular frequency dependence of (**a**) the storage (filled symbols) and loss (open symbols) shear moduli and (**b**) complex viscosity for P(OEGMA$_{80}$-*co*-VBTMAC$_{20}$) (blue symbols) and P(OEGMA$_{60}$-*co*-VBTMAC$_{40}$) (red symbols) DHBCs and their respective homopolymers POEGMA (gray symbols) and PVBTMAC (black symbols) at a temperature of 293 K. Lines with slopes 1 and 2 are also shown in (**a**). The solid and dashed lines in (**b**) represent fits by Equation (1) and Equation (2), respectively.

Initially, the dynamic frequency sweep reveals that the PVBTMAC homopolymer exhibits a strong nature and elastic response ($G' >> G'' \sim \omega^0$). In going to the statistical copolymer containing 40 wt.% VBTMAC, a viscoelastic response, characterized by comparable and frequency-dependent storage (G') and loss (G'') moduli, can be observed. By further decreasing the VBTMAC content to 20 wt.%, a liquid-like response ($G' \sim \omega^2$ and $G'' \sim \omega^1$) is evident. In the latter copolymer, the complex viscosity ($\eta^* = \sqrt{G'^2 + G''^2}/\omega$), exhibits a plateau at lower angular frequencies, from which the zero shear viscosity, η_0, can be extracted. For extracting the zero shear viscosity, the measured data were fitted/modeled with two different empirical models, as detailed in the experimental section. The fitting parameters are provided in Table 1. In addition, the dynamic frequency data of the statistical copolymers are compared and contrasted with those found at PVBTMAC homopolymer under iso-T_g temperatures (i.e., T_g + 30 K) in Figure S4. The iso-T_g comparison reflects the changes in viscoelastic response from elastic to viscoelastic and liquid response by decreasing the VBTMAC content, quantitatively verifying the observations at ambient temperature.

Table 1. Zero shear viscosity and fitting parameters of CY and MC models.

VBTMAC (wt.%)	η_0 **(Pa·s)**	λ **(s)**	n	α
		Carreau–Yasuda (CY) model		
20 (sequential)	$(3 \pm 1) \times 10^7$	580 ± 10	0.11 ± 0.02	2 (fixed)
20 (statistical)	$11{,}970 \pm 30$	0.43 ± 0.01	0.59 ± 0.01	0.91 ± 0.03
40 (statistical)	$(2.68 \pm 0.08) \times 10^7$	3.8 ± 0.3	0.38 ± 0.02	2.0 ± 0.4
100	$(2 \pm 2) \times 10^{11}$	2000 ± 1000	0.015 ± 0.002	0.99 ± 0.2
		Modified Cross (MC) model		
20 (sequential)	$(3 \pm 1) \times 10^7$	1800 ± 900	0.11 ± 0.02	-
20 (statistical)	$11{,}970 \pm 30$	0.43 ± 0.01	0.61 ± 0.01	-
40 (statistical)	$(3.13 \pm 0.08) \times 10^7$	2.65 ± 0.4	0.25 ± 0.04	-
100	$(2.47 \pm 0.09) \times 10^{11}$	1490 ± 60	0.015 ± 0.002	-

Concerning the statistical copolymers, the relaxation time, λ, increases by increasing the concentration of VBTMAC glassy blocks. The extracted values of the power law index, n, imply a shear-thinning behavior of the copolymer melts that becomes stronger by increasing VBTMAC content. For the copolymer with 40 wt.% VBTMAC and the PVBTMAC homopolymer, the values of zero shear viscosity extracted from the MC model are slightly higher than the ones obtained from the CY model.

The effect of block arrangement along the backbone can be discussed with respect to Figure 3 and Table 1, comparing the dynamic frequency sweep data of the sequential and statistical copolymer with a VBTMAC content of 20 wt.%.

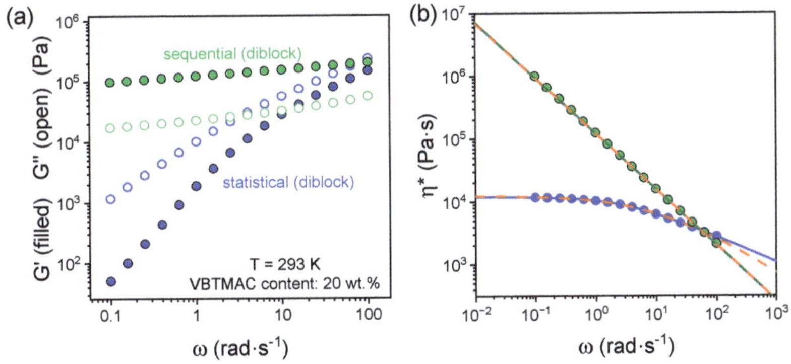

Figure 3. Angular frequency dependence of (**a**) the storage (filled symbols) and loss (open symbols) shear moduli and (**b**) complex viscosity for the sequential (green) and statistical (blue) DHBCs with 20 wt.% of VBTMAC. The solid and dashed lines in (**b**) represent fits by Equations (1) and (2), respectively.

The sequential arrangement of VBTMAC glassy blocks yields an elastic response ($G' >> G''$ and G', $G' \sim \omega^0$), compared to the liquid-like response of the statistical copolymer, indicating the presence of larger and continuous bulk-like VBTMAC regions, despite the fact that both specimens are in the disordered state, as evidenced in our previous study [40]. Parenthetically, a reduced change in heat capacity for T_g^{inter} was evidenced for the sequential copolymer compared to the statistical, verifying the aforementioned observations [40]. The determination of the zero shear viscosity is challenging and inaccurate for the sequential copolymer, due to its predominantly elastic response, even at elevated temperatures (see Figures S5 and S6), hindering the low-frequency plateau in the viscosity data. To mitigate this challenge, the viscosity data were fitted with the empirical models of Equations (1) and (2), yielding a higher value of zero shear viscosity by approximately three orders of magnitude compared to the corresponding statistical copolymer. This indicates that the block arrangement along the copolymer backbone plays a crucial role in tailoring the viscoelastic response of the studied DHBCs.

The extracted values for zero shear viscosity are summarized in Figure 4 at ambient temperature.

The zero shear viscosity values can be dictated by changing (i) the copolymer's composition and (ii) the block arrangement along the backbone. This is advantageous for the use of these copolymers in drug delivery applications. The dynamic frequency sweeps for the statistical copolymers are presented at different temperatures in Figure S7, along with the temperature dependence of the extracted power law index. The latter increases by increasing temperature, implying less shear thinning behavior, as evident from the flattening of the viscosity curves. The shear-thinning behavior of the studied polymeric formulations can be tailored by varying (i) temperature, (ii) composition, and (iii) block arrangement along the polymer backbone, highlighting their potential for biomedical applications.

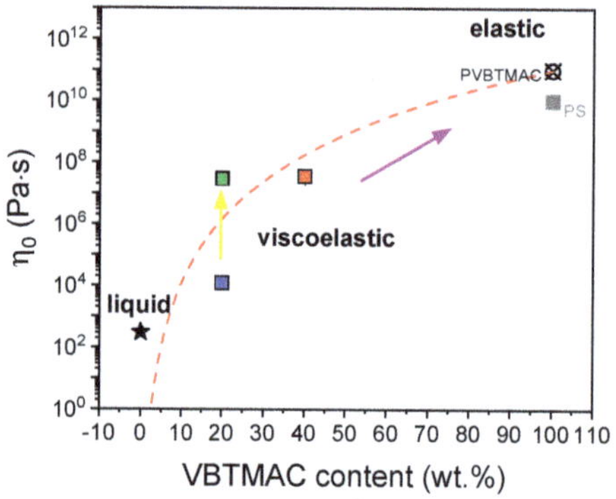

Figure 4. Zero shear viscosity, η_0, as a function of the VBTMAC weight fraction in DHBCs and the corresponding parent homopolymers at ambient temperature. The zero shear viscosity for polystyrene (PS) is also included for comparison and taken from Ref. [47]. The yellow and purple arrows indicate the effect of block arrangement across the backbone and the impact of VBTMAC content on the zero shear viscosity, respectively. The red dashed line is a guide for the eye.

Detailed information about the viscoelastic response of the studied copolymers can be obtained through the construction of master curves by employing the *tTs* principle, which involves the horizontal shifting of dynamic frequency sweeps, as shown in Figure 5. The original data of the storage and loss moduli are presented in Figures S8 and S9.

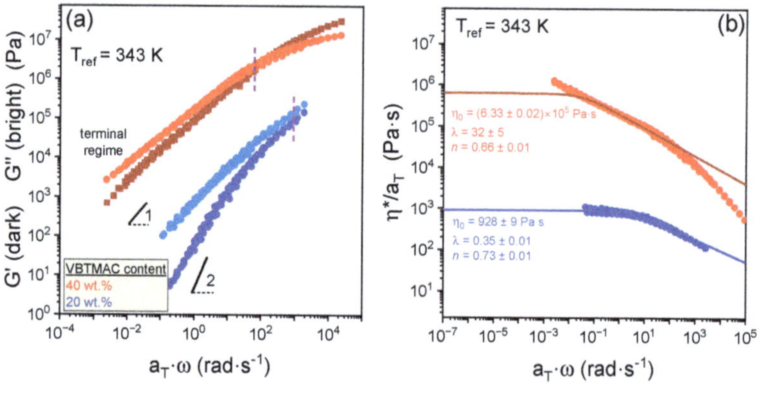

Figure 5. Superimposed master curves of (**a**) the storage (dark symbols) and loss (bright symbols) shear moduli and (**b**) complex viscosity constructed by employing the principle of *tTs* for the statistical DHBCs with 20 wt.% (blue symbols) and 40 wt.% (red symbols) VBTMAC content, as indicated. Lines with slopes 1 and 2 are also shown in (**a**). The solid lines represent fits by Equation (2) in (**b**).

For the copolymer containing 20 wt.% VBTMAC, a well-defined terminal regime is observed, confirming its liquid- or melt-like behavior. The $G'(\omega)$ does not follow the ideal terminal relaxation ($G' \sim \omega^2$), reflecting mainly the ionic interactions arising from the charged VBTMAC side groups. These ionic interactions persist into the melt state, influencing its terminal relaxation. In contrast, the copolymer with 40 wt.% VBTMAC maintains a viscoelastic response ($G' \sim \omega$) even at high temperatures (i.e., low frequencies), highlighting the influence of the glassy VBTMAC segments and the presence of strong ionic interactions. As depicted in Figures 5 and S2, the absence of large rubbery plateaus is attributable to the low molar masses and the short side chain lengths that distinctly reduce the entanglements. Figure 5b presents the normalized complex viscosity, adjusted using horizontal shift factors, as a function of normalized angular frequency. This representation of the master curves for the statistical copolymers confirms the viscosity variation with copolymer composition. Simultaneously, the extracted values of the power law index, by employing the MC model, indicate that both statistical copolymers display comparable levels of shear-thinning behavior. The temperature dependencies of the extracted horizontal shift factors for the statistical DHBCs and the PVBTMAC homopolymer are depicted in Figure 6. The original data of the frequency sweeps for the PVBTMAC homopolymer are presented in Figure S10.

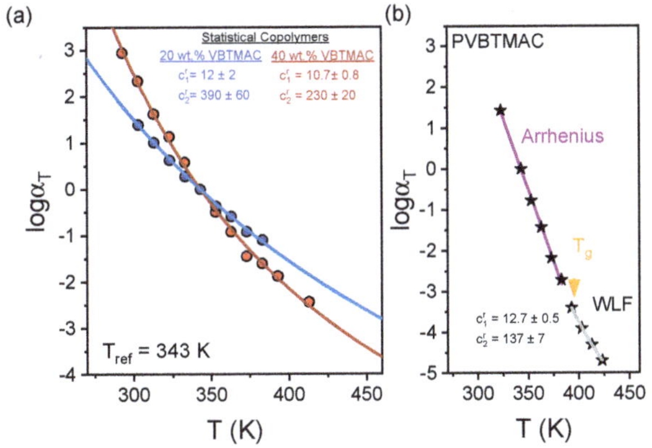

Figure 6. Temperature dependence of the horizontal shift factors for (**a**) statistical DHBCs and (**b**) the PVBTMAC homopolymer.

The results can now be discussed in terms of classical free volume theories of glass formation [43]. As shown in Figure 6, the extracted horizontal shift factors, α_T, were fitted according to the Williamas–Landel–Ferry (WLF) equation [48]:

$$log\alpha_T = -\frac{c_1{}^r(T - T^r)}{c_2{}^r(T - T^r)}, \quad (3)$$

where $c_1{}^r$ and $c_2{}^r$ are empirical parameters at the reference temperature, that is, T_r = 343 K, for the studied statistical DHBCs. For the PVBTMAC homopolymer, a transition from WLF to Arrhenius-type temperature dependence is observed in the vicinity of T_g (~383 K), corresponding to the vitrification of the polystyrene backbone. According to the theory, the empirical parameters can be calculated at T_g: $c_1{}^g = c_1{}^r\, c_2{}^r/(c_2{}^r + T_g - T_r)$ and $c_2{}^g = c_2{}^r + T_g - T_r$, and then the fractional free volume, f(T_g), and the thermal expansion coefficient of free volume, α_f, can be extracted, according to

$$f(T_g) = \frac{1}{2.303 c_1^g}, \quad (4)$$

$$\alpha_f = \frac{f(T_g)}{c_2^g}. \quad (5)$$

Furthermore, the fragility or steepness index, m^*, and apparent activation energy, E_g, can be estimated from [49,50]

$$m^* = \frac{c_1^g T_g}{c_2^g}, \quad (6)$$

$$E_g = 2.303R\frac{c_1^g}{c_2^g}T_g^2. \quad (7)$$

The estimated parameters are summarized in Table 2 and can be discussed with respect to Figures 7 and 8.

Table 2. Rheological T_g and WLF parameters, fractional free volume, thermal expansion coefficient, fragility along with apparent activation energy at T_g for DHBCs and PVBTMAC homopolymer.

Sample Code	T_g (K) *	c_1^g	c_2^g (K)	f(T_g)	α_f (K^{-1})	m^*	E_g (kJ·mol^{-1})
P(OEGMA$_{80}$-*co*-VBTMAC$_{20}$)	290	13.89	337	0.0313	9.29×10^{-5}	12	66.4
P(OEGMA$_{60}$-*co*-VBTMAC$_{40}$)	310	12.49	197	0.0348	1.77×10^{-4}	20	116.7
PVBTMAC	383	9.83	177	0.044	2.5×10^{-4}	21	156.0

* T_g determined from rheology is higher compared to the dielectric one due to the fact that the dynamic frequency sweeps are performed at ω = 10 rad·s^{-1}.

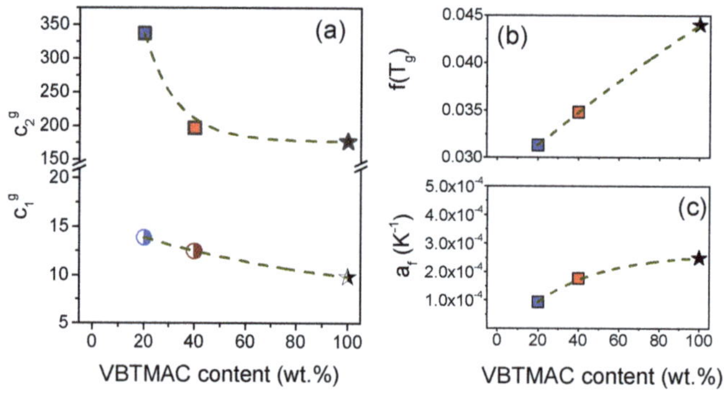

Figure 7. (**a**) WLF coefficients, c_1^g (semi-filled symbols) and c_2^g (filled symbols), (**b**) fractional free volume and (**c**) thermal expansion coefficient of free volume at T_g, determined through rheology, plotted as a function of the VBTMAC content. Dashed lines are guides for the eye.

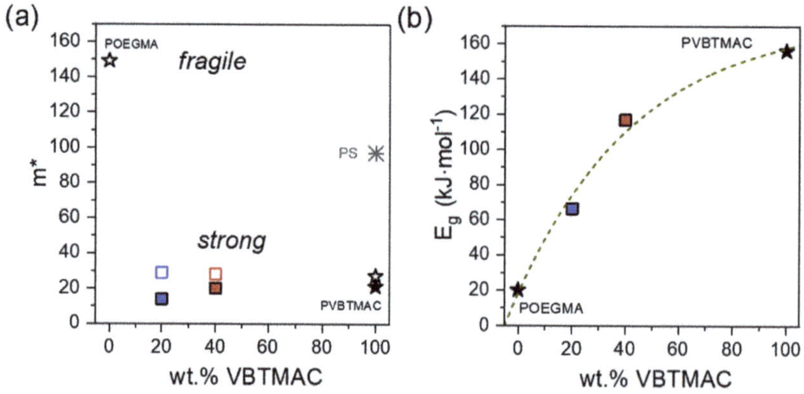

Figure 8. (**a**) Fragility and (**b**) apparent activation energy as a function of the VBTMAC weight fraction for the copolymers (squares) and their parent blocks (stars). Fragility data from dielectric measurements, taken from Ref. [40], are also included along with fragility data for PS taken from Ref. [51].

As depicted in Figure 7, the $c_1{}^g$ and $c_2{}^g$ parameters are dependent on the copolymer composition, leading to differences in the fractional free volume and thermal expansion coefficient according to Equations (4) and (5). Specifically, the fractional free volume increases by increasing the VBTMAC content. The reason for this is most likely that the bulky ionic side chains of the VBTMAC blocks increase the segmental rigidity and steric hindrance, leading to packing frustration. This also results in a slight increment of the thermal expansion coefficient. Importantly, the $c_2{}^g$ of PVBTMAC is approximately three times larger compared to polystyrene (PS) [51], thereby reducing fragility values, as discussed below in relation to Figure 8.

The PVBTMAC homopolymer exhibits one of the lowest fragility values ever reported for common polymers (see Figure S11), reflecting a *superstrong* behavior [52,53]. The rheology-derived values are in line with those found from dielectric measurements [40]. This low value of fragility indicates a less cooperative character of the structural relaxation, meaning that fewer neighboring segments cooperate in the structural relaxation. Specifically, the PVBTMAC homopolymer exhibits distinctly reduced fragility values compared to the linear polystyrene [43,51]. A similar trend has been observed in the family of poly(*p*-alkyl methacrylates), where a change from a "fragile" (m = 92) to a "strong" (m = 36) liquid occurs only by increasing the length of the side chain with the addition of a methylene unit (see Figure S12) [54,55]. Specifically, the PVBTMAC homopolymer exhibits a slightly lower value of fragility than poly(*p*-alkyl methacrylates) with side chain lengths, p, bearing $1 < p < 18$ methylene units [55–57] and the poly(*p*-phenylene) homopolymer [58] bearing side chain length with 8 methylene units, despite the higher T_g of PVBTMAC. Therefore, the superstrong nature of the PVBTMAC homopolymer is probably driven mainly by its side chain length and its high T_g that is mainly dictated by the strong electrostatic interactions taking place along its side chains.

Concerning the statistical copolymers, a similar *superstrong* behavior can be observed. Anticipatedly, as shown in Figure 8b, the apparent activation energy strongly increases by increasing the VBTMAC content, reflecting the stiff nature of VBTMAC segments and the higher amount of energy that is required for their segmental relaxation. This *superstrong* nature of the studied polymeric formulations may be a valuable attribute for their use in biomedical applications.

To sum up, this study offers a comprehensive quantitative analysis of the viscoelastic behavior of both sequential and statistical DHBCs. The findings demonstrate that the viscoelastic response of these systems can be precisely modulated by altering the copolymer composition and the arrangement of the hydrophilic blocks, as well as by adjusting the environmental temperature. Variations in block architecture (e.g., sequential versus statistical copolymers) lead to distinct molecular interactions and microphase behavior, which in turn govern the macroscopic rheological properties such as storage moduli (G') and loss moduli (G''), and thus the viscosity.

These tunable viscoelastic properties are critical for drug delivery applications, where formulation performance often depends on the mechanical adaptability, injectability and controlled release capability of the carrier matrix. For instance, a copolymer with a high shear moduli may provide better mechanical integrity at the site of delivery, while a reduced viscosity or gel-like response at physiological temperatures can enable sustained drug release. The ability to systematically tailor these rheological characteristics makes DHBCs versatile components for designing smart and responsive drug delivery systems.

4. Conclusions

Herein, we have investigated the mechanical properties of block and random DHBCs using rheological measurements. We report that the mechanical properties are governed mainly by the interfacial T_g. Depending on the VBTMAC content and block arrangement across the backbone, the macroscopic mechanical properties change from elastic-like to liquid-like. Specifically, the PVBTMAC homopolymer exhibits an elastic response up to high temperatures, reflecting its strong nature. A viscoelastic response is evident for the statistical copolymer with 40 wt.% VBTMAC. By further decreasing the VBTMAC to 20 wt.%, the statistical copolymer exhibits a liquid-like response. Overall, the zero shear viscosity, determined by fitting the measured data with two empirical models (i.e., the Carreau–Yasuda and modified Cross models), exponentially increases by increasing the VBTMAC content. This behavior reflects the influence of the glassy VBTMAC segments and the presence of strong ionic interactions arising from the charged VBTMAC side groups. Furthermore, for the 20 wt.% VBTMAC content, the change in block arrangement of the backbone from statistical to sequential results in a transition from liquid-like to elastic-like behavior, accompanied by an increase in zero shear viscosity.

In terms of classical free volume theories [43], we report that the copolymers exhibit reduced fragility reminiscent of that found in the PVBTMAC parent block. Importantly, the PVBTMAC homopolymer exhibits 5-fold lower fragility than the one found for polystyrene, reflecting its *superstrong* nature. The above observations are in line with previously reported dielectric results [40]. A *superstrong* behavior is also evident for the statistical copolymers, reflecting the impact of the glassy VBTMAC segments and their miscible state. To conclude, this study illustrates the tunability of the viscosity and shear-thinning behavior through systematic variation in (i) the brush copolymer composition, (ii) the block sequence along the polymer backbone, and (iii) the temperature, which furthermore demonstrates the potential of the investigated copolymers for advanced biomedical applications such as drug delivery.

Supplementary Materials: The following supporting information can be downloaded at https://www.mdpi.com/article/10.3390/polym17131857/s1, Figure S1: Strain sweep test; Figure S2: Temperature ramp of shear storage and loss modulus, along with the loss factor curves; Figure S3: Angular frequency sweeps for the studied statistical copolymers and their parent homopolymers; Figure S4: Angular frequency sweeps at iso-T_g temperatures; Figure S5: temperature ramps for diblock and statistical copolymer with 20 wt.% of VBTMAC; Figure S6: Angular frequency sweeps for diblock copolymer at two temperatures; Figure S7: Complex viscosity curves and the extracted power law index at different temperatures; Figures S8 and S9: Original dynamic frequency sweep data for the statistical copolymers; Figure S10: Original data for the PVBTMAC homopolymer; Figures S11 and S12: Comparison of fragility values with the literature-reported data.

Author Contributions: Conceptualization, A.P.; methodology, A.P. and A.C.; validation, A.P., S.P. and J.S.; formal analysis, A.P.; investigation, A.P.; resources, A.C.; data curation, A.P.; writing—original draft preparation, A.P.; writing—review and editing, A.P., A.C., S.P. and J.S.; visualization, A.P.; supervision, S.P. and J.S.; funding acquisition, J.S. All authors have read and agreed to the published version of the manuscript.

Funding: A.P. and J.S. were financially supported by the Area of Advanced Materials Science at the Chalmers University of Technology, with internal project number: C 2024-0296.

Institutional Review Board Statement: Not applicable.

Data Availability Statement: Data will be made available on request.

Conflicts of Interest: The authors declare no conflicts of interest.

Abbreviations

The following abbreviations are used in this manuscript:

DHBCs	double hydrophilic block copolymers
POEGMA	poly[oligo(ethylene glycol) methacrylate]
PVBTMAC	poly(vinyl benzyl trimethylammonium chloride)
SAOS	small amplitude oscillatory shear
T_g^{inter}	interfacial glass transition temperature
RAFT	reversible addition–fragmentation chain transfer
T_g	glass transition temperature
LVR	linear viscoelastic regime
CY	Carreau–Yasuda model
MC	modified Cross model
$\lvert G^* \rvert$	complex shear moduli
G'	shear storage moduli
G''	shear loss moduli
η_0	zero shear viscosity
tTs	time–temperature superposition
WLF	Williamas–Landel–Ferry
α_T	horizontal shift factor
$f(T_g)$	fractional free volume at T_g
α_f	thermal expansion coefficient of free volume at T_g
m^*	fragility
E_g	apparent activation energy
PS	polystyrene

References

1. Bates, F.S.; Fredrickson, G.H. Block copolymers—Designer soft materials. *Phys. Today* **1999**, *52*, 32–38. [CrossRef]
2. Hadjichristidis, N.; Pispas, S.; Floudas, G. *Block Copolymers: Synthetic Strategies, Physical Properties, and Applications*; John Wiley & Sons: Hoboken, NJ, USA, 2003.
3. Tsukahara, Y.; Kohjiya, S.; Tsutsumi, K.; Okamoto, Y. On the intrinsic viscosity of poly (macromonomer) s. *Macromolecules* **1994**, *27*, 1662–1664. [CrossRef]
4. Neugebauer, D.; Zhang, Y.; Pakula, T.; Sheiko, S.S.; Matyjaszewski, K. Densely-grafted and double-grafted PEO brushes via ATRP. A route to soft elastomers. *Macromolecules* **2003**, *36*, 6746–6755. [CrossRef]
5. Vlassopoulos, D.; Fytas, G.; Loppinet, B.; Isel, F.; Lutz, P.; Benoit, H. Polymacromonomers: Structure and dynamics in nondilute solutions, melts, and mixtures. *Macromolecules* **2000**, *33*, 5960–5969. [CrossRef]
6. Kapnistos, M.; Vlassopoulos, D.; Roovers, J.; Leal, L. Linear rheology of architecturally complex macromolecules: Comb polymers with linear backbones. *Macromolecules* **2005**, *38*, 7852–7862. [CrossRef]
7. Hu, M.; Xia, Y.; McKenna, G.B.; Kornfield, J.A.; Grubbs, R.H. Linear rheological response of a series of densely branched brush polymers. *Macromolecules* **2011**, *44*, 6935–6943. [CrossRef]
8. Daniel, W.F.; Burdyńska, J.; Vatankhah-Varnoosfaderani, M.; Matyjaszewski, K.; Paturej, J.; Rubinstein, M.; Dobrynin, A.V.; Sheiko, S.S. Solvent-free, supersoft and superelastic bottlebrush melts and networks. *Nat. Mater.* **2016**, *15*, 183–189. [CrossRef]
9. Qian, Z.; Chen, D.; McKenna, G.B. Re-visiting the "consequences of grafting density on the linear viscoelastic behavior of graft polymers". *Polymer* **2020**, *186*, 121992. [CrossRef]
10. Jang, J.; Leo, C.M.; Santiago, P.; Kennemur, J.G. Unraveling the Linear-to-Bottlebrush Transition by Linear Viscoelastic Response to Increasing Side Chain Length. *Macromolecules* **2025**, *58*, 2437–2449. [CrossRef]
11. Daniels, D.; McLeish, T.; Crosby, B.; Young, R.; Fernyhough, C. Molecular rheology of comb polymer melts. 1. Linear viscoelastic response. *Macromolecules* **2001**, *34*, 7025–7033. [CrossRef]
12. Lee, J.H.; Fetters, L.J.; Archer, L.A. Stress relaxation of branched polymers. *Macromolecules* **2005**, *38*, 10763–10771. [CrossRef]
13. Roovers, J.; Graessley, W. Melt rheology of some model comb polystyrenes. *Macromolecules* **1981**, *14*, 766–773. [CrossRef]
14. Saha, D.; Witt, C.L.; Fatima, R.; Uchiyama, T.; Pande, V.; Song, D.-P.; Fei, H.-F.; Yavitt, B.M.; Watkins, J.J. Opportunities in Bottlebrush Block Copolymers for Advanced Materials. *ACS Nano* **2024**, *19*, 1884–1910. [CrossRef]

15. Cui, S.; Zhang, B.; Shen, L.; Bates, F.S.; Lodge, T.P. Core–shell gyroid in ABC bottlebrush block terpolymers. *J. Am. Chem. Soc.* **2022**, *144*, 21719–21727. [CrossRef]
16. Song, Q.; Dong, Q.; Liang, R.; Xue, Y.; Zhong, M.; Li, W. Hierarchical self-assembly of ABC-type bottlebrush copolymers. *Macromolecules* **2023**, *56*, 5470–5481. [CrossRef]
17. Detappe, A.; Nguyen, H.V.-T.; Jiang, Y.; Agius, M.P.; Wang, W.; Mathieu, C.; Su, N.K.; Kristufek, S.L.; Lundberg, D.J.; Bhagchandani, S.; et al. Molecular bottlebrush prodrugs as mono-and triplex combination therapies for multiple myeloma. *Nat. Nanotechnol.* **2023**, *18*, 184–192. [CrossRef] [PubMed]
18. Pan, T.; Dutta, S.; Kamble, Y.; Patel, B.B.; Wade, M.A.; Rogers, S.A.; Diao, Y.; Guironnet, D.; Sing, C.E. Materials design of highly branched bottlebrush polymers at the intersection of modeling, synthesis, processing, and characterization. *Chem. Mater.* **2022**, *34*, 1990–2024. [CrossRef]
19. Lapkriengkri, I.; Albanese, K.R.; Rhode, A.; Cunniff, A.; Pitenis, A.A.; Chabinyc, M.L.; Bates, C.M. Chemical Botany: Bottlebrush Polymers in Materials Science. *Annu. Rev. Mater. Res.* **2024**, *54*, 27–46. [CrossRef]
20. Tu, S.; Choudhury, C.K.; Luzinov, I.; Kuksenok, O. Recent advances towards applications of molecular bottlebrushes and their conjugates. *Curr. Opin. Solid State Mater. Sci.* **2019**, *23*, 50–61. [CrossRef]
21. Zardalidis, G.; Pipertzis, A.; Mountrichas, G.; Pispas, S.; Mezger, M.; Floudas, G. Effect of polymer architecture on the ionic conductivity. Densely grafted poly (ethylene oxide) brushes doped with LiTf. *Macromolecules* **2016**, *49*, 2679–2687. [CrossRef]
22. Pipertzis, A.; Kafetzi, M.; Giaouzi, D.; Pispas, S.; Floudas, G.A. Grafted Copolymer Electrolytes Based on the Poly (acrylic acid-*co*-oligo ethylene glycol acrylate) (P (AA-*co*-OEGA)) Ion-Conducting and Mechanically Robust Block. *ACS Appl. Polym. Mater.* **2022**, *4*, 7070–7080. [CrossRef]
23. Rolland, J.; Brassinne, J.; Bourgeois, J.-P.; Poggi, E.; Vlad, A.; Gohy, J.-F. Chemically anchored liquid-PEO based block copolymer electrolytes for solid-state lithium-ion batteries. *J. Mater. Chem. A* **2014**, *2*, 11839–11846. [CrossRef]
24. Shim, J.; Bates, F.S.; Lodge, T.P. Superlattice by charged block copolymer self-assembly. *Nat. Commun.* **2019**, *10*, 2108. [CrossRef] [PubMed]
25. Zhang, B.; Zheng, C.; Sims, M.B.; Bates, F.S.; Lodge, T.P. Influence of charge fraction on the phase behavior of symmetric single-ion conducting diblock copolymers. *ACS Macro Lett.* **2021**, *10*, 1035–1040. [CrossRef]
26. Kataoka, K.; Harada, A.; Nagasaki, Y. Block copolymer micelles for drug delivery: Design, characterization and biological significance. *Adv. Drug Deliv. Rev.* **2012**, *64*, 37–48. [CrossRef]
27. Thang, N.H.; Chien, T.B.; Cuong, D.X. Polymer-based hydrogels applied in drug delivery: An overview. *Gels* **2023**, *9*, 523. [CrossRef]
28. Mai, Y.; Eisenberg, A. Self-assembly of block copolymers. *Chem. Soc. Rev.* **2012**, *41*, 5969–5985. [CrossRef]
29. Yan, C.; Altunbas, A.; Yucel, T.; Nagarkar, R.P.; Schneider, J.P.; Pochan, D.J. Injectable solid hydrogel: Mechanism of shear-thinning and immediate recovery of injectable β-hairpin peptide hydrogels. *Soft Matter* **2010**, *6*, 5143–5156. [CrossRef]
30. Chen, M.H.; Wang, L.L.; Chung, J.J.; Kim, Y.-H.; Atluri, P.; Burdick, J.A. Methods to assess shear-thinning hydrogels for application as injectable biomaterials. *ACS Biomater. Sci. Eng.* **2017**, *3*, 3146–3160. [CrossRef]
31. Altunbas, A.; Lee, S.J.; Rajasekaran, S.A.; Schneider, J.P.; Pochan, D.J. Encapsulation of curcumin in self-assembling peptide hydrogels as injectable drug delivery vehicles. *Biomaterials* **2011**, *32*, 5906–5914. [CrossRef]
32. Prince, D.A.; Villamagna, I.J.; Hopkins, C.C.; de Bruyn, J.R.; Gillies, E.R. Effect of drug loading on the properties of temperature-responsive polyester–poly (ethylene glycol)–polyester hydrogels. *Polym. Int.* **2019**, *68*, 1074–1083. [CrossRef]
33. Singh, A.; Dowdall, N.; Hoare, T. Poly (Oligo (Ethylene Glycol) Methacrylate)-Based Polymers in Biomedical Applications: Preparation and Applications. *Biomacromolecules* **2025**. [CrossRef] [PubMed]
34. Taneja, H.; Salodkar, S.M.; Parmar, A.S.; Chaudhary, S. Hydrogel based 3D printing: Bio ink for tissue engineering. *J. Mol. Liq.* **2022**, *367*, 120390. [CrossRef]
35. Pispas, S. Double hydrophilic block copolymers of sodium (2-sulfamate-3-carboxylate) isoprene and ethylene oxide. *J. Polym. Sci. A Polym. Chem.* **2006**, *44*, 606–613. [CrossRef]
36. Mountrichas, G.; Pispas, S. Novel double hydrophilic block copolymers based on poly (*p*-hydroxystyrene) derivatives and poly (ethylene oxide). *J. Polym. Sci. A Polym. Chem.* **2007**, *45*, 5790–5799. [CrossRef]
37. Chroni, A.; Forys, A.; Sentoukas, T.; Trzebicka, B.; Pispas, S. Poly [(vinyl benzyl trimethylammonium chloride)]-based nanoparticulate copolymer structures encapsulating insulin. *Eur. Polym. J.* **2022**, *169*, 111158. [CrossRef]
38. Chroni, A.; Forys, A.; Trzebicka, B.; Alemayehu, A.; Tyrpekl, V.; Pispas, S. Poly [oligo (ethylene glycol) methacrylate]-*b*-poly [(vinyl benzyl trimethylammonium chloride)] Based Multifunctional Hybrid Nanostructures Encapsulating Magnetic Nanoparticles and DNA. *Polymers* **2020**, *12*, 1283. [CrossRef]
39. Al-Tahami, K.; Singh, J. Smart polymer based delivery systems for peptides and proteins. *Recent Pat. Drug Deliv. Formul.* **2007**, *1*, 65–71. [CrossRef]

40. Pipertzis, A.; Chroni, A.; Pispas, S.; Swenson, J. Molecular Dynamics and Self-Assembly in Double Hydrophilic Block and Random Copolymers. *J. Phys. Chem. B* **2024**, *128*, 11267–11276. [CrossRef]
41. Moad, G.; Rizzardo, E.; Thang, S.H. Radical addition–fragmentation chemistry in polymer synthesis. *Polymer* **2008**, *49*, 1079–1131. [CrossRef]
42. Skandalis, A.; Sentoukas, T.; Selianitis, D.; Balafouti, A.; Pispas, S. Using RAFT Polymerization Methodologies to Create Branched and Nanogel-Type Copolymers. *Materials* **2024**, *17*, 1947. [CrossRef]
43. Ferry, J.D. *Viscoelastic Properties of Polymers*; John Wiley & Sons: Hoboken, NJ, USA, 1980.
44. Carreau, P.J. Rheological equations from molecular network theories. *Trans. Soc. Rheol.* **1972**, *16*, 99–127. [CrossRef]
45. Yasuda, K. Investigation of the Analogies Between Viscometric and Linear Viscoelastic Properties of Polystyrene Fluids. Ph.D. Thesis, Massachusetts Institute of Technology, Cambridge, MA, USA, 1979.
46. Cross, M.M. Rheology of non-Newtonian fluids: A new flow equation for pseudoplastic systems. *J. Colloid Sci.* **1965**, *20*, 417–437. [CrossRef]
47. Wagner, M.H.; Narimissa, E.; Poh, L.; Shahid, T. Modeling elongational viscosity and brittle fracture of polystyrene solutions. *Rheol. Acta* **2021**, *60*, 385–396. [CrossRef]
48. Plazek, D.J.; Ngai, K.L. Correlation of polymer segmental chain dynamics with temperature-dependent time-scale shifts. *Macromolecules* **1991**, *24*, 1222–1224. [CrossRef]
49. Qin, Q.; McKenna, G.B. Correlation between dynamic fragility and glass transition temperature for different classes of glass forming liquids. *J. Non Cryst. Solids* **2006**, *352*, 2977–2985. [CrossRef]
50. Angell, C. Spectroscopy simulation and scattering, and the medium range order problem in glass. *J. Non Cryst. Solids* **1985**, *73*, 1–17. [CrossRef]
51. Pipertzis, A.; Hossain, M.D.; Monteiro, M.J.; Floudas, G. Segmental dynamics in multicyclic polystyrenes. *Macromolecules* **2018**, *51*, 1488–1497. [CrossRef]
52. Huang, D.; McKenna, G.B. New insights into the fragility dilemma in liquids. *J. Chem. Phys.* **2001**, *114*, 5621–5630. [CrossRef]
53. Böhmer, R.; Ngai, K.L.; Angell, C.A.; Plazek, D.J. Nonexponential relaxations in strong and fragile glass formers. *J. Chem. Phys.* **1993**, *99*, 4201–4209. [CrossRef]
54. Kunal, K.; Robertson, C.G.; Pawlus, S.; Hahn, S.F.; Sokolov, A.P. Role of chemical structure in fragility of polymers: A qualitative picture. *Macromolecules* **2008**, *41*, 7232–7238. [CrossRef]
55. Floudas, G.; Štepánek, P. Structure and dynamics of poly (*n*-decyl methacrylate) below and above the glass transition. *Macromolecules* **1998**, *31*, 6951–6957. [CrossRef]
56. Pipertzis, A.; Hess, A.; Weis, P.; Papamokos, G.; Koynov, K.; Wu, S.; Floudas, G. Multiple segmental processes in polymers with cis and trans stereoregular configurations. *ACS Macro Lett.* **2018**, *7*, 11–15. [CrossRef] [PubMed]
57. Pipertzis, A.; Skandalis, A.; Pispas, S.; Floudas, G. Nanophase Segregation Drives Heterogeneous Dynamics in Amphiphilic PLMA-b-POEGMA Block-Copolymers with Densely Grafted Architecture. *Macromol. Chem. Phys.* **2024**, *225*, 2400180. [CrossRef]
58. Mierzwa, M.; Floudas, G.; Neidhöfer, M.; Graf, R.; Spiess, H.W.; Meyer, W.H.; Wegner, G. Constrained dynamics in supramolecular structures of poly (*p*-phenylenes) with ethylene oxide side chains: A combined dielectric and nuclear magnetic resonance investigation. *J. Chem. Phys.* **2002**, *117*, 6289–6299. [CrossRef]

polymers

MDPI

Article

Self-Assembly of Amphiphilic Comb-like Copolymers into Micelles and Vesicles in Solution

Qiaoyue Chen [1,†], Kun Tian [2,†], Ruiqi Zhu [1], Mingming Ding [1,3,*] and Zhanwen Xu [2,*]

1 Xinjiang Laboratory of Phase Transitions and Microstructures in Condensed Matters, College of Physical Science and Technology, Yili Normal University, Yining 835000, China; chenqy210701@163.com (Q.C.); rqzhu135@163.com (R.Z.)

2 State Key Laboratory of Molecular Engineering of Polymers, Key Laboratory of Computational Physical Sciences, Department of Macromolecular Science, Fudan University, Shanghai 200433, China; 23210440020@m.fudan.edu.cn

3 School of Chemical Engineering and Light Industry, Guangdong University of Technology, Guangzhou 510006, China

* Correspondence: mmding@gdut.edu.cn (M.D.); xuzhanwen@fudan.edu.cn (Z.X.)

† These authors contributed equally to this work.

Abstract: Combining Brownian dynamics simulations and self-consistent field theory, we demonstrate that stable assembled structures, such as vesicles, toroidal micelles, bowl-like micelles, sheet-like micelles, non-spherical vesicles, and cylindrical micelles, are dependent on the molecular parameters of amphiphilic comb-like copolymers. Importantly, we find that vesicle formation involves two intermediate states, sheet-like and bowl-like micelles, and the difference in their free energies is minimal, which illustrates the coexisting phase between them. Moreover, the assembled vesicles can be modulated in the membrane thickness with overall size, unchanged only by adjusting the backbone length. We also demonstrate the coexistence of toroidal and cylindrical micelles because neither structure has a significant advantage over the other in free energy. Our work points out how to obtain different morphologies by adjusting the molecular parameters of amphiphilic comb-like copolymers, instilling confidence in their potential for stable drug encapsulation and enhanced targeted drug delivery.

Keywords: amphiphilic comb-like copolymers; micelles; vesicles; Brownian dynamics

1. Introduction

Copolymer self-assembly has been a topic of great interest because of its ability to form functional micelles or vesicles, which, in turn, determine the potential applications of copolymers [1–5]. For example, spherical micelles and vesicles exhibit stable drug encapsulation and enhanced targeted drug delivery [6–10]. Toroidal micelles have applications in delivery systems that mimic bacteriophages and vertebrate DNA rings [11,12]. In biomedical sensing, self-assembled length-controlled cylindrical micelle brushes are used to enhance the signal recognition capability of detectors [13].

The self-assembled morphologies of copolymers in solution are influenced by numerous factors, with molecular structure being the most intrinsic. To date, the self-assembling mechanisms of linear block copolymers [7,14,15], comb-like copolymers [16–19], hyperbranched copolymers [20,21], and other topologically structured copolymers continue to be subjects of ongoing exploration in both simulation and experimental studies [22]. Among them, the self-assembly of non-linear polymers, such as graft copolymers, emerges as a more promising component in self-assembled materials [23–30]. This is attributed to their lower critical micelle concentration [31] and higher interfacial stability [32], making them withstand rigorous environments such as high pressure and heat, and strong alkalis and acids [33,34].

Given this, many researchers have used different graft molecular models to explore their self-assembly morphologies. For instance, Wang et al. employed dissipative particle dynamics (DPD) simulations to investigate the self-assembly of amphiphilic comb-like copolymers, specifically using the $PA_6(B_2)_3$ and $QC_6(D_2)_3$ models (the subscript numbers 6, 2, and 3, respectively, represent the backbone length, side-chain length, and grafting number), which results in monolayer vesicles [35]. Lin et al. combined self-consistent field theory (SCFT) with DPD simulations to study the A_9-graft-$(B_6)_3$ model, revealing the formation of segmented, multi-layered composite vesicles [10]. Liu et al. focused on the spontaneous formation and fusion of onion-like vesicles using DPD simulations with the $A_6(B_2)_3$ model [19]. Additionally, Khokhlov et al. used Monte Carlo (MC) simulations to examine the conformational transitions of a single amphiphilic comb-like copolymer, observing structural changes from spherical to elongated ellipsoids and pearl-necklace configurations as the grafting density increased [36]. Sheng and Tsao et al. separately employed comb-like graft copolymers composed of a backbone of poly((N-vinylcarbazole)-co-(4-vinylbenzyl chloride)) (P(NVK-co-VBC)) and side chains of poly(((2-dimethylamino)ethyl methacrylate)-co-(tert-butyl acrylate)) (P(DMAEMA-co-tBA)) to regulate the formation of vesicles from unilamellar to multilamellar structures using the DPD method. They found that variations in grafting density can modulate the thickness and the number of layers in the hydrophobic region [37,38]. These studies highlight the critical role of molecular model selection in determining the self-assembly morphologies of amphiphilic comb-like copolymers. However, a significant challenge remains: the diversity of molecular models makes direct comparisons between different studies difficult. This limitation hinders the ability to establish a precise relationship between molecular parameters and the resulting self-assembled morphologies.

In the current study, we systematically investigate the relationship between the molecular parameters and self-assembled morphologies of amphiphilic comb-like chains in solution by employing Brownian dynamics (BD) simulations. The objective is to control the self-assembly structures by tuning these molecular parameters for diverse applications. The study revealed that various self-assembled morphologies could be observed, including toroid micelles, bowl-like micelles, vesicles, sheet-like micelles, cylindrical micelles, and non-spherical vesicles, by adjusting the length of the backbone and side chains, and the grafting density. In addition, we reveal the transition mechanisms between different morphologies by calculating the free energy of different self-assembled morphologies using SCFT. The remainder of this article is organized as follows. Section 2 provides the model and simulation details. Section 3 constructs the morphological phase diagram, elucidates the self-assembly mechanism, and provides the relationship between the molecular parameters and self-assembled morphologies. Section 4 presents the conclusion and future outlook.

2. Model and Methods

BD simulations: We conduct BD simulations to investigate the self-assembly dynamics of amphiphilic comb-like chains. As demonstrated in our previous work and related simulations, the coarse-grained model is employed, which can reasonably capture the essential interactions and structural features with the reduced unit system [31,32,39–45]. As illustrated in Figure 1, the amphiphilic comb-like chains are modeled as bead-spring chains: The hydrophilic side chains B (red) are randomly grafted onto the hydrophobic backbone A (yellow). We denote amphiphilic comb-like chains as A_x-comb-$(B_y)_n$, where x and y represent the lengths of the backbone and side chains, respectively, and n denotes the grafting number. All bonded interactions, including the adjacent beads of the backbone and the side chains, and the grafting points between the beads of the backbone and the side chains, are calculated by the finite extensible nonlinear elastic (FENE) potential [46], i.e.,

$$U_{FENE}(r) = -0.5\kappa R_0^2 \ln[1 - (r/R_0)^2] \quad (1)$$

where κ denotes the spring constant and R_0 is the maximum separation distance for the bead–bead distance r.

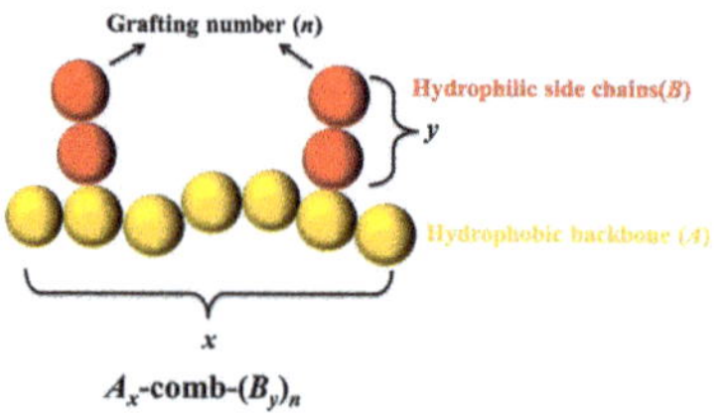

Figure 1. Coarse-grained model of amphiphilic comb-like copolymers: A_x-comb-$(B_y)_n$, composed of a hydrophobic backbone with a length of x (represented by A, yellow) and n hydrophilic side chains with a length of y (represented by B, red), where $n \leq x$.

The non-bonded interactions between two beads accounted for the excluded-volume interactions described through the Lennard-Jones (LJ) potential, i.e.,

$$U_{LJ}(r) = 4\epsilon[(\sigma/r)^{12} - (\sigma/r)^6] \quad (2)$$

Here, ϵ and σ represent the energy and length parameters, respectively. Implicit solvents are used in the simulation system. For the interactions between hydrophobic beads, we specify a cutoff distance $r_c = 2.5\sigma$ for the LJ potential, corresponding to a poor solvent condition [14,47]. For the other pairwise beads, we set the cutoff to $r_c = 2^{1/6}\sigma$, corresponding to a good solvent condition [48]. Additionally, both LJ potentials are shifted separately to ensure $U_{\mathrm{LJ}} = 0$ at r_c.

We employ BD simulations, where the motion for bead i is given by [49]

$$\dot{\mathbf{r}}_i(t) = -\frac{1}{\xi}\nabla U_i + \frac{1}{\xi}\mathbf{f}_i(t) \quad (3)$$

U_i represents the potentials from all FENE and LJ interactions as discussed, and ξ denotes the friction coefficient experienced by a bead in the solvent. The random force $\mathbf{f}_i(t)$ is generated based on the thermal bath model using a Gaussian-distributed random number generator, satisfying the fluctuation–dissipation theorem: $\langle \mathbf{f}_i(t)\mathbf{f}_j(t')\rangle = 2\xi k_B T \delta_{ij}\delta(t-t')\mathbf{I}$, where k_B is Boltzmann's constant, T is the temperature, and $\mathbf{I}$ denotes the unit tensor. Equation (3) neglects inertial effects and can be efficiently integrated using the Euler method. In this work, we disregard hydrodynamic interactions between different beads, which enables us to compute the steps for a longer time with reduced computational effort.

We adopt length, mass, and energy units as σ, m, and ϵ, respectively. Consequently, the unit of time can be described as $\tau = (m\sigma^2/\epsilon)^{1/2}$. We set the temperature as $k_BT = 1.0$, the time step as $\Delta t = 10^{-4}$, and the friction coefficient as $\xi = 0.5$. Furthermore, we fix the parameters of the FENE potential as $\kappa = 30$ and $R_0 = 1.5$ to prevent chain crossing [50].

We conduct simulations within a $20\sigma \times 20\sigma \times 20\sigma$ cubic box with periodic boundary conditions. All samples are fixed with the same volume fraction 0.15 and are simulated with 3.5–5.0 $\times 10^8$ time steps. In addition, independent simulations are carried out 5–10 times with different random seeds.

SCFT: In the SCFT calculations, we consider an incompressible blending system composed of AB comb-like copolymers and solvent molecules of short A homopolymer. Under the approximations of the mean-field treatment and Gaussian-chain model, the free energy in the canonical ensemble can be expressed as

$$\begin{aligned}\frac{NF}{\rho_0 V k_B T} &= -\phi \ln \frac{Q_C}{\phi} - \frac{1-\phi}{\gamma} \ln \frac{Q_H}{1-\phi} \\ &+ \frac{1}{V}\int d\mathbf{r}\{\chi N \phi_A(\mathbf{r})\phi_B(\mathbf{r}) - \omega_A(\mathbf{r})\phi_A(\mathbf{r}) - \omega_B(\mathbf{r})\phi_B(\mathbf{r}) \\ &- \eta(\mathbf{r})[1-\phi_A(\mathbf{r}) - \phi_B(\mathbf{r})]\}\end{aligned} \tag{4}$$

where $\omega_K(\mathbf{r})$ is the mean-field conjugating to the volume fraction $\phi_K(\mathbf{r})$ (K = A or B). The spatial function $\eta(\mathbf{r})$ is a Lagrange multiplier used to enforce the incompressibility condition, $\phi_A(\mathbf{r}) + \phi_B(\mathbf{r}) = 1$. The two quantities Q_C and Q_H are the partition functions of single copolymer chain and single homopolymer chain interacting with the mean fields of $\omega_A(\mathbf{r})$ and $\omega_B(\mathbf{r})$, respectively.

The A_1-comb-$(B_1)_1$, which can be seen as the repeating units of A_x-comb-$(B_y)_n$ at $y = n = 1$ in the BD simulation, is modelled as an AB_2 star copolymer with $f_A = 0.5$ and N segments in the SCFT. In other words, $0.5N$ segments of the comb-like copolymer in the SCFT correspond to 1 bead of the copolymer model in the BD. By setting the number of total segments for the solvent to be $0.2N$ and the interaction parameter between A and B components $\chi_{AB}N$ to be 15 (χ_{AB} denotes the Flory–Huggins parameter), we observed the order–disorder transitions and self-assembled structures in the SCFT similar to that in the BD. More details of the SCFT can be found in references [51,52] or the Supporting Information.

3. Results and Discussion

As shown in Figure 2, we initially fix $y = 1$, that is A_x-comb-$(B_1)_n$, to explore the possible self-assembling structures. The results reveal that when $n = x$, the system cannot form a stable structure and is in the dispersed phase. At this time, each bead on the backbone is grafted with a side chain. As we reported earlier, at this time, the hydrophobic backbone chain is surrounded by the hydrophilic side chains to form a better encapsulation layer, that is to say, the formation of a single molecule micelle [40], which leads to the difficulty of further self-assembly with other chains. As n decreases, we begin to be able to observe various structures. For example, first, one can self-assemble to form cylindrical micelles; however, when x is small, the coexistence of cylindrical micelles and toroidal micelles can also be observed. The next larger percentage is sheet-like micelles. In addition, when $n < 4$, the structures obtained are more abundant; for example, the coexistence of sheet-like micelles and vesicles, and non-spherical vesicles and cylindrical micelles can be obtained when x is small.

We then fix $n = 2$, that is, A_x-comb-$(B_2)_2$, to investigate the self-assembling structures with two side chains. Our results indicate that most stable self-assembled structures cannot be formed. This implies that the desired self-assembled structure will be formed only if the molecular parameters are tuned to a suitable range. We find that only sheet-like micelles and cylindrical micelles can be formed with small y. When x is larger, it is in the coexistence phase of sheet-like micelles and cylindrical micelles. As reported in the literature, stabilized micelles and vesicles with different shapes have very many promising applications [6–10].

Therefore, we proceed to further analyze the self-assembly process of toroidal micelles, cylindrical micelles, and vesicles, as well as the free energy of their structures.

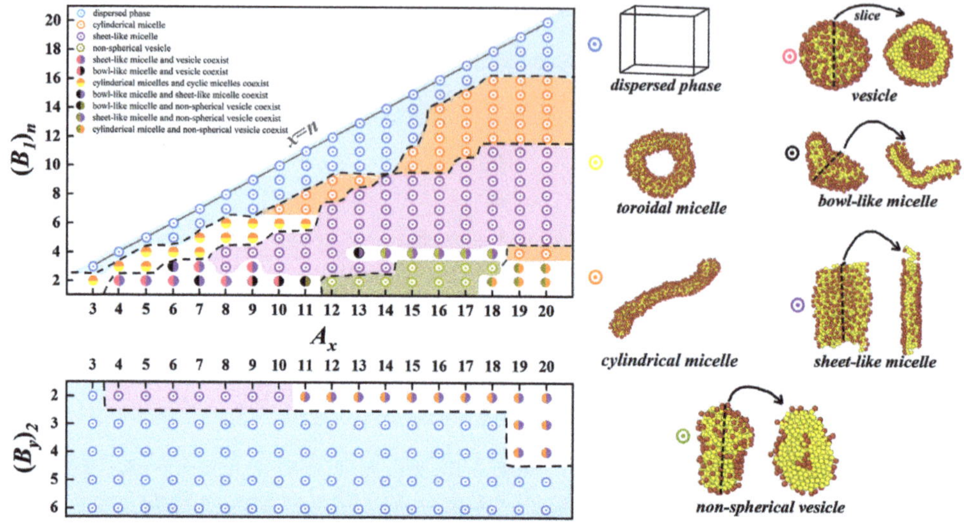

Figure 2. Phase diagrams of A_x-comb-$(B_1)_n$ and A_x-comb-$(B_y)_2$, respectively, where the blue region represents the dispersed phase, the yellow region denotes toroidal micelle, the orange region indicates cylindrical micelle, the pink region represents vesicle, the black region signifies bowl-like micelle, the purple region denotes sheet-like micelle, and the gray region indicates non-spherical vesicle. The corresponding simulation snapshots for each region are indicated on the right, where the arrows point to the cross-sectional slice of the structures. The two colored circles represent the coexistence state of two structures simulated under the same starting conditions.

We trace the dynamic process of vesicle formation, as shown in Figure 3a. The chains rapidly aggregate into numerous small clusters at the initial stage and assemble into sheet-like micelles. At this stage, two distinct scenarios emerge, one where the system always maintains the sheet-like structure, accompanying occasional minor bending. The other is the gradual bending of the sheet-like micelles, transitioning into bowl-like structures. Similarly, the system either remains in the bowl-like micelles or the bowl's opening gradually narrows until completely closed, forming a spherical vesicle. This suggests that vesicle formation involves two intermediate states, sheet-like and bowl-like micelles, rather than direct formation. This closely resembles the findings of Gianneschi et al., who utilized in situ liquid-cell transmission electron microscopy to observe the formation of bicontinuous micelle structures during the transition of amphiphilic block copolymer micelles into higher-order vesicle structures [53].

As shown in Figure 2, the simulation results indicate the coexistence of vesicles, bowl-like micelles, and sheet-like micelles. To justify this coexistence of states, we calculate their formation and the corresponding free energies using SCFT, as shown in Figure 3b. Consistently, we observe the coexistence of vesicles and bowl-like micelles as well as sheet-like micelles, except that $A_6(B_1)_2$ and $A_7(B_1)_2$ failure to form the sheet-like micelles. Although the results of the coexisting phases do not fully agree with the simulation results of BD, the results show that the difference in the free energy of the different phases is minimal for the same molecular parameters, which confirms that this coexisting phase is reasonable. It also shows that this system's potential barrier for vesicle formation is low, which opens up new possibilities for self-assembly to form vesicles.

(a)

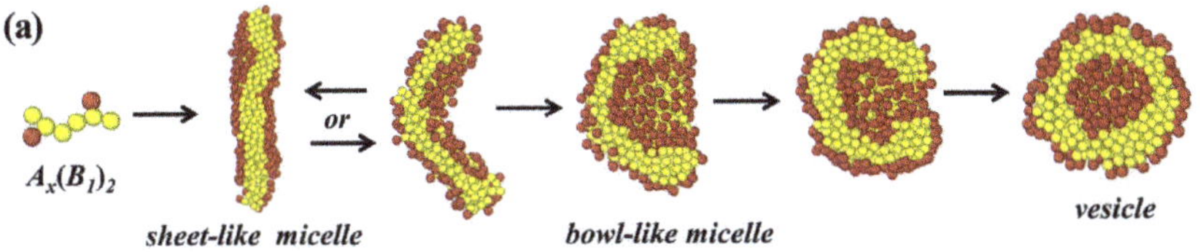

(b)	*conformations*	$A_6(B_1)_2$	$A_7(B_1)_2$	$A_8(B_1)_2$	$A_9(B_1)_2$
sheet-like micelle	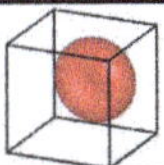			15.9900	15.7394
bowl-like micelle	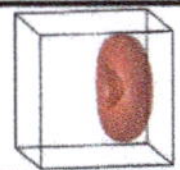	16.4324	16.1990	15.9954	15.8185
vesicle	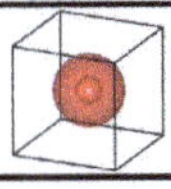	16.3606	16.1201	15.9150	15.7394

Figure 3. (**a**) Self-assembly roadmap of amphiphilic comb-like copolymers with the molecular parameters $A_x(B_1)_2$, which indicates that they first form sheet-like micelles, then the structure bends and forms bowl-like micelles, or may revert to sheet-like micelles, and finally the structure closes to form vesicles. (**b**) Free energies calculated by SCFT for the sheet-like micelles, the bowl-like micelles, and the vesicles correspond representatively to (**a**).

The control of vesicle size and membrane thickness has been a widely studied topic [8,10,35,54,55]. In Figure 4a, we statistically analyze the density distribution of self-assembled vesicles with molecular parameters of A_x-comb-$(B_1)_2$, where $x = 4$, 8, 10, and 14, respectively. Here, r represents the distance from the vesicle center, and $\rho(r)$ denotes the average density of the vesicle at $r + \Delta r$ on the spherical shell. Specifically, the vesicle's center of mass is first determined. Then, the number of hydrophilic beads at different radial distances within concentric spherical shells of equal thickness is counted to obtain the distribution profile. We observe two peaks with increasing r, except $A_14(B_1)_2$, which further demonstrates the formation of vesicles. When $x = 14$, corresponding to non-spherical vesicles, the density distribution no longer follows the pattern of two peaks. As x increases, the first peak continuously shifts towards the vesicle's center. However, the second peak is almost unchanged, suggesting that the membrane thickness is increasing while the overall dimensions remain constant, which matches the calculated values. This is a promising method for controlling vesicle size to modulate membrane thickness, which infers that the assembled vesicles formed by comb-like copolymers with larger gap lengths (i.e., fewer grafting points) possess thicker hydrophobic membranes and smaller cavities.

To clarify the molecular mechanism of such size-invariant but membrane-thickness-variable vesicles, we give amplified molecular packing, as illustrated in Figure 4b. When $x = 4$, due to the high graft density of hydrophilic side chains, most hydrophobic backbones exhibit a straighter shape, and the average distance from backbone end to end is approximately 4. However, when $x = 10$, with side chains spaced farther apart, the backbone tends to bend, and the average distance from the backbone end to end is approximately 2.36. As hydrophobic backbones tend to aggregate, thicker hydrophobic membranes are formed. Furthermore, when increased to "overloaded" lengths, such as $x = 14$, the hydrophobic portion of the packing is not constrained, resulting in a non-spherical vesicle.

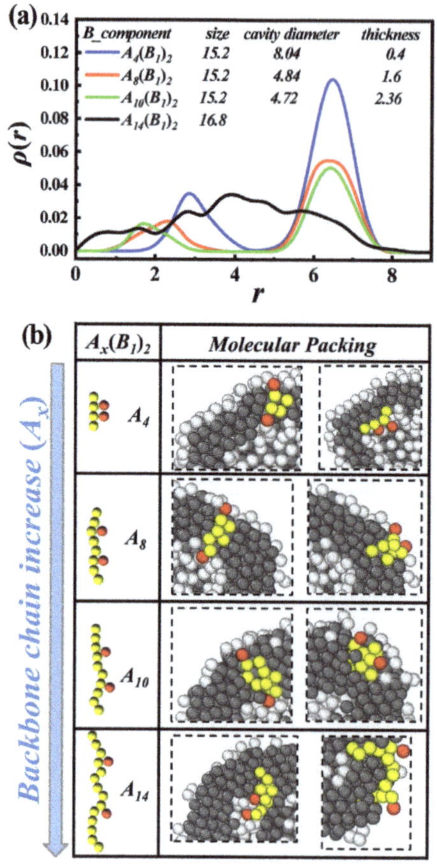

Figure 4. (**a**) Average density [$\rho(r)$] of the hydrophilic beads (B) as a function of the distance r from the center of the assembled vesicles, which are composed of $A_x(B_1)_2$ and $x = 4, 8, 10$, and 14, respectively. The overall size of the vesicles, the cavity diameters, and the membrane thicknesses are also listed. (**b**) Amplified molecular packing of the vesicles corresponds to the cases of (**a**).

Figure 2 shows a large coexistence zone for toroidal and cylindrical micelles. To explore their formation mechanism and to verify the rationality of their coexistence, we calculate the radius of gyration during their formation respectively and calculate the free energy of their steady states using SCFT, as shown in Figure 5. The comb-like chains rapidly aggregate at the initial simulation. Partial aggregates gradually become straight, which may be attributed to the rearrangement of comb-like chains, causing the formation of cylindrical micelles. Differing from that of cylindrical micelles, the middle of the aggregate gradually broke open, eventually forming toroidal micelles. As depicted in Figure 5a,b, when the structure of the toroidal and cylindrical micelles is gradually stabilized, their radii of gyration are also only stable, accompanied by small fluctuations. We then calculate their free energies using SCFT in Figure 5c and find their values almost identical. This suggests that under the same molecular parameter conditions, neither structure has a significant advantage over the other, and thus, coexisting phases can easily occur.

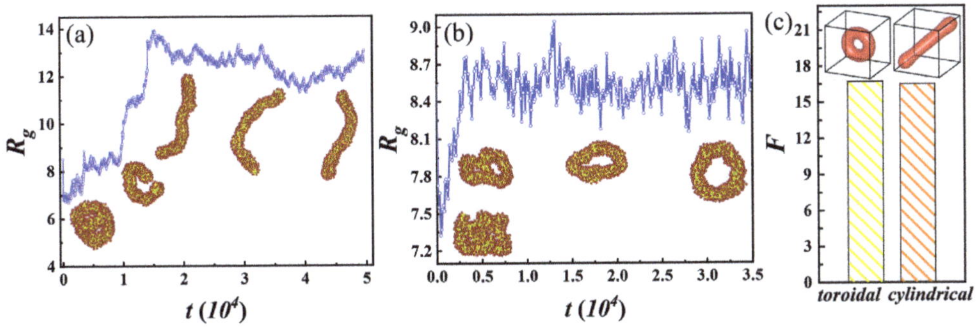

Figure 5. Gyration radius (R_g) during the formation of cylindrical micelles (**a**) and toroidal micelles (**b**) as a function of simulation time (t). The insets show representative simulation snapshots captured during the self-assembly process. (**c**) Free energy (F) for toroidal and cylindrical micelles, where the insets exhibit stable states.

4. Conclusions

In this work, we systematically investigate the relationship between the molecular parameters and self-assembled morphologies of amphiphilic comb-like chains A_x-comb-$(B_y)_n$ in solution by employing BD and SCFT. The results indicate that the adjustments in x and n lead to different arrangements of chains, resulting in stable states such as vesicles, toroidal micelles, bowl-like micelles, sheet-like micelles, non-spherical vesicles, and cylindrical micelles. Importantly, we find that vesicle formation involves two intermediate states, sheet-like and bowl-like micelles, and the difference in their free energies is minimal, which illustrates the coexisting phase between them. Moreover, the assembled vesicles can be modulated in the membrane thickness with overall size unchanged by only adjusting the backbone length. We also demonstrate the coexistence of toroidal and cylindrical micelles because neither structure has a significant advantage over the other in free energy. Our work shows how to obtain different morphologies by adjusting A_x-comb-$(B_y)_n$, especially the size-invariant but membrane-thickness-variable vesicles and toroidal and cylindrical micelles, which provides insight into their applications such as stable drug encapsulation and enhanced targeted drug delivery. Of course, the interaction strength of hydrophobic beads is another critical factor influencing the outcomes, alongside molecular parameters. The synergistic effects of these factors warrant further investigation. Additionally, the availability of efficient sampling methods to obtain the distribution of coexisting phases is a question worth exploring. Furthermore, future studies could extend to environmentally responsive dynamic assembly, investigating phenomena such as temperature-gradient-induced dynamic phase transitions.

Supplementary Materials: The following supporting information can be downloaded at: https://www.mdpi.com/article/10.3390/polym17131870/s1, SCFT Method.

Author Contributions: Methodology, Q.C., K.T., R.Z. and Z.X.; Software, K.T.; Formal analysis, K.T. and R.Z.; Investigation, Q.C. and K.T.; Data curation, Q.C.; Writing—original draft, Q.C. and R.Z.; Writing—review & editing, M.D. and Z.X.; Supervision, M.D. and Z.X.; Project administration, M.D.; Funding acquisition, Q.C. and M.D. All authors have read and agreed to the published version of the manuscript.

Funding: This work is financially supported by the Research Program for Universities in Xinjiang Uygur Autonomous Region (No. XJEDU2024P69) and the National Natural Science Foundation of China (No. 22373025).

Institutional Review Board Statement: Not applicable.

Data Availability Statement: The original contributions presented in this study are included in the article/Supplementary Material. Further inquiries can be directed to the corresponding author.

Conflicts of Interest: The authors declare no conflict of interest.

References

1. Mei, G.; Zheng, Y.; Fu, Y.; Huo, M. Polymerization-induced self-assembly of random bottlebrush copolymers. *Polym. Chem.* **2022**, *13*, 5389–5396. [CrossRef]
2. MacFarlane, L.; Zhao, C.; Cai, J.; Qiu, H.; Manners, I. Emerging applications for living crystallization-driven self-assembly. *Chem. Sci.* **2021**, *12*, 4661–4682. [CrossRef]
3. Wen, W.; Huang, T.; Guan, S.; Zhao, Y.; Chen, A. Self-assembly of single chain janus nanoparticles with tunable liquid crystalline properties from stilbene-containing block copolymers. *Macromolecules* **2019**, *52*, 2956–2964. [CrossRef]
4. Shi, H.; Zhao, Y.; Dong, X.; Zhou, Y.; Wang, D. Frustrated crystallisation and hierarchical self-assembly behaviour of comb-like polymers. *Chem. Soc. Rev.* **2013**, *42*, 2075–2099. [CrossRef]
5. Sugihara, S.; Blanazs, A.; Armes, S.P.; Ryan, A.J.; Lewis, A.L. Aqueous dispersion polymerization: A new paradigm for in situ block copolymer self-assembly in concentrated solution. *J. Am. Chem. Soc.* **2011**, *133*, 15707–15713. [CrossRef] [PubMed]
6. Rösler, A.; Vandermeulen, G.W.; Klok, H.A. Advanced drug delivery devices via self-assembly of amphiphilic block copolymers. *Adv. Drug Deliv. Rev.* **2001**, *53*, 95–108. [CrossRef]
7. Ding, M.; Hou, L.; Duan, X.; Shi, T.; Li, W.; Shi, A.C. Translocation of micelles through a nanochannelNanochannel. *Macromolecules* **2022**, *55*, 6487–6492. [CrossRef]
8. Nishimura, T.; Sasaki, Y.; Akiyoshi, K. Biotransporting self-assembled nanofactories using polymer vesicles with molecular permeability for enzyme prodrug cancer therapy. *Adv. Mater.* **2017**, *29*, 1702406. [CrossRef]
9. Nguyen, H.N.; Ezzat, M.; Huang, C.J. Lysolipid-inspired amphiphilic polymer nanostructures: Implications for drug delivery. *ACS Appl. Nano Mater.* **2022**, *5*, 107–112. [CrossRef]
10. Wang, L.; Jiang, T.; Lin, J. Self-assembly of graft copolymers in backbone-selective solvents: A route toward stable hierarchical vesicles. *RSC Adv.* **2013**, *3*, 19481–19491. [CrossRef]
11. van den Broek, B.; Noom, M.C.; van Mameren, J.; Battle, C.; MacKintosh, F.C.; Wuite, G.J.L. Visualizing the formation and collapse of DNA toroids. *Biophys. J.* **2010**, *98*, 1902–1910. [CrossRef] [PubMed]
12. Xu, P.; Gao, L.; Cai, C.; Lin, J.; Wang, L.; Tian, X. Polymeric toroidal self-assemblies: Diverse formation mechanisms and functions. *Adv. Funct* **2021**, *32*, 2106036. [CrossRef]
13. Zhang, Y.; Yang, L.; Sun, Y.; Lin, G.; Manners, I.; Qiu, H. Surface-Initiated Living Self-Assembly of Polythiophene-Based Conjugated Block Copolymer into Erect Micellar Brushes. *Angew. Chem. Int. Ed.* **2024**, *63*, e202315740. [CrossRef]
14. Ding, M.; Chen, Q.; Duan, X.; Shi, T. Flow-driven translocation of a diblock copolymer through a nanopore. *J. Phys. Chem. B* **2019**, *123*, 8848–8852. [CrossRef] [PubMed]
15. Shi, Q.; Wu, J.; Chen, H.; Xu, X.; Yang, Y.B.; Ding, M. Inertial migration of polymer micelles in a square microchannel. *Soft Matter* **2024**, *20*, 1760–1766. [CrossRef] [PubMed]
16. Qi, H.; Liu, D.; Ding, M. Flow-driven translocation of strung micelles formed by core–shell bottle-brush copolymers and the accompanying morphological transformation through a nanochannel. *Macromolecules* **2024**, *57*, 2786–2791. [CrossRef]
17. Phan, Q.T.; Zhang, H.; Pham, D.A.; Rabanel, J.M.; Filippini, A.; Boffito, D.; Banquy, X. Multicompartment micro- and nanoparticles using supramolecular assembly of core–shell bottlebrush polymers. *ACS Macro Lett.* **2023**, *12*, 1589–1594. [CrossRef]
18. Qi, H.; Zhou, H.W.; Duan, C.; Li, W.H.; Ding, M.M. Self-assembled conformations of a core-shell comb-like chain with adjustable architectural parameters. *Chin. J. Polym. Sci.* **2023**, *41*, 1439–1446. [CrossRef]
19. Liu, Y.T.; Li, Y.R.; Wang, X. Spontaneous onion shape vesicle formation and fusion of comb-like block copolymers studied by dissipative particle dynamics. *RSC Adv.* **2017**, *7*, 5130–5135. [CrossRef]
20. Selianitis, D.; Pispas, S. Multi-responsive poly(oligo(ethylene glycol)methyl methacrylate)-co-poly(2-(diisopropylamino)ethyl methacrylate) hyperbranched copolymers via reversible addition fragmentation chain transfer polymerization. *Polym. Chem.* **2021**, *12*, 6582–6593. [CrossRef]
21. Qi, M.; Li, K.; Zheng, Y.; Rasheed, T.; Zhou, Y. Hyperbranched multiarm copolymers with a ucst phase transition: topological effect and the mechanism. *Langmuir* **2018**, *34*, 3058–3067. [CrossRef] [PubMed]
22. Ozawa, N.; Lee, J.H.; Akiba, I.; Nishimura, T. Exploring the effects of glyco-copolymer architectures on the solution self-assembly of amphiphilic thermoresponsive linear, star, and cyclic polymers. *Polym. Chem.* **2023**, *14*, 3834–3842. [CrossRef]

23. Mukkamala, R.S.; Hore, M.J.A. Simulation and analysis of molecular bottlebrush dynamics in dilute solutions. *Macromolecules* **2024**, *57*, 445–455. [CrossRef]
24. Mohammadi, E.; Joshi, S.Y.; Deshmukh, S.A. A review of computational studies of bottlebrush polymers. *Comp. Mater. Sci.* **2021**, *199*, 110720. [CrossRef]
25. Li, Z.; Tang, M.; Liang, S.; Zhang, M.; Biesold, G.M.; He, Y.; Hao, S.M.; Choi, W.; Liu, Y.; Peng, J.; et al. Bottlebrush polymers: From controlled synthesis, self-assembly, properties to applications. *Prog. Polym. Sci.* **2021**, *116*, 101387. [CrossRef]
26. Zhao, B. Shape-changing bottlebrush polymers. *J. Phys. Chem. B* **2021**, *125*, 6373–6389. [CrossRef] [PubMed]
27. Xie, G.; Martinez, M.R.; Olszewski, M.; Sheiko, S.S.; Matyjaszewski, K. Molecular bottlebrushes as novel materials. *Biomacromolecules* **2019**, *20*, 27–54. [CrossRef] [PubMed]
28. Rzayev, J. Molecular bottlebrushes: New opportunities in nanomaterials fabrication. *ACS Macro Lett.* **2012**, *1*, 1146–1149. [CrossRef]
29. Kent, E.W.; Lewoczko, E.M.; Zhao, B. Effect of buffer anions on pearl-necklace morphology of tertiary amine-containing binary heterografted linear molecular bottlebrushes in acidic aqueous buffers. *Langmuir* **2020**, *36*, 13320–13330. [CrossRef]
30. Nese, A.; Li, Y.; Averick, S.; Kwak, Y.; Konkolewicz, D.; Sheiko, S.S.; Matyjaszewski, K. Synthesis of amphiphilic poly(n-vinylpyrrolidone)-b-poly(vinyl acetate) molecular bottlebrushes. *ACS Macro Lett.* **2012**, *1*, 227–231. [CrossRef]
31. Lyubimov, I.; Wessels, M.G.; Jayaraman, A. Molecular dynamics simulation and PRISM theory study of assembly in solutions of amphiphilic bottlebrush block copolymers. *Macromolecules* **2018**, *51*, 7586–7599. [CrossRef]
32. He, Y.; Li, L.; Ding, M.; Li, W. Flow-driven translocation of comb-like copolymer micelles through a nanochannel. *Soft Matter* **2023**, *19*, 9166–9172. [CrossRef]
33. Zhang, Y.; Ren, T.; Gou, J.; Zhang, L.; Tao, X.; Tian, B.; Tian, P.; Yu, D.; Song, J.; Liu, X.; et al. Strategies for improving the payload of small molecular drugs in polymeric micelles. *J. Control. Release* **2017**, *261*, 352–366. [CrossRef] [PubMed]
34. De Nicola, A.; Hezaveh, S.; Zhao, Y.; Kawakatsu, T.; Roccatano, D.; Milano, G. Micellar drug nanocarriers and biomembranes: How do they interact? *Phys. Chem. Chem. Phys.* **2014**, *16*, 5093–5105. [CrossRef]
35. Liu, Y.T.; Li, Y.R.; Wang, X. Dynamic evolution of a vesicle formed by comb-like block copolymer-tethered nanoparticles: A dissipative particle dynamics simulation study. *Phys. Chem. Chem. Phys.* **2017**, *19*, 27313–27319. [CrossRef]
36. Kramarenko, E.Y.; Pevnaya, O.S.; Khokhlov, A.R. Stoichiometric polyelectrolyte complexes as comb copolymers. *J. Chem. Phys.* **2005**, *122*, 084902. [CrossRef] [PubMed]
37. Chang, H.Y.; Lin, Y.L.; Sheng, Y.J.; Tsao, H.K. Structural characteristics and fusion pathways of onion-like multilayered polymersome formed by amphiphilic comb-like graft copolymers. *Macromolecules* **2013**, *46*, 5644–5656. [CrossRef]
38. Li, M.; Li, G.L.; Zhang, Z.; Li, J.; Neoh, K.G.; Kang, E.T. Self-assembly of pH-responsive and fluorescent comb-like amphiphilic copolymers in aqueous media. *Polymer* **2010**, *51*, 3377–3386. [CrossRef]
39. Tang, Z.; Pan, X.; Zhou, H.; Li, L.; Ding, M. Conformation of a comb-like chain free in solution and confined in a nanochannel: From linear to bottlebrush structure. *Macromolecules* **2022**, *55*, 8668–8675. [CrossRef]
40. Hao, P.; Mai, X.; Chen, Q.; Ding, M. Conformation of an amphiphilic comb-like copolymer in a selective solvent. *Chin. J. Polym. Sci.* **2023**, *41*, 1386–1391. [CrossRef]
41. Yethiraj, A. A Monte Carlo simulation study of branched polymers. *J. Chem. Phys.* **2006**, *125*, 204901. [CrossRef] [PubMed]
42. Hsu, H.P.; Paul, W.; Rathgeber, S.; Binder, K. Characteristic length scales and radial monomer density profiles of molecular bottle-brushes: Simulation and experiment. *Macromolecules* **2010**, *43*, 1592–1601. [CrossRef]
43. Wessels, M.G.; Jayaraman, A. Molecular dynamics simulation study of linear, bottlebrush, and star-like amphiphilic block polymer assembly in solution. *Soft Matter* **2019**, *15*, 3987–3998. [CrossRef]
44. Bejagam, K.K.; Singh, S.K.; Ahn, R.; Deshmukh, S.A. Unraveling the conformations of backbone and side chains in thermosensitive bottlebrush polymers. *Macromolecules* **2019**, *52*, 9398–9408. [CrossRef]
45. Dutta, S.; Wade, M.A.; Walsh, D.J.; Guironnet, D.; Rogers, S.A.; Sing, C.E. Dilute solution structure of bottlebrush polymers. *Soft Matter* **2019**, *15*, 2928–2941. [CrossRef]
46. Kremer, K.; Grest, G.S. Dynamics of entangled linear polymer melts: A molecular-dynamics simulation. *J. Chem. Phys.* **1990**, *92*, 5057–5086. [CrossRef]
47. Reddy, G.; Yethiraj, A. Implicit and explicit solvent models for the simulation of dilute polymer solutions. *Macromolecules* **2006**, *39*, 8536–8542. [CrossRef]
48. Weeks, J.D.; Chandler, D.; Andersen, H.C. Role of repulsive forces in determining the equilibrium structure of simple liquids. *J. Chem. Phys.* **1971**, *54*, 5237–5247. [CrossRef]
49. Ermak, D.L.; McCammon, J.A. Brownian dynamics with hydrodynamic interactions. *J. Chem. Phys.* **1978**, *69*, 1352–1360. [CrossRef]

50. Kumar, S.; Larson, R.G. Brownian dynamics simulations of flexible polymers with spring–spring repulsions. *J. Chem. Phys.* **2001**, *114*, 6937–6941. [CrossRef]
51. Zhang, L.; Lin, J.; Lin, S. Effect of molecular architecture on phase behavior of graft copolymers. *J. Phys. Chem. B* **2008**, *112*, 9720–9728. [CrossRef] [PubMed]
52. Duan, C.; Li, W.; Qiu, F.; Shi, A.C. lanet-satellite micellar superstructures formed by abcb terpolymers in solution. *ACS Macro Lett.* **2017**, *6*, 257–261. [CrossRef] [PubMed]
53. Parent, L.R.; Bakalis, E.; Ramirez-Hernandez, A.; Kammeyer, J.K.; Park, C.; de Pablo, J.J.; Zerbetto, F.; Patterson, J.P.; Gianneschi, N.C. Directly Observing Micelle Fusion and Growth in Solution by Liquid-Cell Transmission Electron Microscopy. *J. Am. Chem. Soc.* **2017**, *139*, 17140–17151. [CrossRef] [PubMed]
54. Kang, S.W.; Li, Y.; Park, J.H.; Lee, D.S. pH-triggered unimer/vesicle-transformable and biodegradable polymersomes based on PEG-b-PCL–grafted poly (β-amino ester) for anti-cancer drug delivery. *Polymer* **2013**, *54*, 102–110. [CrossRef]
55. Tseng, Y.C.; Chang, H.Y.; Sheng, Y.J.; Tsao, H.K. Atypical vesicles and membranes with monolayer and multilayer structures formed by graft copolymers with diblock side-chains: Nonlamellar structures and curvature-enhanced permeability. *Soft Matter* **2022**, *18*, 7559–7568. [CrossRef]

polymers

MDPI

Article

Amphiphilic Diblock Copolymers of Poly(N-vinyl pyrrolidone) and Poly(vinyl esters) Bearing N-Alkyl Side Chains for the Encapsulation of Curcumin and Indomethacin

Nikolaos V. Plachouras, Aikaterini-Maria Gkolemi, Alexandros Argyropoulos, Athanasios Bouzoukas, Theodosia-Panagiota Papazoglou, Nikoletta Roka and Marinos Pitsikalis *

Industrial Chemistry Laboratory, Department of Chemistry, National and Kapodistrian University of Athens, Panepistimiopolis Zografou, 15771 Athens, Greece; plachouras.v.nikolaos@gmail.com (N.V.P.); katerinagolemi228@gmail.com (A.-M.G.); argiropoulosal@gmail.com (A.A.); thanasisbuzukas@gmail.com (A.B.); pnpapazoglou@gmail.com (T.-P.P.); nikolettaroka@yahoo.gr (N.R.)
* Correspondence: pitsikalis@chem.uoa.gr; Tel.: +30-210-727-4768

Abstract: The self-assembly behavior of a series of amphiphilic diblock copolymers, each consisting of a hydrophilic poly(N-vinyl pyrrolidone) (PNVP) block and a hydrophobic block derived from n-alkyl vinyl esters, namely poly(vinyl butyrate) (PVBu), poly(vinyl decanoate) (PVDc), and poly(vinyl stearate) (PVSt), in aqueous solutions was investigated. Dynamic and static light scattering (DLS and SLS) techniques were employed to monitor the micellization behavior. In addition, the self-assembled structures were observed with Transmission Electron Microscopy (TEM). The effect of the nature of the hydrophobic block, the copolymer composition and the copolymer molecular weight on the self-assembly properties was thoroughly examined. The encapsulation of curcumin and indomethacin within the dry cores of the micellar structures was conducted in aqueous solutions for all block copolymers at various curcumin/indomethacin-to-polymer mass ratios. UV-Vis spectroscopy was used to evaluate the drug-loading capacity and efficiency (%DLC and %DLE). In several cases, the encapsulation of both hydrophobic drugs was found to be nearly quantitative. Combined with the observed stability of the micellar structures, these findings suggest that the block copolymers demonstrate significant potential as carriers for drug delivery applications.

Keywords: amphiphilic block copolymers; poly(N-vinyl pyrrolidone); poly(n-alkyl vinyl esters); self-assembly; polymeric micelles; drug encapsulation; curcumin; indomethacin; light scattering

1. Introduction

A major obstacle in the efficient development of pharmaceuticals is still their limited solubility in aqueous media. Despite efforts to address this issue, over 40% of licensed medications and up to 90% of investigational substances exhibit low solubility in water, according to estimates [1]. The proliferation of barely water-soluble drug candidates, which is attributed to the strategic application of combinatorial chemistry and the routine implementation of high-throughput screening in non-aqueous systems [2,3], is typically accompanied by a range of associated problems. Their poor water solubility lowers their pharmacological potential by restricting dissolution and absorption, promoting aggregation or premature precipitation in aqueous environments that may result in recrystallization and variable pharmacokinetics, and facilitating rapid systemic clearance through tissue distribution, plasma protein binding, or accelerated elimination. Furthermore, their clinical activity is further limited by intrinsic chemical and oxidative instability before they reach the target site [4–7].

An effective strategy to overcome the aforementioned limitations is the employment of biocompatible amphiphilic block copolymers, capitalizing on their spontaneous self-assembling into micelles in aqueous solutions. The hydrophobic core of the micelles can encapsulate therapeutic molecules that would otherwise be insoluble in water, enhancing their apparent solubility and preventing the degradation of the active ingredient [8,9]. Moreover, the hydrophilic shell of the micelles confers colloidal stability, extending their circulation half-life and biodistribution. The tunable architecture of block copolymers enables high drug loading and controlled release profiles. This can be achieved by adding functional groups to target specific tissues or tumors or by including pH- or redox-responsive functionalities [10–12]. Finally, the broad spectrum of available loading methods, including thin-film hydration, solvent evaporation, dialysis and co-solvent methods, facilitates optimization of micellar formulation for a wide variety of hydrophobic drugs [13].

Curcumin is a highly hydrophobic molecule with extremely low water solubility, which limits its bioavailability. It is considered a natural ingredient that has sparked scientific interest due to its numerous advantages for human health, as it is believed to have anti-inflammatory, antioxidant, antibacterial, antiviral, antifungal, antidiabetic, and anticancer properties [14]. Another practically water-insoluble molecule with particularly interesting pharmacological properties is indomethacin. It belongs to the class of nonsteroidal anti-inflammatory drugs (NSAIDs) and has been clinically employed for the treatment of acute pain, rheumatoid arthritis, ankylosing spondylitis, osteoarthritis, bursitis, gouty arthritis, patent ductus arteriosus, and certain types of migraine [15]. Both curcumin and indomethacin can easily be traced by UV-Vis spectroscopy since they exhibit a maximum absorption at 429 nm [16] and 318 nm [17], respectively. The chemical structures of both curcumin and indomethacin are shown in Scheme 1.

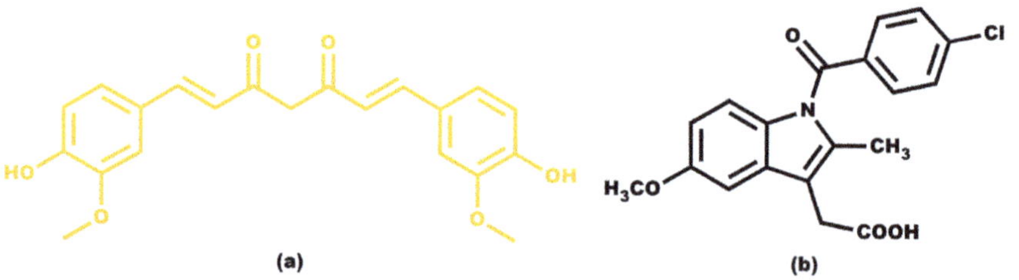

Scheme 1. (**a**) Chemical structure of the hydrophobic drug compound curcumin; (**b**) chemical structure of the hydrophobic drug compound indomethacin.

The micellization behavior of a series of amphiphilic diblock copolymers, each consisting of a hydrophilic poly(N-vinyl pyrrolidone) (PNVP) block and a hydrophobic block derived from n-alkyl vinyl esters, namely poly(vinyl butyrate) (PVBu), poly(vinyl decanoate), (PVDc) and poly(vinyl stearate) (PVSt), in aqueous solutions was investigated. The chemical structures of each different family of amphiphilic diblock copolymers are given in Scheme 2. Poly(N-vinyl pyrrolidone) possesses distinctive properties, including biocompatibility and non-toxicity, which have led to its extensive industrial utilization in the pharmaceutical and biomedical sectors, as well as others [18]. In fact, recent studies suggest PNVP as the optimal choice for replacing poly(ethylene oxide) (PEO), which is currently the most widely used hydrophilic material employed in the aforementioned sectors [19,20]. Poly(vinyl esters) (PVEs) represent another significant class of polymeric materials, which exhibit either hydrophilic or hydrophobic characteristics, depending on the nature of the side group. These materials are of pharmacological significance due to their capability to serve as precursors for the production of poly(vinyl alcohol) (PVA) through the hydrolysis of their ester group [21,22].

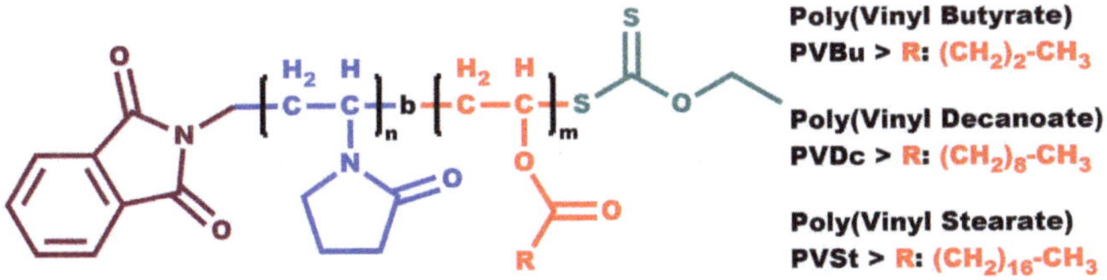

Scheme 2. Chemical structures of the amphiphilic diblock copolymers consisting of a hydrophilic PNVP block and a hydrophobic block derived from PVBu, PVDc and PVSt.

The micellization behavior of amphiphilic block copolymers is governed by a delicate balance of molecular and environmental parameters. The nature of the hydrophobic block is a crucial molecular factor. Flexible and amorphous hydrophobic blocks frequently result in equilibrium micelles, while more rigid and crystallizable blocks favor highly ordered aggregates due to restricted chain mobility [23–25]. The length of the alkyl side chain within the hydrophobic block has been demonstrated to further modulate core packing and stability. It has been shown that longer alkyl chains markedly reduce the critical micelle concentration (CMC) and promote the formation of larger, more stable aggregates, as the increased hydrophobic interactions drive self-assembly [26–29]. The composition of the diblock copolymer also determines the resulting morphology. Specifically, the hydrophilic-to-hydrophobic ratio affects the formed micellar structure. Higher percentage of the hydrophilic block promote spherical micelles, intermediate ratios yield cylindrical or worm-like micelles, and higher percentages of the hydrophobic block lead to the formation of vesicular or lamellar structures [30–32]. Finally, the molecular weight of the amphiphilic block copolymers affects both thermodynamics and kinetics of micellization behavior. Higher molecular weights, especially of the hydrophobic block, lower the CMC and enable the formation of well-defined, stable micelles with improved thermodynamic stability [33–35]. The targets of the present work are summarized as follows. (a) A study of the micellization behavior of the PNV-b-PVEs block copolymers in aqueous solutions. Water is a selective solvent for the PNVP blocks [36–38]. The self-assembly behavior of the same series of diblock copolymers in tetrahydrofuran (THF) has already been presented [39]. Parameters, such as the nature of the PVEs block, the composition and the molecular weight of the block copolymers, are expected to play a crucial role in defining the self-assembly behavior of the amphiphilic block copolymers in water. (b) Application of the supramolecular structures as vehicles for the encapsulation of hydrophobic compounds. Two hydrophobic drugs, curcumin and indomethacin, will be tested for encapsulation into the core of the micellar structures that are formed. The drug loading capacity and efficiency (%DLC and %DLE) values [40] will be calculated. The influence of the nature of the hydrophobic block along with the molecular characteristics of the block copolymers will be examined.

2. Materials and Methods

2.1. Synthesis of Diblock Copolymers PNVP-b-PVBu, PNVP-b-PVDc and PNVP-b-PVSt

The synthesis of the PNVP-based amphiphilic diblock copolymers has already been published [41]. In summary, the synthesis of the copolymers was accomplished through the implementation of the reversible addition-fragmentation chain transfer (RAFT) radical polymerization technique, in conjunction with the sequential addition of monomers. In each instance, the PNVP block was synthesized first and, after purification and characterization, was used as a macro chain transfer agent (macro-CTA) for the polymerization of each PVE block. For each of the three distinct copolymer classes, designated PNVP-b-PVBu, PNVP-b-PVDc, and PNVP-b-PVSt, a total of four or five different copolymers with varying molecular weights and molar ratios of each block were synthesized in order to investigate the effect of these parameters on the micellization behavior of the products in aqueous solutions. The molecular characteristics of the PNVP-b-PVEs diblock copolymers are provided in Table 1, whereas more details regarding the synthesis and characterization are provided in the Supplementary Material File S1 of the Supporting Information Section (SIS).

Table 1. Molecular characteristics of the PNVP-b-PVEs amphiphilic diblock copolymers.

Sample	PNVP [1]		PNVP-b-PVEs [1]		NVP [2]	PVEs [2]
	M_n $\times 10^3$ (Da)	Đ	M_n $\times 10^3$ (Da)	Đ	Molar Ratio (%)	Molar Ratio (%)
PNVP-b-PVBu #1	8.5	1.30	16.0	1.90	22	78
PNVP-b-PVBu #2	28.0	1.27	32.0	1.32	84	16
PNVP-b-PVBu #3	8.9	1.35	17.5	1.40	57	43
PNVP-b-PVBu #4	8.9	1.35	15.5	1.54	48	52
PNVP-b-PVBu #5	8.5	1.30	13.0	1.55	56	44
PNVP-b-PVDc #1	8.5	1.30	12.5	1.31	63	37
PNVP-b-PVDc #2	5.5	1.47	12.5	1.60	38	62
PNVP-b-PVDc #3	8.5	1.30	11.0	1.45	56	44
PNVP-b-PVDc #4	9.5	1.36	10.5	1.36	93	7
PNVP-b-PVSt #1	8.5	1.30	10.5	1.44	78	22
PNVP-b-PVSt #2	7.5	1.30	10.4	1.51	61	39
PNVP-b-PVSt #3	8.1	1.30	10.9	1.37	85	15
PNVP-b-PVSt #4	8.1	1.30	12.5	1.22	83	17
PNVP-b-PVSt #5	5.5	1.47	9.0	1.62	40	60

[1] Derived by size exclusion chromatography (SEC) in $CHCl_3$. [2] Derived by ^1H-NMR in $CDCl_3$.

2.2. Dynamic and Static Light Scattering (DLS and SLS) Techniques

2.2.1. Preparation of Stock Solutions for the DLS and SLS Measurements

The solvent exchange method was selected for the preparation of stock solutions for each of the distinct samples in each of the three different types of diblock copolymers. Specifically, 20 mg of each sample was initially dissolved in 5 mL of dry THF in a vial. The THF had been dried overnight over sodium and distilled prior to use. The samples were left in THF overnight to ensure complete dissolution of the diblock copolymers. Thereafter, 20 mL of extra pure water was added to each vial. Subsequently, the solutions were heated at 65 °C to guarantee full evaporation of the THF, thereby enabling the self-assembly of the amphiphilic diblock copolymers in water. The initial concentrated solution of each sample, with a concentration of about 10^{-3} g/mL, was further diluted to different lower concentrations. For the SLS measurements, the stock solutions of each sample were prepared via the exact same procedure. The examined concentrations for all samples ranged from 0.8×10^{-4} up to 1.5×10^{-3} g/mL.

2.2.2. Instrumentation and Methods for the DLS and SLS Measurements

Both DLS and SLS measurements were performed on a NanoBrook Omni system from Brookhaven Instruments operating at λ = 640 nm. A JDS Uniphase 22 mW He–Ne laser was used as the light source.

For the DLS experiments, correlation functions were collected at 25 °C at a 90° scattering angle and analyzed by the cumulant method and the CONTIN software [42]. Measurements were carried out ten times for each concentration of each sample and were averaged. Before each measurement, all solutions were filtered using a 25 mm PVDF hydrophilic welded syringe filter with 0.45 μm pore size and directly introduced into the scattering cell. SLS experiments were conducted at the same temperature and scattering angle.

The intensity time autocorrelation function, $g_2(q,t)$, was determined following standard procedures, where q is the scattering vector, t the time, and t_0 the lag time. The scattering vector was defined as $q = \frac{4\pi n_0}{\lambda}\sin\left(\frac{\theta}{2}\right)$, with θ being the observation angle, n_0 the refractive index of the solvent, and λ the wavelength of the incident light.

As mentioned, the autocorrelation functions were analyzed using the cumulant method. The normalized electric field autocorrelation function can be expressed as $g_1(t) = \int_0^\infty G(\Gamma)e^{-\Gamma t}d\Gamma$, where $G(\Gamma)$ is the distribution of decay rates and Γ the average decay rate. In the cumulant expansion, this is written as $g_1(t) = \exp\left(-\Gamma t + \frac{\mu_2}{2!}t^2 - \frac{\mu_3}{3!}t^3 + \cdots\right)$, where Γ is the average decay rate of the correlation function and μ_2 is the second cumulant describing the variance of the distribution. The polydispersity index (PDI) was calculated as $PDI = \frac{\mu_2}{\Gamma^2}$, providing a quantitative measure of the distribution width of the relaxation times. From the mean decay rate, the apparent translational diffusion coefficients were obtained. Extrapolation to zero concentration yielded $D_{0,app}$, while the apparent hydrodynamic radii, R_h^0, were obtained from the Stokes–Einstein relation: $R_h^0 = \frac{kT}{6\pi\eta_S D_{0,app}}$, where k is the Boltzmann constant, T the absolute temperature, and η_s the solvent viscosity. The concentration dependence of the diffusion coefficient was described by $D = D_{0,app}(1 + k_D c)$, where k_D denotes the diffusion interaction parameter. Characteristic autocorrelation functions from the self-organizing copolymer solution are given in Figure S1 of the SIS.

Furthermore, k_D is related to the second virial coefficient A_2 according to $k_D = 2A_2M - k_f - u$, where M is the molecular weight, k_f the concentration dependence of the friction coefficient, and u the polymer's partial specific volume.

SLS data were processed according to the Debye formalism, employing the equation: $\frac{Kc}{\Delta R_\theta} = \frac{1}{M_w} + 2A_2c$ where the optical constant K is an optical constant incorporating the refractive index increment dn/dc, and ΔR_θ represents the excess Rayleigh ratio relative to the solvent. Refractive index increments measurements were performed on a Chromatix KMX-16 refractometer from Milton Roy, operating at 633 nm and calibrated with aqueous NaCl solutions. These analyses enabled the determination of molecular weight and second virial coefficient values.

2.3. *Transmission Electron Microscopy (TEM) Measurements*

Transmission electron microscopy (TEM) experiments were conducted employing a Hitachi model HT7800. For the preparation of the samples, carbon-coated copper TEM grids on 200 mesh (ProSciTech Pty. Ltd., Thurnigowa, Australia) were used. An aliquot (5 μL) of the sample was pipetted onto each grid. The solution was allowed to be adsorbed for 2 min. The grids were then manually blotted using Whatman 541 filter paper to remove the excess liquid on the grids. The grids were then left to air-dry at room temperature for approximately 20 min. No staining protocol was applied to the grid preparation for all samples.

2.4. Encapsulation of Curcumin and Indomethacin

2.4.1. Preparation of Stock Solutions for Drug Encapsulation

The stock solutions employed for the study of the encapsulation of curcumin and indomethacin within the dry core of the micellar structures formed by the amphiphilic diblock copolymers in water were prepared with the same methodology as those utilized for the DLS and SLS measurements. The only difference is the following: the initial solution of the diblock copolymer, dissolved in THF, was equally distributed between separate vials. Before the addition of only 5 mL of extra pure water on this occasion, a solution of the drug in THF was introduced into these vials at varying drug-to-polymer mass ratios. The vials were then left overnight to ensure full interaction between the drug and the polymer. Subsequently, the remaining steps of the procedure were identical to those previously mentioned.

At this point, it would be beneficial to provide a concrete illustration concerning concentrations. Approximately 20 mg of a sample IS dissolved in 20 mL of dry THF. This solution is equally divided into 5 mL proportions in four vials. To these vials, a solution of indomethacin in THF is introduced at concentrations of 10, 15, and 20% *w*/*w* indomethacin or curcumin/polymer. The solution that remains without the administration of a drug is utilized as a blank for the UV-Vis measurements, as described below. The precise concentrations of polymers and drugs are provided, along with the results obtained in the Results and Discussion section.

2.4.2. Drug Loading Capacity and Efficiency (%DLC and %DLE) Calculation

It has previously been established that both curcumin and indomethacin absorb at 429 and 320 nm, respectively. Consequently, UV-Vis spectroscopy was employed for their quantitative determination in the solutions in which their encapsulation was attempted. The experiments were conducted on a Lamda 650 spectrophotometer from Perkin Elmer, operating within a wavelength range of 250 to 800 nm, at 25 °C, using a 3 mL quartz cell.

Initially, two separate calibration curves were created from UV-Vis spectroscopy, measuring the absorbance of curcumin and indomethacin solutions in THF with varying concentrations. Using the corresponding calibration curve and measuring the absorbance of the polymer solutions with the encapsulated drugs, two critical parameters, namely drug loading capacity (%DLC) and drug loading efficiency (%DLE), were evaluated via the following equations:

$$Drug\ Loading\ Capacity\ (DLC)\% = \frac{mass\ of\ loaded\ drug}{mass\ of\ polymer} * 100 \tag{1}$$

$$Drug\ Loading\ Efficiency\ (DLE)\ \% = \frac{mass\ of\ loaded\ drug}{mass\ of\ drug\ in\ feed} * 100 \tag{2}$$

3. Results and Discussion

3.1. Self-Assembly Behavior of the Amphiphilic Diblock Copolymers in Aqueous Solutions

3.1.1. DLS Results for the PNVP-b-PVBu Samples

As illustrated in Table 1, sample PNVP-b-PVBu #2 exhibits the highest percentage of the PNVP block and the lowest percentage of the PVBu block among the other PNVP-b-PVBu samples. This suggests that the micellar structures formed from this sample in water are expected to have a large shell and a small core, thus falling under the classification of the star-like micelles. In contrast, sample PNVP-b-PVBu #1 represents the opposite extreme, containing the highest percentage of the PVBu block and the lowest percentage of the PNVP block. Consequently, the expected morphology of the micellar structures formed from this sample is reversed, featuring a comparatively thin shell and an expanded core, which places it in the category of crew-cut micelles. The remaining three samples have a roughly equivalent PNVP block yet vary from each other in terms of the percentage of their hydrophobic PVBu block.

The experimental results of the DLS analysis for the PNVP-b-PVBu samples are summarized in Table 2, and representative DLS plots are displayed in Figure 1.

Table 2. DLS results for the PNVP-b-PVBu samples.

Sample	D_0 $\times 10^8$ (cm^2/s)	K_D	R_{h0} (nm)	$R_{h1,0}$ (nm)	$R_{h2,0}$ (nm)
PNVP-b-PVBu #1	2.653	−86.11	92.4	-	-
PNVP-b-PVBu #2	4.813	177.46	50.9	30.8	127.6
PNVP-b-PVBu #3	4.392	−174.02	55.8	23.1	104.6
PNVP-b-PVBu #4	3.645	−149.42	67.3	29.5	99.2
PNVP-b-PVBu #5	6.198	362.45	39.6	20.3	115.7

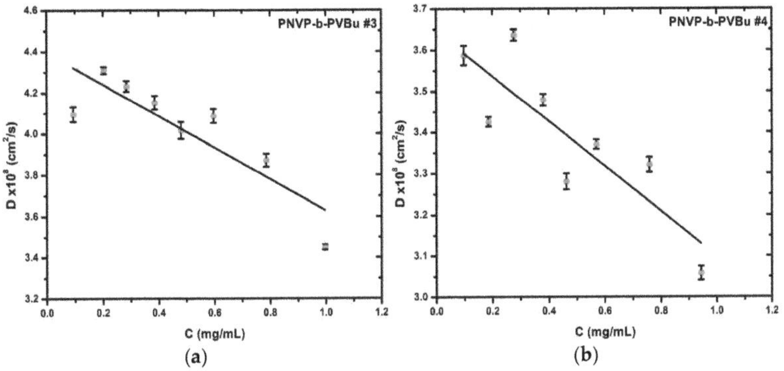

Figure 1. (**a**) Apparent diffusion coefficient vs. concentration plot of sample PNVP-b-PVBu #3; (**b**) Apparent diffusion coefficient vs. concentration plot of sample PNVP-b-PVBu #4.

The R_{h0} values refer to the hydrodynamic radius at infinite dilution coming from the cumulant analysis and the plots of diffusion coefficient versus concentration. In case of the existence of two populations, the R_{h1} and R_{h2} values refer to the results given by the CONTIN algorithm. Sample PNVP-b-PVBu #1, which contains the highest percentage of the PVBu block, according to both CONTIN analysis and the results demonstrated in Table 2, exhibits only a single population of particles in aqueous solutions with $R_{h0} = 92.4$ nm, and a polydispersity index of $0.1 < PDI < 0.2$. This finding suggests that the supramolecular structures formed are relatively uniform, most notably when considering the relatively broad polydispersity, Đ = 1.90, of the initial diblock copolymer PNVP-b-PVBu #1. This result, in conjunction with the large size of detected structures, indicates the presence of either micellar aggregates or vesicles.

In contrast with the sample PNVP-b-PVBu #1, the remaining samples with the lower percentage of the hydrophobic PVBu block, as revealed by CONTIN analysis and whose results are presented in Table 2, exhibit two distinct populations with dimensions ranging from 20 to 30 nm and 100 to 130 nm, respectively. Taking into consideration the low molecular weights of the diblock copolymers, it can be concluded that there is an equilibrium between micelles and either micellar aggregates or vesicles. In each sample, the overall PDI attains relatively high values, due to the aforementioned equilibrium; however, each individual population exhibits a notable uniform size, a finding that is in alignment with the observations derived from sample PNVP-b-PVBu #1.

In the case of the sample PNVP-b-PVBu #2, a substantial enhancement in the fraction of micellar aggregates or vesicles was observed, with an increase from 30% at low concentrations up to 55% at high concentrations. Upon increasing concentration, the micellar structures approach one another, thereby facilitating interactions mediated by the hydrophilic PNVP block. This phenomenon may ultimately lead to either the formation of micellar aggregates or to the generation of vesicles, accompanied by rearrangement of the hydrophilic and hydrophobic segments of the copolymers.

In the case of the remaining samples, namely PNVP-b-PVBu #3, #4 and #5, which contain a higher percentage of the hydrophobic PVBu block ranging from 43% to 52%, the predominance of micellar aggregates or vesicles over the simple micellar structures is clearly evident, as their percentage ranges from 65% to 80% according to CONTIN analysis. Additionally, the variation of this fraction with respect to concentration is minimal. The increased percentage of the hydrophobic PVBu block results in more stable and compact micellar cores, and in conjunction with the small PNVP water-soluble block, it facilitates micelle–micelle interactions, ultimately favoring the formation of micellar aggregates or vesicles. The linearity of the plots of diffusion coefficients with concentration (Figure 1) is not perfect, since these points are coming from the cumulant analysis and reflect the equilibrium between micelles and larger supramolecular structures. Upon increasing the concentration, the equilibrium shifts in favor of the larger aggregates.

The k_D value for the sample PNVP-b-PVBu #1 is negative, indicating that hydrodynamic interactions are predominant over thermodynamic ones. The negative k_D value further suggests that the molecular weight of the supramolecular structures is not particularly high, thereby implying that the degree of aggregation is not very high. This finding aligns more closely with the presence of micellar aggregates or vesicles in solutions, which are defined by an elevated size yet comparatively low molecular weight, as opposed to micellar aggregates.

A similar behavior is also observed in the individual distribution profiles corresponding to the two particle populations detected in the other samples by CONTIN analysis. Two characteristic CONTIN plots of sample PNVP-b-PVBu #1, which exhibits one population, and of sample #2 with two populations are given in Figure 2. Additional CONTIN plots are given in Figures S2 and S3 in the Supporting Information Section (SIS).

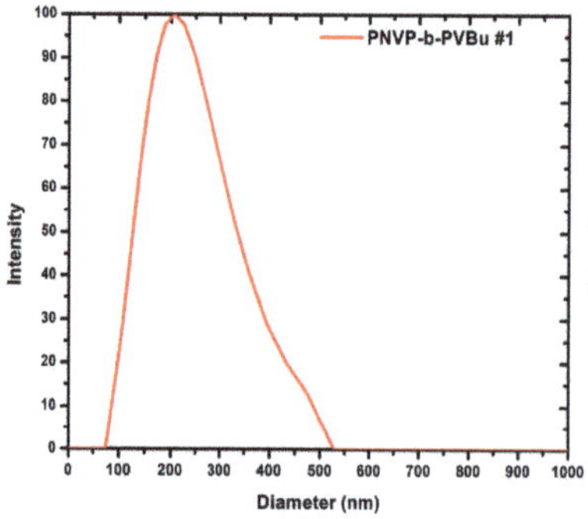

(a)

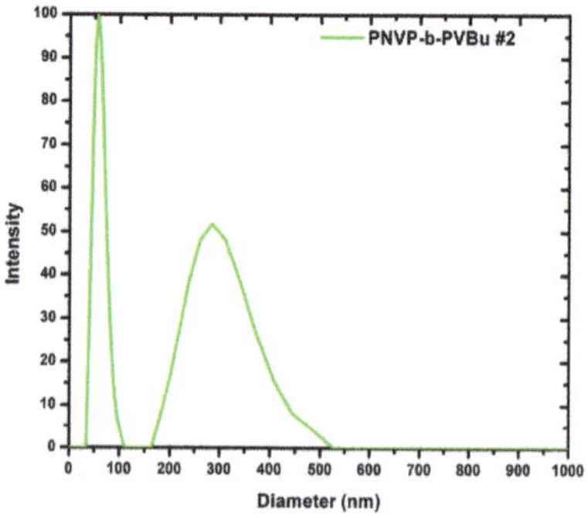

(b)

Figure 2. (**a**) CONTIN plot of sample PNVP-b-PVBu #1 at C = 0.59 mg/mL; (**b**) CONTIN plot of sample PNVP-b-PVBu #2 at C = 0.289 mg/mL.

3.1.2. DLS Results for the PNVP-b-PVDc Samples

As can be seen from the data in Table 1, all PNVP-b-PVDc samples display remarkably similar molecular weights. Consequently, the results are directly comparable in terms of molecular weight and rely exclusively on the composition of the copolymers. The results of the DLS analysis are summarized in Table 3, and characteristic DLS plots are featured in Figure 3.

Table 3. DLS results for the PNVP-b-PVDc samples.

Sample	D_0 $\times 10^8$ (cm^2/s)	K_D	R_{h0} (nm)	$R_{h1,0}$ (nm)	$R_{h2,0}$ (nm)
PNVP-b-PVDc #1	5.84	−71	42.0	24.2	91.8
PNVP-b-PVDc #2	2.88	−502	85.0	15.6	95.4
PNVP-b-PVDc #3	3.93	−73	62.4	35.6	131.1
PNVP-b-PVDc #4	6.62	981	37.0	15.4	136.4

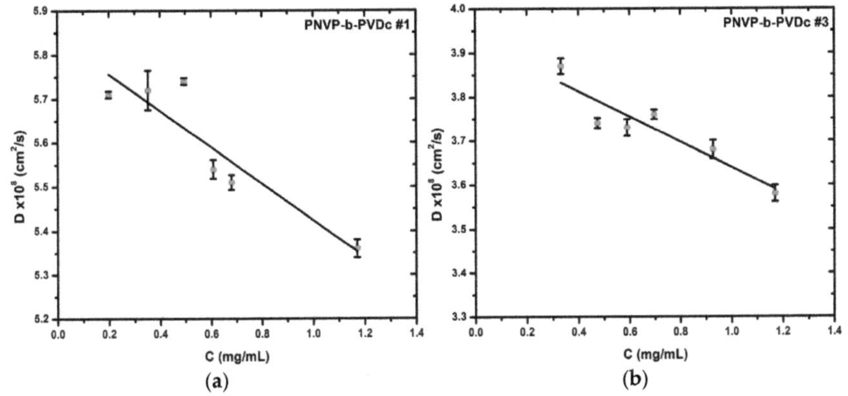

Figure 3. (**a**) Apparent diffusion coefficient vs. concentration plot of sample PNVP-b-PVDc #1; (**b**) apparent diffusion coefficient vs. concentration plot of sample PNVP-b-PVDc #3.

The initial observation, which is directly related to the composition of the samples, is that the overall hydrodynamic radii of the supramolecular structures formed in aqueous solution increase with increasing percentage of the hydrophobic PVDc block.

In addition, the CONTIN analysis reveals the presence of two distinct populations, a behavior consistent with that observed for the PNVP-b-PVBu samples, with the exception of PNVP-b-PVBu #1. Plots of the hydrodynamic radius of the two populations are displayed in Figure 4. The linearity of the plots is not perfect, since the system is in dynamic equilibrium, and in addition, the polydispersity of the structures is significant. The relative dimensions of the two populations further indicate that there is an equilibrium between micelles and micellar aggregates or vesicles. In samples PNVP-b-PVDc #1, #2, and #3, the larger structures are predominant. Increasing the concentration of the solutions used for the DLS measurements reduces the contribution of simple micelles and shifts the equilibrium further toward the larger aggregates. The only exception to this is sample PNVP-b-PVDc #4, where the smaller micellar structure prevails, due to the fact that the percentage of the hydrophobic PVDc block in this sample is only 7%, as shown in Table 1. Representative CONTIN analysis plots are given in Figure 5 for the samples PNVP-b-PVDc #1 and #4. The CONTIN plots of the remaining two PNVP-b-PVDc samples are provided in Figure S4 in SIS.

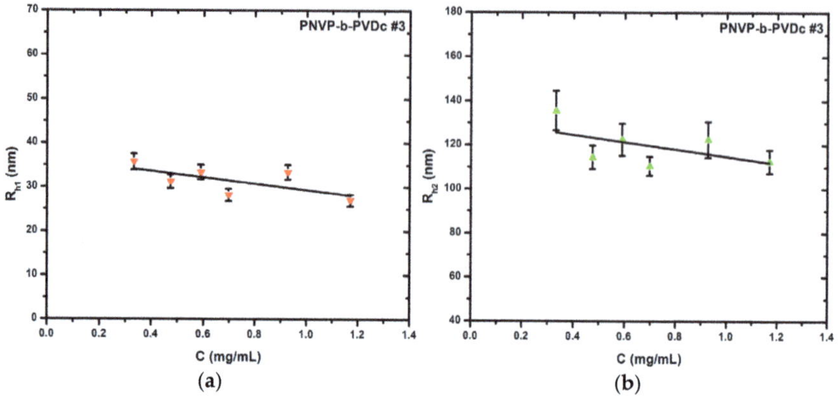

Figure 4. (**a**) R_{h1} vs. C plot of sample PNVP-b-PVDc #3 in H_2O; (**b**) R_{h2} vs. C plot of sample PNVP-b-PVDc #3 in H_2O.

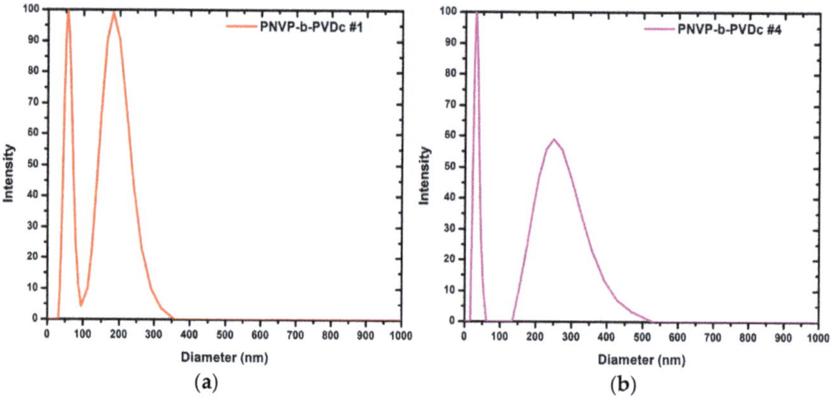

Figure 5. (**a**) CONTIN plot of sample PNVP-b-PVDc #1 at C = 0.352 mg/mL; (**b**) CONTIN plot of sample PNVP-b-PVDc #4 at C = 0.925 mg/mL.

3.1.3. DLS and SLS Results for the PNVP-b-PVSt Samples

As observed in the case of the amphiphilic block copolymers containing PVDc, in the last family of samples, namely PNVP-b-PVSt, all copolymers exhibit comparable molecular weights. Therefore, the results, presented in Table 4, again depend only on the composition of the copolymers. Representative DLS plots are also provided in Figure 6.

In contrast to all previously discussed systems, and with the sole exception of sample PNVP-b-PVBu #1, CONTIN analysis of the PNVP-b-PVSt samples reveals the presence of a single population across all concentrations of the stock solutions used for the measurements. CONTIN analysis plots are given for the sample PNVP-b-PVSt #1 and PNVP-b-PVSt #2 in Figure 7. The CONTIN analysis plots of samples PNVP-b-PVSt #3, PNVP-b-PVSt #4 and PNVP-b-PVSt #5 are included in SIS in Figures S5 and S6. Taking into account the measured hydrodynamic radii (Table 4) and the relatively low molecular weights of the free chains (Table 1), it can be concluded that the structures formed in aqueous environment correspond either to micelles with a high degree of aggregation, to micellar aggregates, or to stable vesicular assemblies.

Table 4. DLS results for the PNVP-b-PVSt samples.

Sample	D_0 $\times 10^8$ (cm^2/s)	K_D	R_{h0} (nm)
PNVP-b-PVSt #1	4.37	65.8	56.1
PNVP-b-PVSt #2	3.73	69.3	65.7
PNVP-b-PVSt #3	4.52	693.0	54.2
PNVP-b-PVSt #4	4.35	187.8	56.4
PNVP-b-PVSt #5	2.54	69.0	96.6

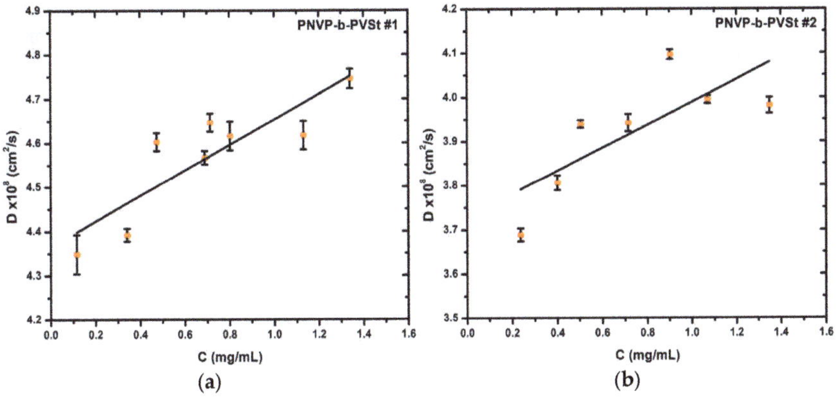

Figure 6. (**a**) Apparent diffusion coefficient vs. concentration plot of sample PNVP-b-PVSt #1; (**b**) apparent diffusion coefficient vs. concentration plot of sample PNVP-b-PVSt #2.

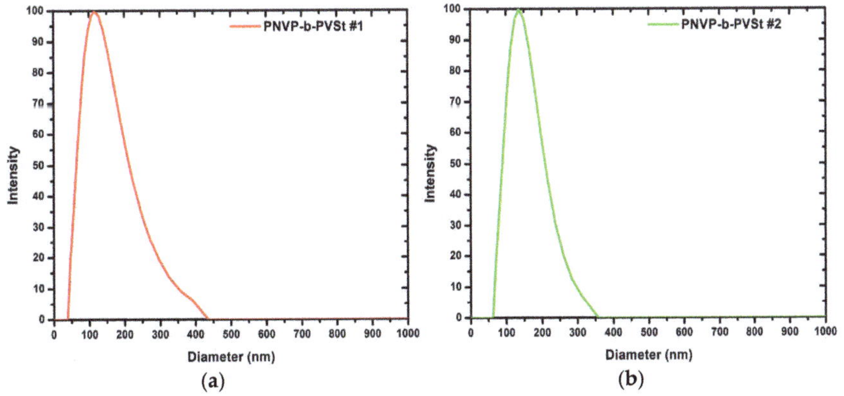

Figure 7. (**a**) CONTIN plot of sample PNVP-b-PVSt #1 at C = 0.115 mg/mL; (**b**) CONTIN plot of sample PNVP-b-PVSt #2 at C = 0.236 mg/mL.

In accordance with the behavior previously observed in the PVBu- and PVDc-based amphiphilic block copolymers, the hydrodynamic radius of the supramolecular structures formed in PNVP-b-PVSt samples increases upon increasing the percentage of the hydrophobic PVSt block, most likely due to a higher degree of aggregation.

The values of the PDI for all samples exceed 0.2, indicating that the supramolecular structures formed in aqueous solution exhibit a broad size distribution. It is evident that this phenomenon can be attributed, in part, to the polydispersity of the initial copolymers.

However, it is primarily attributable to the aggregation process itself. More specifically, the hydrophobic interactions within the core-forming material in the PNVP-b-PVSt copolymers are stronger compared to the two other families of copolymers. The long alkyl side chains of PVSt are able to form crystalline domains, leading to the formation of compact micellar cores.

In all PNVP-b-PVSt samples, the k_D values are found to be relatively low but positive due to the huge M_w of the aggregates, thus indicating that thermodynamic interactions predominate over hydrodynamic ones. As a direct consequence, and due to the reduced interactions of the polymeric system with the solvent, the corresponding A_2 values persist at sufficiently low levels.

As already mentioned, the PNVP-b-PVSt samples are the only ones that consistently exhibit a single population. Therefore, SLS measurements for these samples were carried out as well. The detailed results are presented in Table 5, which, for convenience, also includes the information from Table 1 regarding the molar ratio of the hydrophobic PVSt block. Furthermore, characteristic SLS plots are presented in Figure 8.

Table 5. SLS results for the PNVP-b-PVSt samples.

Sample	PVSt [1] Molar Ratio (%)	$(M_w)_{micc}$ [2] (Daltons)	$(A_2)_{micc}$ [2] (cm^3mol/g^2)	Nw
PNVP-b-PVSt #1	22	4.02×10^7	2.15×10^{-5}	3828
PNVP-b-PVSt #2	39	1.18×10^8	1.65×10^{-5}	11,346
PNVP-b-PVSt #3	15	5.98×10^7	1.85×10^{-5}	5486
PNVP-b-PVSt #4	17	4.77×10^7	1.64×10^{-5}	3816
PNVP-b-PVSt #5	60	4.35×10^8	1.87×10^{-6}	48,333

[1] Derived by ^{1}H-NMR in $CDCl_3$. [2] Derived by SLS.

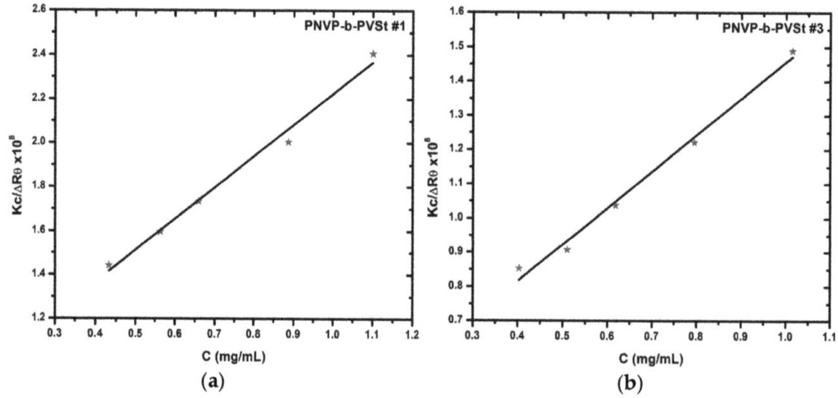

Figure 8. (**a**) SLS plot of sample PNVP-b-PVSt #1; (**b**) SLS plot of sample PNVP-b-PVSt #3.

In agreement with the DLS results, the structures formed upon self-assembly of these amphiphilic block copolymers in aqueous solution are micelles characterized by an exceptionally high degree of aggregation. In general, the degree of aggregation was governed by both the block copolymer composition and the degree of polymerization. Since all samples share nearly identical molecular characteristics, the degree of association is dictated exclusively by the copolymer composition.

More specifically, as the percentage of the hydrophilic PNVP block decreases, corresponding to an increase in the percentage of the hydrophobic PVSt block, the degree of association value, Nw, rises dramatically. For instance, in PNVP-b-PVSt samples #1, #2, and #3, where the PNVP content ranges between 78 and 85%, Nw is already very high, from approximately 3800 up to 5500. Upon reduction in the percentage of the PNVP block to 61% in sample #4, Nw increases to about 11,000, and in sample #5, containing only 40% PNVP, Nw reaches an exceptionally high value of over 48,000. This behavior reflects the formation of a highly compact and rigid hydrophobic core, effectively minimizing the unfavorable hydrophobic–water interfacial area.

3.2. Self-Assembly Behaviour of the PNVP-b-PVEs Block Copolymers in Aqueous Solutions by TEM Imaging

TEM images were collected on carbon-coated copper grids. Efforts were made to stain the specimens with uranyl acetate to improve the contrast of the images. However, this procedure was not successful, since uranium strongly interacted through complexation with the nitrogen atoms of the pyrrolidone rings, producing a black surface without any possibility of observing any micellar structure on the grids. Therefore, the images were obtained without staining. Characteristic images from the micellar solutions in THF are given in Figures 9–14.

A characteristic image from sample PNVP-b-PVBu#1 is given in Figure 9. As was revealed from the DLS measurements, single micellar, relatively spherical structures were revealed by TEM imaging. In addition, the distribution of sizes is not very broad in agreement with CONTIN analysis. It is characteristic that the size of the micellar structures is lower than that measured by DLS, although in most cases the opposite behaviour is obtained [43,44], due to the expansion of the structure on the surface of the grid in the absence of any solvent. This result is associated with the hydrophilic character of the PNVP chains, which is the corona-forming block, and the hydrophobic character of the carbon-coated copper grids. In this case, the PNVP chains shrink to avoid contact with the hydrophobic surface.

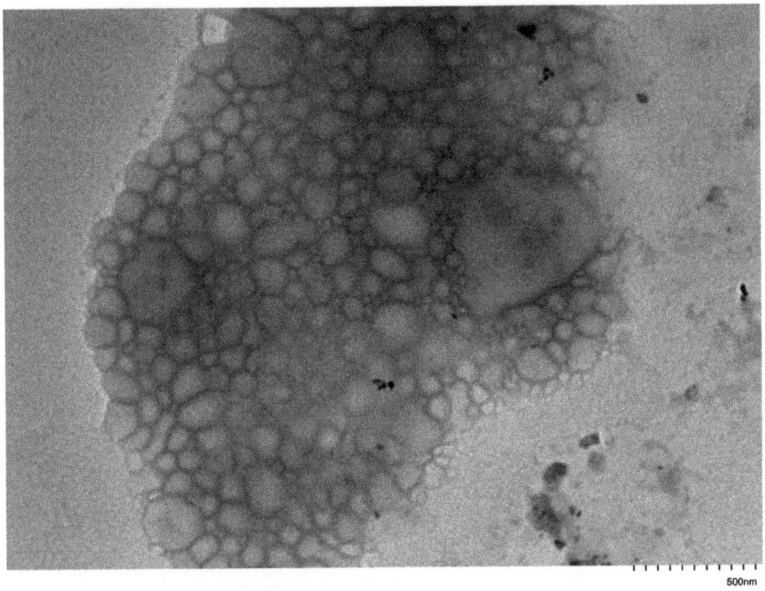

Figure 9. TEM image from the sample PNVP-b-PVBu #1.

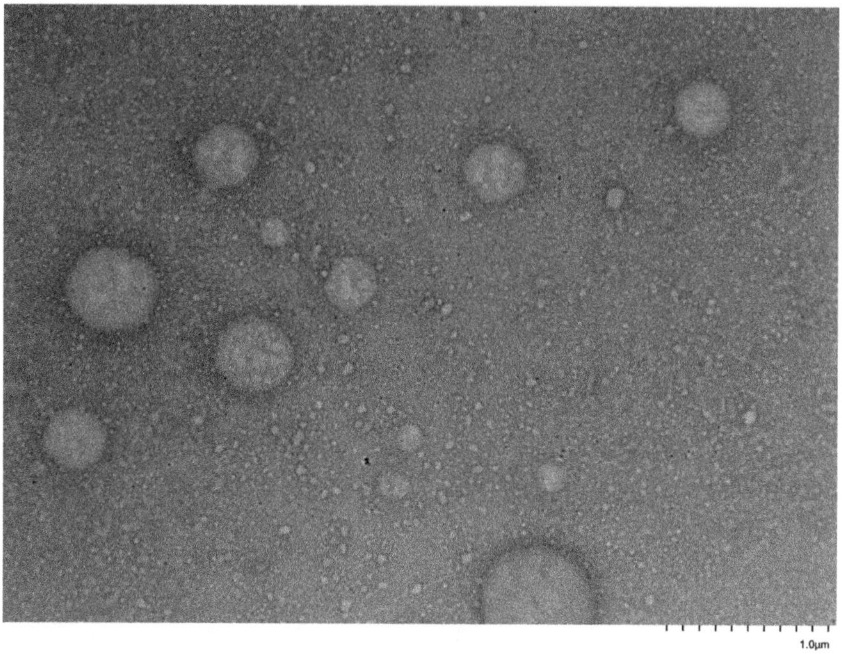

Figure 10. TEM image from the sample PNVP-b-PVBu #2.

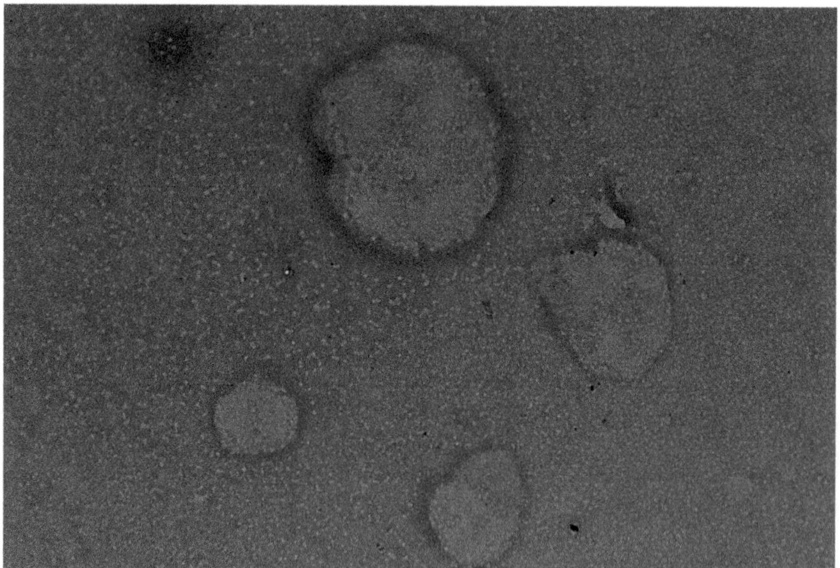

Figure 11. TEM image from the sample PNVP-b-PVDc#1.

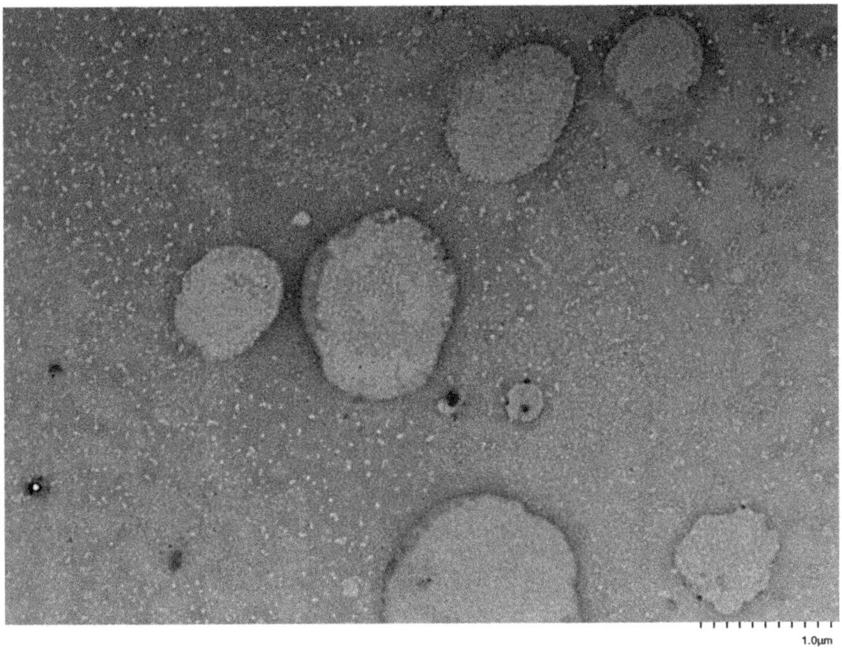

Figure 12. TEM image from the sample PNVP-b-PVDc#4.

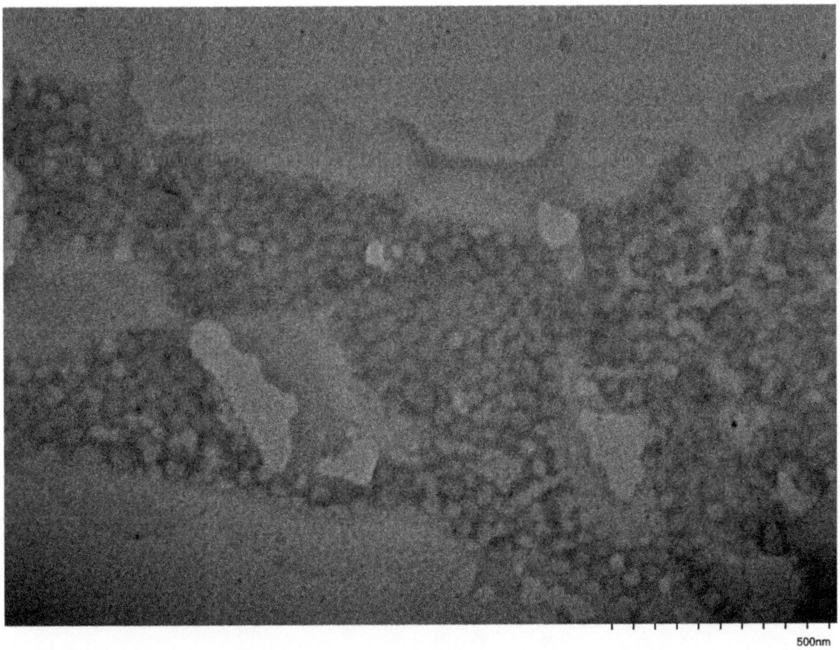

Figure 13. TEM image from the sample PNVP-b-PVSt#3.

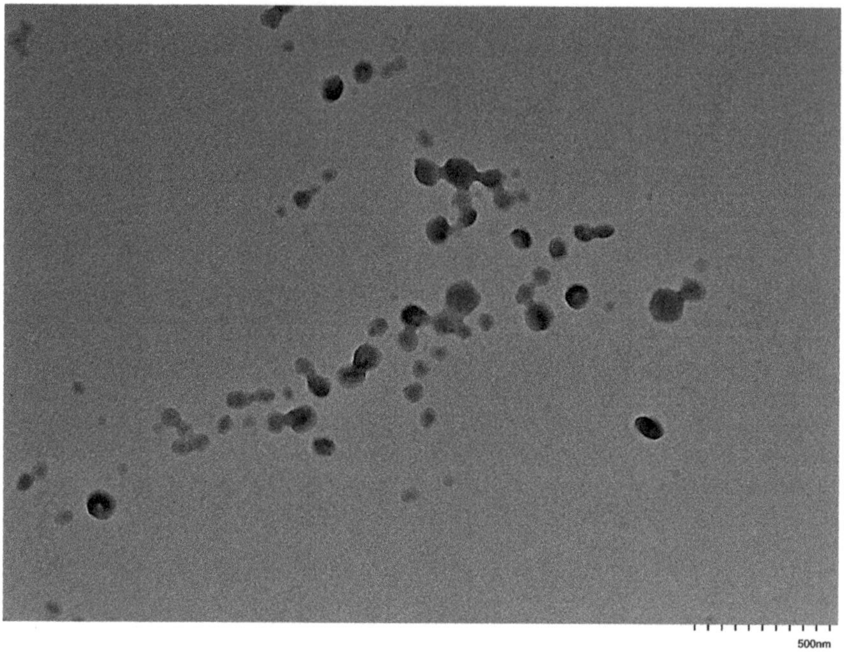

Figure 14. TEM image from the sample PNVP-b-PVSt#5.

The DLS measurements were also confirmed with sample PNVP-b-PVBu#2. A characteristic image is given in Figure 10. It is clear that in this case, there is an equilibrium between large supramolecular structures with small micelles. The large structures are polydisperse and seem to be constructed as assemblies of multimolecular micelles. Therefore, the scenario of the existence of vesicles from the DLS measurements seems to be abandoned in favour of the presence of micellar aggregates.

Similar conclusions were also drawn for samples PNVP-b-PVDc#1 and #4, as shown by the images given in Figures 11 and 12, respectively.

Finally, for samples PNVP-b-PVSt#3 and #5, the DLS conclusion for the presence of single populations of micelles is confirmed, as shown in Figures 13 and 14, respectively.

3.3. Drug Encapsulation Results

3.3.1. Encapsulation of Curcumin

The results of curcumin encapsulation within the dry cores of the micellar structures formed by the self-assembly of the PNVP-b-PVBu amphiphilic block copolymers in aqueous solution are summarized in Table 6. UV-Vis spectra of PNVP-b-PVBu samples with all concentrations of encapsulated curcumin are given in Figures S7 and S8 in SIS.

As demonstrated in Table 6, the encapsulation efficiency is found to be noteworthy, with the DLE values attaining a maximum of 80%. As might have been anticipated, sample PNVP-b-PVBu #1, which contains the highest amount of the hydrophobic PVBu block and exhibits a single population of micellar aggregates, as evidenced by the DLS results in Table 2, shows the highest encapsulation capacity.

Among the three remaining samples, PNVP-b-PVBu #2 may demonstrate a slightly higher encapsulation performance due to the larger hydrodynamic radius of its second micellar population; however, this sample, along with samples PNVP-b-PVBu #3 and #4, displays a very similar behavior overall. This observation is consistent with the DLS results presented in Table 2, given the coexistence of micelles with clusters or micellar aggregates in all three cases, as previously discussed.

Table 6. Results of curcumin encapsulation within the dry core of the micellar structures formed in aqueous solutions by the PNVP-b-PVBu samples.

Sample	Cur. [1]	A [2] (a.u.)	Pol. [3] (μg/mL)	DIF [4] (μg/mL)	LD [5] (μg/mL)	DLC (%)	DLE (%)
PNVP-b-PVBu #1	1%	1.85	481.20	12.60	9.19	1.91	73.01
	2%	1.78	670.09	14.59	8.83	1.32	60.54
	3%	2.85	653.62	17.93	14.33	2.19	79.90
	4%	2.74	643.69	21.22	13.76	2.14	64.85
PNVP-b-PVBu #2	1%	0.45	543.50	6.23	2.00	0.37	32.08
	2%	1.57	604.45	14.17	7.75	1.28	54.72
	3%	1.85	599.34	17.42	9.19	1.53	52.75
	4%	2.55	621.73	21.75	12.79	2.06	58.80
PNVP-b-PVBu #3	1%	0.38	583.23	6.33	1.64	0.28	25.90
	2%	1.84	650.11	14.12	9.14	1.41	64.73
	3%	1.97	637.42	17.22	9.81	1.54	56.94
	4%	2.34	630.40	21.68	11.71	1.86	54.01
PNVP-b-PVBu #4	1%	0.37	537.80	5.97	1.59	0.30	26.61
	2%	1.94	632.52	14.38	9.65	1.53	67.13
	3%	1.93	619.94	17.38	9.60	1.55	55.23
	4%	2.89	638.81	22.02	14.53	2.27	65.99

[1] Curcumin concentration. [2] Absorbance. [3] Concentration of the polymer stock solutions in H_2O. [4] Concentration of the drug in feed. [5] Concentration of the loaded drug.

Overall, the observed behavior is in accordance with the anticipated trend, i.e., the percentage of curcumin encapsulation, increases as the feed concentration increases, as demonstrated by the increase in DLC values. However, in the case of DLE, after an initial rise, DLE values appear to level off, reaching a plateau with nearly constant values.

The corresponding results for the PNVP-b-PVDc samples are presented in Table 7, and the respective UV-Vis spectra are featured in Figures S9 and S10 in SIS. The values of DLC and DLE are satisfactory and quite comparable to those of the PNVP-b-PVBu samples, which is consistent with the similar core sizes of the micellar structures formed through their self-assembly in aqueous environments, as also indicated by the hydrodynamic radii of the second micellar populations, summarized in Tables 2 and 3, for the PNVP-b-PVBu samples and the PNVP-b-PVDc samples, respectively. Nevertheless, the increase in the hydrophobic segment length of the n-alkyl side chain in PVDc compared to that of PVBu results in a rise in the DLE values, most likely due to stronger interactions with curcumin.

More specifically, in the case of the PNVP-b-PVDc amphiphilic block copolymers, for samples #2 and #3, the DLE values increase with increasing concentration of curcumin in the feed. On the other hand, for samples #3 and #4, as the concentration of curcumin in the feed increases from C_3 to C_4, a slight decrease in DLE values is observed.

This phenomenon was also noted to a smaller extent in the PNVP-b-PVBu samples and may be attributed to the fact that at higher concentrations, curcumin tends to form aggregates and precipitate. Such behavior is not uncommon for organic dyes, where molecular aggregation in solution frequently occurs, notably in the form of H-aggregates and J-aggregates, particularly at elevated dye concentrations [45–48]. It is characteristic that dye aggregates do not exhibit the same absorption maximum wavelength as the individual dye molecules, due to slight modifications of the conjugated double-bond system within the aggregate compared to the isolated chromophore. Consequently, in some cases, anomalous absorption behavior is observed in the visible region as a function of the dye concentration in the measured solution.

Table 7. Results of curcumin encapsulation within the dry core of the micellar structures formed in aqueous solutions by the PNVP-b-PVDc samples.

Sample	Cur. [1]	A [2] (a.u.)	Pol. [3] (μg/mL)	DIF [4] (μg/mL)	LD [5] (μg/mL)	DLC (%)	DLE (%)
PNVP-b-PVDc #1	1%	0.61	684.13	6.18	4.82	0.71	78.05
	2%	2.12	746.70	13.79	2.82	0.38	20.46
	3%	2.64	765.59	17.73	10.58	1.38	59.64
	4%	1.92	767.62	21.11	13.25	1.73	62.77
PNVP-b-PVDc #2	1%	0.73	601.64	6.37	9.55	1.59	49.88
	2%	2.17	668.63	13.92	3.44	0.51	24.69
	3%	2.68	676.55	17.69	10.83	1.60	61.25
	4%	1.69	641.88	20.09	13.45	2.10	66.96
PNVP-b-PVDc #3	1%	0.93	615.39	6.91	8.37	1.36	21.18
	2%	2.08	619.55	14.25	4.46	0.72	31.33
	3%	2.52	592.96	17.34	10.37	1.75	59.81
	4%	2.19	624.96	21.81	12.63	2.02	57.93
PNVP-b-PVDc #4	1%	0.86	560.35	66.07	10.94	1.95	65.54
	2%	1.96	552.45	13.03	4.11	0.74	31.50
	3%	2.21	546.12	16.24	9.76	1.79	60.06
	4%	2.81	544.62	19.41	11.04	2.03	56.87

[1] Curcumin concentration. [2] Absorbance. [3] Concentration of the polymer stock solutions in H_2O. [4] Concentration of the drug in feed. [5] Concentration of the loaded drug.

The investigation of curcumin encapsulation within the micellar structures of the PNVP-b-PVSt amphiphilic block copolymer samples yielded promising results, as summarized in Table 8. The characteristic UV-Vis spectra are all displayed in Figures S11 and S12 in SIS. The hydrophobic drug was successfully incorporated to a satisfactory extent, with all samples exhibiting DLE values above 40%. Similarly to the case of the PNVP-b-PVBu samples, the DLE initially tends to increase until it reaches a plateau with nearly constant values. However, for the PNVP-b-PVSt samples, there is a more specific explanation. As previously mentioned, the long alkyl side chains of PVSt are capable of forming crystalline domains [49], which reduce the capacity of the compact core, thereby limiting the available space for the encapsulation of higher amounts of curcumin.

Table 8. Results of curcumin encapsulation within the dry core of the micellar structures formed in aqueous solutions by the PNVP-b-PVSt samples.

Sample	Cur. [1]	A [2] (a.u.)	Pol. [3] (μg/mL)	DIF [4] (μg/mL)	LD [5] (μg/mL)	DLC (%)	DLE (%)
PNVP-b-PVSt #1	1%	0.63	553.69	6.20	2.92	0.53	47.17
	2%	1.48	570.94	13.26	7.29	1.28	54.98
	3%	1.61	522.77	14.91	7.96	1.52	53.36
	4%	2.24	567.39	19.22	11.19	1.97	58.25
PNVP-b-PVSt #2	1%	0.67	567.63	6.25	3.13	0.55	50.04
	2%	1.74	571.86	12.87	8.63	1.51	67.01
	3%	1.95	580.85	16.02	9.70	1.67	60.59
	4%	2.38	577.08	19.90	11.91	2.06	59.87

Table 8. *Cont.*

Sample	Cur. [1]	A [2] (a.u.)	Pol. [3] (μg/mL)	DIF [4] (μg/mL)	LD [5] (μg/mL)	DLC (%)	DLE (%)
PNVP-b-PVSt #3	1%	0.68	554.27	6.15	3.18	0.57	51.70
	2%	1.57	574.55	12.91	7.75	1.35	60.06
	3%	1.86	574.47	16.01	9.24	1.61	57.74
	4%	2.51	591.43	19.06	12.58	2.13	65.99
PNVP-b-PVSt #4	1%	0.48	554.60	5.42	2.15	0.39	39.72
	2%	1.48	560.06	11.74	7.29	1.30	62.12
	3%	2.04	558.56	14.81	10.17	1.82	68.64
	4%	2.19	566.32	18.06	10.94	1.93	60.55

[1] Curcumin concentration. [2] Absorbance. [3] Concentration of the polymer stock solutions in H_2O. [4] Concentration of the drug in feed. [5] Concentration of the loaded drug.

3.3.2. Encapsulation of Indomethacin

In addition to the encapsulation of curcumin, the encapsulation of the hydrophobic drug indomethacin within the dry cores of the micellar structures formed by the self-assembly of the PNVP-b-PVEs amphiphilic block copolymers in aqueous solution was also investigated. The results of the encapsulation of indomethacin into the PNVP-b-PVBu samples are presented in Table 9. As mentioned, UV-Vis spectroscopy was utilized for the quantitative determination of indomethacin in the solutions in which its encapsulation was attempted; therefore, the UV-Vis spectra of indomethacin with various concentrations and the calibration curve for indomethacin at 320 nm are given in SIS Figure S13. The UV-Vis spectra of sample PNVP-b-PVBu #2 and PNVP-b-PVBu #5, respectively, with all three different concentrations of encapsulated indomethacin combined, are provided in Figure 15, while the UV-Vis spectra of the remaining PNVP-b-PVBu samples are included in Figures S14 and S15 in SIS.

Table 9. Results of indomethacin encapsulation within the dry core of the micellar structures formed in aqueous solutions by the PNVP-b-PVBu samples.

Sample	Ind. [1] (*w/w*)	A [2] (a.u.)	Pol. [3] (μg/mL)	DIF [4] (μg/mL)	LD [5] (μg/mL)	DLC (%)	DLE (%)
PNVP-b-PVBu #1	10%	0.32	178.34	17.16	12.53	7.02	73.00
	15%	0.69	169.51	24.48	28.93	17.07	118.17
	20%	0.80	168.85	34.39	33.98	20.13	98.83
PNVP-b-PVBu #2	10%	0.27	168.70	15.97	10.56	6.26	66.16
	15%	0.37	164.15	25.65	14.93	9.09	58.21
	20%	0.45	166.63	33.01	18.24	10.95	55.26
PNVP-b-PVBu #3	10%	0.32	180.86	16.18	12.55	6.94	77.58
	15%	0.40	182.12	25.69	16.12	8.85	62.76
	20%	0.61	187.77	34.82	25.70	13.69	73.82
PNVP-b-PVBu #4	10%	0.30	171.48	16.54	11.90	6.94	71.95
	15%	0.32	173.91	23.46	42.45	7.16	53.07
	20%	0.56	173.42	32.31	23.32	13.45	72.18
PNVP-b-PVBu #5	10%	0.30	165.14	15.93	11.75	7.11	73.73
	15%	0.43	165.03	24.82	17.41	10.55	70.13
	20%	0.61	164.20	35.00	25.46	15.51	72.75

[1] Concentration of the indomethacin solutions in THF (*w/w* indomethacin/polymer). [2] Absorbance. [3] Concentration of the polymer stock solutions in H_2O. [4] Concentration of the drug in feed. [5] Concentration of the loaded drug.

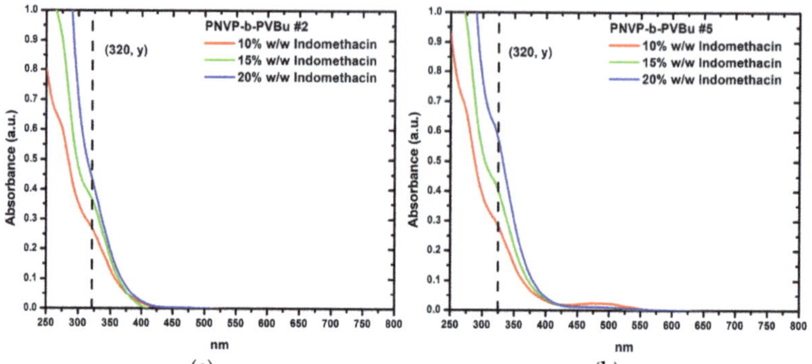

(**a**) (**b**)

Figure 15. (**a**) UV-Vis spectra of PNVP-b-PVBu #2 with all concentrations of encapsulated indomethacin; (**b**) UV-Vis spectra of PNVP-b-PVBu #5 with all concentrations of encapsulated indomethacin.

Sample PNVP-b-PVBu #1, which, as shown in its composition in Table 1, contains the highest percentage of the hydrophobic PVBu block, displays the highest DLC and DLE values for the encapsulation of indomethacin, consistent with the trend previously observed for curcumin. Remarkably, the addition of up to 20% *w*/*w* indomethacin leads to nearly quantitative encapsulation. In contrast to sample #1, PNVP-b-PVBu #2 shows the lowest DLC and DLE values, precisely because, as indicated in Table 1, it contains the lowest percentage of the hydrophobic PVBu block.

The three remaining samples, PNVP-b-PVBu #3, #4 and #5, which possess similar compositions and intermediate percentages of the hydrophobic PVBu block relative to the two extreme cases of samples #1 and #2, exhibit comparable behaviors and DLC/DLE values, falling between those observed for samples #1 and #2, as expected.

The results regarding the encapsulation of indomethacin within the PNVP-b-PVDc samples are summarized in Table 10, and characteristic UV-Vis spectra are given in Figure 16. Additional UV-Vis spectra are provided in Figure S16 in the SIS.

Table 10. Results of indomethacin encapsulation within the dry core of the micellar structures formed in aqueous solutions by the PNVP-b-PVDc samples.

Sample	Ind. [1] (*w*/*w*)	A [2] (a.u.)	Pol. [3] (μg/mL)	DIF [4] (μg/mL)	LD [5] (μg/mL)	DLC (%)	DLE (%)
PNVP-b-PVDc #1	10%	0.24	73.25	17.63	9.18	12.53	52.05
	15%	0.32	94.67	16.09	12.57	13.28	78.09
	20%	0.49	98.63	19.59	20.06	20.34	102.38
PNVP-b-PVDc #2	10%	0.39	95.43	10.15	15.66	16.41	154.22
	15%	0.55	98.59	15.65	22.65	22.98	144.74
	20%	0.69	101.96	19.60	29.14	28.58	148.66
PNVP-b-PVDc #3	10%	0.20	107.78	10.36	7.46	6.92	72.02
	15%	0.33	109.01	17.25	12.88	11.82	74.97
	20%	0.48	105.61	19.48	19.90	18.84	102.16
PNVP-b-PVDc #4	10%	0.26	97.04	10.24	10.00	10.31	97.64
	15%	0.31	94.03	16.07	11.98	12.75	74.59
	20%	0.52	92.95	19.92	21.67	23.32	108.77

[1] Concentration of the indomethacin solutions in THF (*w*/*w* indomethacin/polymer). [2] Absorbance. [3] Concentration of the polymer stock solutions in H_2O. [4] Concentration of the drug in feed. [5] Concentration of the loaded drug.

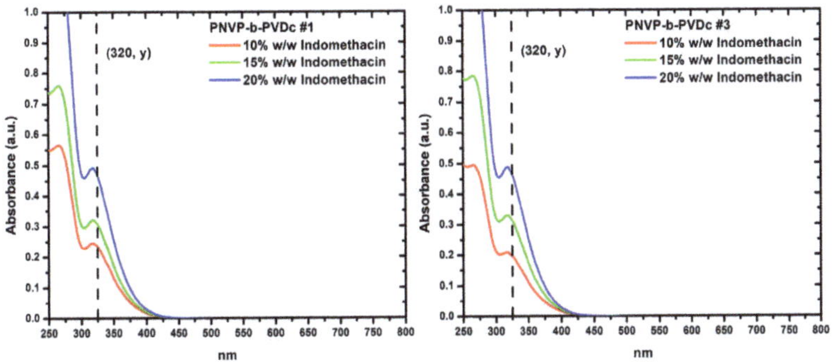

(a) (b)

Figure 16. (**a**) UV-Vis spectra of PNVP-b-PVDc #1 with all concentrations of encapsulated indomethacin (**b**) UV-Vis spectra of PNVP-b-PVDc #3 with all concentrations of encapsulated indomethacin.

The encapsulation results of indomethacin in the PNVP-b-PVDc samples are remarkably high. The DLC values closely match the respective feed concentrations of indomethacin, while the DLE demonstrates that in the vast majority of samples, the incorporation of indomethacin into the dry micellar core is essentially quantitative. These findings highlight that indomethacin is more efficiently accommodated within the micelles compared to curcumin. In the case of PNVP-b-PVDc, the encapsulation efficiency surpasses that of the PNVP-b-PVBu samples, which can be attributed to the longer n-alkyl hydrophobic side chains of PVDc that promote stronger hydrophobic interactions with the hydrophobic drug molecules.

Finally, the encapsulation performance of indomethacin within the PNVP-b-PVSt samples is highlighted in Table 11, accompanied by UV-Vis spectra in Figure 17. Additional UV-Vis spectra are given in Figures S17 and S18 in the SIS.

Table 11. Results of indomethacin encapsulation within the dry core of the micellar structures formed in aqueous solutions by the PNVP-b-PVSt samples.

Sample	Ind. [1] (*w/w*)	A [2] (a.u.)	Pol. [3] (μg/mL)	DIF [4] (μg/mL)	LD [5] (μg/mL)	DLC (%)	DLE (%)
PNVP-b-PVSt #1	10%	0.39	165.90	15.95	15.66	9.44	98.19
	15%	0.46	164.14	23.65	18.75	11.42	79.26
	20%	0.55	166.71	38.53	22.68	13.60	58.87
PNVP-b-PVSt #2	10%	0.73	32.93	3.27	31.06	9.43	94.9
	15%	0.36	166.27	24.02	14.17	8.52	58.98
	20%	0.65	166.17	33.13	27.49	16.54	82.98
PNVP-b-PVSt #3	10%	0.45	165.82	18.98	18.22	10.99	96.01
	15%	0.55	181.78	27.56	22.82	12.55	82.80
	20%	0.62	168.32	33.06	25.75	15.30	77.90
PNVP-b-PVSt #4	10%	0.41	172.92	16.89	16.71	9.66	98.93
	15%	0.51	167.90	24.12	21.07	12.55	87.33
	20%	0.55	171.99	35.92	22.74	13.22	63.30
PNVP-b-PVSt #5	10%	0.84	157.27	16.47	35.96	22.86	218.31
	15%	0.84	152.17	25.01	35.64	23.42	142.47
	20%	0.90	150.52	33.06	38.35	25.48	116.01

[1] Concentration of the indomethacin solutions in THF (*w*/*w* indomethacin/polymer). [2] Absorbance. [3] Concentration of the polymer stock solutions in H_2O. [4] Concentration of the drug in feed. [5] Concentration of the loaded drug.

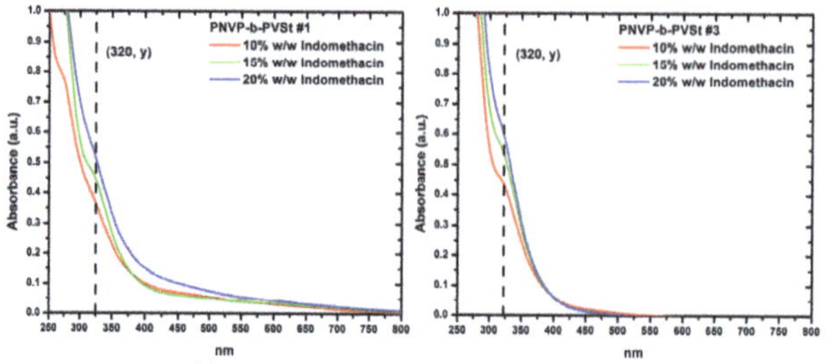

(a) (b)

Figure 17. (**a**) UV-Vis spectra of PNVP-b-PVSt #1 with all concentrations of encapsulated indomethacin (**b**) UV-Vis spectra of PNVP-b-PVSt #3 with all concentrations of encapsulated indomethacin.

All PNVP-b-PVSt samples exhibit consistent behavior. At an indomethacin feed concentration of 10% w/w indomethacin/polymer, encapsulation is nearly quantitative, with DLE values approaching 100%. However, further increases in feed concentration lead to a sharp decline in DLE values. This behavior contrasts with that of PNVP-b-PVBu and PNVP-b-PVDc, yet it can be fully explained by the nature of the micelles. In the PNVP-b-PVDc samples, longer n-alkyl side chains enhance loading capacity and efficiency, compared to PNVP-b-PVBu samples, due to the amorphous micellar core, which provides ample space for drug incorporation. In PNVP-b-PVSt samples; however, the PVSt block crystallizes, significantly reducing the available core volume. Consequently, full encapsulation is achieved only up to 10% w/w, corresponding to the observed DLC values, while beyond this threshold, the core becomes saturated, preventing further indomethacin loading, which is clearly reflected in the steep drop of DLE at higher feed concentrations.

4. Conclusions

Water acts as a selective solvent for the PNVP-b-PVEs amphiphilic block copolymers, enabling their self-assembly into micellar structures upon dissolution. In the cases of PNVP-b-PVBu and PNVP-b-PVDc copolymers, where k_D values are negative, interparticle hydrodynamic interactions prevail, whereas in the case of PNVP-b-PVSt samples, self-assembly in aqueous media is a thermodynamically driven process, as indicated by the positive k_D values. The investigation of the self-assembly of PNVP-b-PVEs, which differ in the length of their n- alkyl side chains, further demonstrated the impact of this parameter on their micellization behavior. Specifically, PNVP-b-PVBu and PNVP-b-PVDc samples, bearing short- and medium-length side chains, respectively, exhibited an equilibrium between two populations of micelles and micelle aggregates, whereas in the PNVP-b-PVSt samples, with the longest aliphatic side chain, only a single population was detected. Within each copolymer family, the effect of composition on the morphology of the formed structures was also evident. In all cases, increasing the percentage of the hydrophobic PVE block resulted in an enlargement of the micellar core size and, consequently, an increase in the hydrodynamic radius. Encapsulation of hydrophobic bioactive molecules within the micellar cores was successful for both curcumin and indomethacin. These amphiphilic block copolymers proved more effective in encapsulating indomethacin compared to curcumin. Copolymer composition was identified as the key parameter for enhancing loading capacity and efficiency, as higher hydrophobic PVE percentage consistently led to increased DLC and DLE values due to enlargement of the hydrodynamic radius. Comparison among the three PNVP-b-PVEs copolymer families revealed that with PVEs having short alkyl side chains, the DLC and DLE values either increase or at least remain stable upon increasing the amount of the drug, whereas with PVEs with long alkyl side chains, DLC and

DLE decrease in the same sense. This is due to the stronger interaction of the long side chains and the ability to form crystalline domains that restrict the incorporation of higher amounts of drug. In summary, the morphology, size, and stability of the self-assembled nanostructures, combined with the rather high DLC and DLE values, particularly in the case of indomethacin, strongly suggest that the amphiphilic block copolymers investigated in this study can be considered as possible carriers for drug delivery applications.

Supplementary Materials: The following supporting information can be downloaded at: https://www.mdpi.com/article/10.3390/polym17212852/s1, Supplementary Material File S1. Synthesis of the PNVP-b-PVEs block copolymers. Figure S1. Autocorrelation functions from DLS experiments from various block copolymer micellar aqueous solutions. Figure S2. (a) CONTIN plot of sample PNVP-b-PVBu #3 at C = 0.093 mg/mL; (b) CONTIN plot of sample PNVP-b-PVBu #4 at C = 0.758 mg/mL; Figure S3: CONTIN plot of sample PNVP-b-PVBu #5 at C = 0.095 mg/mL; Figure S4: (a) CONTIN plot of sample PNVP-b-PVDc #2 at C = 0.659 mg/mL; (b) CONTIN plot of sample PNVP-b-PVDc #3 at C = 0.59 mg/mL; Figure S5: (a) CONTIN plot of sample PNVP-b-PVSt #3 at C = 0.257 mg/mL; (b) CONTIN plot of sample PNVP-b-PVSt #4 at C = 0.081 mg/mL; Figure S6: CONTIN plot of sample PNVP-b-PVSt #3 at C = 0.054 mg/mL; Figure S7: (a) UV-Vis spectra of PNVP-b-PVBu #1 with all concentrations of encapsulated curcumin; (b) UV-Vis spectra of PNVP-b-PVBu #2 with all concentrations of encapsulated curcumin; Figure S8; (a) UV-Vis spectra of PNVP-b-PVBu #3 with all concentrations of encapsulated curcumin; (b) UV-Vis spectra of PNVP-b-PVBu #4 with all concentrations of encapsulated curcumin; Figure S9; (a) UV-Vis spectra of PNVP-b-PVDc #1 with all concentrations of encapsulated cur-cumin; (b) UV-Vis spectra of PNVP-b-PVDc #2 with all concentrations of encapsulated curcumin; Figure S10: (a) UV-Vis spectra of PNVP-b-PVDc #3 with all concentrations of encapsulated cur-cumin; (b) UV-Vis spectra of PNVP-b-PVDc #4 with all concentrations of encapsulated curcumin; Figure S11: (a) UV-Vis spectra of PNVP-b-PVSt #1 with all concentrations of encapsulated curcumin; (b) UV-Vis spectra of PNVP-b-PVSt #2 with all concentrations of encapsulated curcumin; Figure S12: (a) UV-Vis spectra of PNVP-b-PVSt #3 with all concentrations of encapsulated curcumin; (b) UV-Vis spectra of PNVP-b-PVSt #4 with all concentrations of encapsulated curcumin; Figure S13: (a) UV-Vis spectra of indomethacin with various concentrations; (b) Calibration curve for indomethacin at 320 nm; Figure S14: (a) UV-Vis spectra of PNVP-b-PVBu #1 with all concentrations of encapsulated indomethacin; (b) UV-Vis spectra of PNVP-b-PVBu #3 with all concentrations of encapsulated indomethacin; Figure S15: UV-Vis spectra of PNVP-b-PVBu #4 with all concentrations of encapsulated indomethacin; Figure S16: (a) UV-Vis spectra of PNVP-b-PVDc #2 with all concentrations of encapsulated indomethacin; (b) UV-Vis spectra of PNVP-b-PVDc #4 with all concentrations of encapsulated indomethacin; Figure S17: (a) UV-Vis spectra of PNVP-b-PVSt #2 with all concentrations of encapsulated indomethacin; (b) UV-Vis spectra of PNVP-b-PVSt #4 with all concentrations of encapsulated indomethacin; Figure S18: UV-Vis spectra of PNVP-b-PVSt #5 with all concentrations of encapsulated indomethacin.

Author Contributions: Conceptualization, M.P.; methodology, N.V.P. and N.R.; investigation, N.V.P. (DLS, SLS, indomethacin encapsulation), A.-M.G. (DLS, SLS), A.A. (DLS), A.B. (DLS), T.-P.P. (investigation) and N.R. (curcumin encapsulation); writing—original draft preparation, N.V.P.; writing—review and editing, M.P.; visualization, N.V.P.; supervision, M.P.; project administration, M.P.; funding acquisition, N.R. All authors have read and agreed to the published version of the manuscript.

Funding: This research was co-funded by Greece and the European Union (European Social Fund-ESF) through the Operational Program «Human Resources Development, Education and Lifelong Learning» in the context of the project "Strengthening Human Resources Research Potential via Doctorate Research" (MIS-5000432), implemented by the State Scholarships Foundation (IKY).

Institutional Review Board Statement: Not applicable.

Data Availability Statement: The original contributions presented in this study are included in the article/Supplementary Materials. Further inquiries can be directed to the corresponding author.

Conflicts of Interest: The authors declare no conflicts of interest.

References

1. Xie, B.; Liu, Y.; Li, X.; Yang, P.; He, W. Solubilization techniques used for poorly water-soluble drugs. *Acta Pharm. Sin. B* **2024**, *14*, 4683–4716. [CrossRef]
2. Williams, H.D.; Trevaskis, N.L.; Charman, S.A.; Shanker, R.M.; Charman, W.N.; Pouton, C.W.; Porter, C.J.H. Strategies to Address Low Drug Solubility in Discovery and Development. *Pharmacol. Rev.* **2013**, *65*, 315–499. [CrossRef]
3. Kawabata, Y.; Wada, K.; Nakatani, M.; Yamada, S.; Onoue, S. Formulation Design for Poorly Water-Soluble Drugs Based on Biopharmaceutics Classification System: Basic Approaches and Practical Applications. *Int. J. Pharm.* **2011**, *420*, 1–10. [CrossRef]
4. Bhalani, D.V.; Nutan, B.; Kumar, A.; Chandel, A.K.S. Bioavailability Enhancement Techniques for Poorly Aqueous Soluble Drugs and Therapeutics. *Biomedicines* **2022**, *10*, 2055. [CrossRef]
5. Boyd, B.J.; Bergström, C.A.S.; Vinarov, Z.; Kuentz, M.; Brouwers, J.; Augustijns, P.; Brandl, M.; Bernkop-Schnürch, A.; Shrestha, N.; Préat, V.; et al. Successful Oral Delivery of Poorly Water-Soluble Drugs Both Depends on the Intraluminal Behavior of Drugs and of Appropriate Advanced Drug Delivery Systems. *Eur. J. Pharm. Sci.* **2019**, *137*, 104967. [CrossRef]
6. Stielow, M.; Witczyńska, A.; Kubryń, N.; Fijałkowski, Ł.; Nowaczyk, J.; Nowaczyk, A. The Bioavailability of Drugs—The Current State of Knowledge. *Molecules* **2023**, *28*, 8038. [CrossRef] [PubMed]
7. Kalepu, S.; Nekkanti, V. Insoluble Drug Delivery Strategies: Review of Recent Advances and Business Prospects. *Acta Pharm. Sin. B* **2015**, *5*, 442–453. [CrossRef] [PubMed]
8. Hwang, D.; Ramsey, J.D.; Kabanov, A.V. Polymeric Micelles for the Delivery of Poorly Soluble Drugs: From Nanoformulation to Clinical Approval. *Adv. Drug Deliv. Rev.* **2020**, *156*, 80–118. [CrossRef] [PubMed]
9. Kuperkar, K.; Patel, D.; Atanase, L.I.; Bahadur, P. Amphiphilic Block Copolymers: Their Structures, and Self-Assembly to Polymeric Micelles and Polymersomes as Drug Delivery Vehicles. *Polymers* **2022**, *14*, 4702. [CrossRef]
10. Negut, I.; Bita, B. Polymeric Micellar Systems—A Special Emphasis on "Smart" Drug Delivery. *Pharmaceutics* **2023**, *15*, 976. [CrossRef]
11. Cabral, H.; Miyata, K.; Osada, K.; Kataoka, K. Block Copolymer Micelles in Nanomedicine Applications. *Chem. Rev.* **2018**, *118*, 6844–6892. [CrossRef]
12. Othman, R.S.; Zarei, S.; Haghighat, H.R.; Taromi, A.A.; Khonakdar, H.A. Recent Advances in Smart Polymeric Micelles for Targeted Drug Delivery. *Polym. Adv. Technol.* **2025**, *36*, e70180. [CrossRef]
13. Chroni, A.; Chrysostomou, V.; Skandalis, A.; Pispas, S. Drug Delivery: Hydrophobic Drug Encapsulation into Amphiphilic Block Copolymer Micelles. In *Supramolecules in Drug Discovery and Drug Delivery*; Mavromoustakos, T., Tzakos, A.G., Durdagi, S., Eds.; Methods in Molecular Biology; Humana: New York, NY, USA, 2021; Volume 2207, pp. 71–85. [CrossRef]
14. Zheng, B.; McClements, D.J. Formulation of More Efficacious Curcumin Delivery Systems Using Colloid Science: Enhanced Solubility, Stability, and Bioavailability. *Molecules* **2020**, *25*, 2791. [CrossRef]
15. Pacifici, G.M. Clinical Pharmacology of Indomethacin in Preterm Infants: Implications in Patent Ductus Arteriosus Closure. Pediatr. *Drugs* **2013**, *15*, 363–376. [CrossRef]
16. Kunnumakkara, A.B.; Hegde, M.; Parama, D.; Girisa, S.; Kumar, A.; Daimary, U.D.; Garodia, P.; Yenisetti, S.C.; Oommen, O.V.; Aggarwal, B.B. Role of Turmeric and Curcumin in Prevention and Treatment of Chronic Diseases: Lessons Learned from Clinical Trials. *ACS Pharmacol. Transl. Sci.* **2023**, *6*, 447–518. [CrossRef] [PubMed]
17. Sarkis, N.; Sawan, A. Method Development for Simultaneously Determining Indomethacin and Nicotinamide in New Combination in Oral Dosage Formulations and Co-Amorphous Systems Using Three UV Spectrophotometric Techniques. *Int. J. Anal. Chem.* **2024**, *2024*, 2035824. [CrossRef] [PubMed]
18. Roka, N.; Kokkorogianni, O.; Kontoes-Georgoudakis, P.; Choinopoulos, I.; Pitsikalis, M. Recent Advances in the Synthesis of Complex Macromolecular Architectures Based on Poly(N-vinyl pyrrolidone) and the RAFT Polymerization Technique. *Polymers* **2022**, *14*, 701. [CrossRef]
19. Berger, M.; Toussaint, F.; Ben Djemaa, S.; Laloy, J.; Pendeville, H.; Evrard, B.; Jerôme, C.; Lechanteur, A.; Mottet, D.; Debuigne, A.; et al. Poly(vinyl pyrrolidone) derivatives as PEG alternatives for stealth, non-toxic and less immunogenic siRNA-containing lipoplex delivery. *J. Control. Release* **2023**, *361*, 87–101. [CrossRef]
20. Liu, Y.; Luo, X.; Xu, X.; Gao, N.; Liu, X. Preparation, characterization and in vivo pharmacokinetic study of PVP-modified oleanolic acid liposomes. *Int. J. Pharm.* **2017**, *517*, 1–7. [CrossRef] [PubMed]
21. Rinno, H. Poly(vinyl esters). In *Ullmann's Encyclopedia of Industrial Chemistry*; Wiley-VCH Verlag GmbH & Co. KGaA: Weinheim, Germany, 2000; Volume 28, pp. 469–479. [CrossRef]
22. Yamada, K.; Nakano, T.; Okamoto, Y. Synthesis of syndiotactic poly(vinyl alcohol) from fluorine-containing vinyl esters. *Polym. J.* **1998**, *42*, 9679–9686. [CrossRef]
23. Riess, G. Micellization of block copolymers. *Prog. Polym. Sci.* **2003**, *28*, 1107–1170. [CrossRef]

24. Mai, Y.; Eisenberg, A. Self-assembly of block copolymers. *Chem. Soc. Rev.* **2012**, *41*, 5969–5985. [CrossRef]
25. Blanazs, A.; Armes, S.P.; Ryan, A.J. Self-assembled block copolymer aggregates: From micelles to vesicles and their biological applications. *Macromol. Rapid Commun.* **2009**, *30*, 267–277. [CrossRef]
26. Perinelli, D.R.; Cespi, M.; Lorusso, N.; Palmieri, G.F.; Bonacucina, G.; Blasi, P. Surfactant Self-Assembling and Critical Micelle Concentration: One Approach Fits All? *Langmuir* **2020**, *36*, 5745–5753. [CrossRef]
27. Li, H.; Hu, D.; Liang, F.; Huang, X.; Zhu, Q. Influence Factors on the Critical Micelle Concentration Determination Using Pyrene as a Probe and a Simple Method of Preparing Samples. *R. Soc. Open Sci.* **2020**, *7*, 192092. [CrossRef]
28. Azeri, Ö.; Schönfeld, D.; Noirez, L.; Grillo, I.; Canetta, E.; Dejeu, J. Structural control in micelles of alkyl acrylate–acrylate copolymers via alkyl chain length and block length. *Colloid Polym. Sci.* **2020**, *298*, 829–840. [CrossRef]
29. Li, Y.; Gao, Q.; Ruan, Z. Effects of side-chain lengths on the structure and properties of anhydrides modified starch micelles: Experimental and DPD simulation studies. *Carbohydr. Polym.* **2024**, *322*, 122451. [CrossRef] [PubMed]
30. Discher, D.E.; Eisenberg, A. Polymer vesicles. *Science* **2002**, *297*, 967–973. [CrossRef] [PubMed]
31. Bates, F.S.; Fredrickson, G.H. Block copolymer thermodynamics: Theory and experiment. *Annu. Rev. Phys. Chem.* **1990**, *41*, 525–557. [CrossRef]
32. Wang, J.; Si, J.; Zhu, J.; Nie, Y.; Lu, Y.; Chen, Y. Effect of block ratios on self-assembly morphologies of poly(ε-caprolactone)-block-poly(tert-butyl acrylate) block copolymer. *Polymer* **2023**, *283*, 126292. [CrossRef]
33. Lu, Y.; Zhang, E.; Yang, J.; Cao, Z. Strategies to Improve Micelle Stability for Drug Delivery. *Nano Res.* **2018**, *11*, 4985–4998. [CrossRef]
34. Perumal, S.; Atchudan, R.; Lee, W. A Review of Polymeric Micelles and Their Applications. *Polymers* **2022**, *14*, 2510. [CrossRef]
35. Sandoval, R.W.; Williams, D.E.; Kim, J.; Roth, C.B.; Torkelson, J.M. Critical micelle concentrations of block and gradient copolymers in homopolymer: Effects of sequence distribution, composition, and molecular weight. *J. Polym. Sci. Part B Polym. Phys.* **2008**, *46*, 2673–2687. [CrossRef]
36. Roka, N.; Pitsikalis, M. Synthesis and Micellization Behavior of Amphiphilic Block Copolymers of Poly(N-Vinyl Pyrrolidone) and Poly(Benzyl Methacrylate): Block versus Statistical Copolymers. *Polymers* **2023**, *15*, 2225. [CrossRef] [PubMed]
37. Roka, N.; Pitsikalis, M. Synthesis, Characterization, and Self-Assembly Behavior of Block Copolymers of N-Vinyl Pyrrolidone with n-Alkyl Methacrylates. *Polymers* **2025**, *17*, 1122. [CrossRef] [PubMed]
38. Fokaidis-Psyllas, A.; Kokkorogianni, O.; Pitsikalis, M. Statistical Copolymers of N-Vinylpyrrolidone and Phenoxyethyl Methacrylate via RAFT Polymerization: Monomer Reactivity Ratios, Thermal Properties, Kinetics of Thermal Decomposition and Self-Assembly Behavior in Selective Solvents. *Polym. Bull.* **2025**, *82*, 9161–9179. [CrossRef]
39. Roka, N.; Skiadas, V.-C.; Kolovou, A.; Papazoglou, T.-P.; Pitsikalis, M. Micellization Studies of Block Copolymers of Poly(N-Vinyl Pyrrolidone) and Poly(Vinyl Esters) Bearing N-Alkyl Side-Groups in Tetrahydrofuran. *Polymers* **2025**, *17*, 2842.
40. Kontoes-Georgoudakis, P.; Plachouras, N.V.; Kokkorogianni, O.; Pitsikalis, M. Amphiphilic Block Copolymers of Poly(N-Vinyl Pyrrolidone) and Poly(Isobornyl Methacrylate): Synthesis, Characterization and Micellization Behaviour in Selective Solvents. *Eur. Polym. J.* **2024**, *208*, 112873. [CrossRef]
41. Roka, N.; Papazoglou, T.-P.; Pitsikalis, M. Block Copolymers of Poly(N-Vinyl Pyrrolidone) and Poly(Vinyl Esters) Bearing n-alkyl Side Groups via Reversible Addition-Fragmentation Chain-Transfer Polymerization: Synthesis, Characterization, and Thermal Properties. *Polymers* **2024**, *16*, 2447. [CrossRef]
42. Provencher, S.W. CONTIN: A general purpose constrained regularization program for inverting noisy linear algebraic and integral equations. *Comput. Phys. Commun.* **1982**, *27*, 229–242. [CrossRef]
43. Ghezzi, M.; Pescina, S.; Padula, C.; Santi, P.; Del Favero, E.; Cantù, L.; Nicoli, S. Polymeric micelles in drug delivery: An insight of the techniques for their characterization and assessment in biorelevant conditions. *J. Control. Release* **2021**, *332*, 312–336. [CrossRef]
44. Li, T.; Nowell, C.J.; Cipolla, D.; Rades, T.; Boyd, B.J. Direct Comparison of Standard Transmission Electron Microscopy and Cryogenic-TEM in Imaging Nanocrystals Inside Liposomes. *Mol. Pharm.* **2019**, *16*, 1775–1781. [CrossRef] [PubMed]
45. Calzaferri, G.; Gfeller, N. Thionine in the Cage of *Zeolite* L. *J. Phys. Chem.* **1992**, *96*, 3428–3435. [CrossRef]
46. Roulia, M.; Vassiliadis, A.A. Clay-Catalyzed Phenomena of Cationic-Dye Aggregation and Hydroxo-Chromium Oligomerization. *Microporous Mesoporous Mater.* **2009**, *122*, 13–19. [CrossRef]
47. Jelley, E.E. Spectral Absorption and Fluorescence of Dyes in the Molecular State. *Nature* **1936**, *138*, 1009–1010. [CrossRef]
48. Kobayashi, T. Excitons in J-Aggregates with Hierarchical Structure. *Supramol. Sci.* **1998**, *5*, 343–347. [CrossRef]
49. Pitsikalis, M.; Siakali-Kioulafa, E.; Hadjichristidis, N. Block Copolymers of Styrene and Stearyl Methacrylate. Synthesis and Micellization Properties in Selective Solvents. *Macromolecules* **2000**, *33*, 5460–5469. [CrossRef]

polymers

MDPI

Article

Morphological Characterization of High Molecular Weight Poly(styrene-*b*-isoprene) or PS-*b*-PI and Its Hydrogenated, Sulfonated Derivatives: An AFM Study

Nikolaos Politakos [1,*], Galder Kortaberria [2] and Apostolos Avgeropoulos [3,*]

[1] Institute of Chemical Biology, National Hellenic Research Foundation, 11635 Athens, Greece
[2] Department of Chemical and Environmental Engineering, University of the Basque Country (UPV/EHU), 20018 Donostia, Spain; galder.cortaberria@ehu.eus
[3] Department of Materials Science Engineering, University of Ioannina, 45110 Ioannina, Greece
* Correspondence: npolit@eie.gr (N.P.); aavger@uoi.gr (A.A.)

Abstract: The surface morphology of high molecular weight poly(styrene-*b*-isoprene) block copolymer was analyzed after chemical modification. Poly(styrene-*b*-isoprene) was converted into poly(styrene-*b*-(ethylene-alt-propylene)) by hydrogenation and into poly(styrene-*b*-sulfonated isoprene) by mild sulfonation of the PI block. Obtained morphologies were examined by atomic force microscopy, analyzing the effect of sample preparation parameters such as solvent (tetrahydrofuran, toluene, and cyclohexane), casting technique (spin casting and drop casting), and annealing temperature [room temperature, 80, 100, and 120 °C]. Significant morphological and topographical changes were found depending on the different parameters. Each modification step introduces new variables that can affect the final structure and properties of the copolymer. Finding the balance between solvent choice, casting technique, and annealing conditions was a difficult task and required extensive experimentation and understanding of the principles of block copolymer self-assembly.

Keywords: hydrogenation; sulfonation; poly(isoprene); anionic polymerization; atomic force microscopy (AFM); self-assembly; nanostructures; chemical modification reactions; molecular characterization; thin films

1. Introduction

New challenges in nanotechnology lead to the demand of innovative materials for different applications. Using novel polymeric systems with new monomers and different architectures is one way to obtain the necessary materials. Polymers can be synthesized with different methods, achieving different architectures and block combinations [1–10]. The capability of block copolymers to self-assemble into different nanostructures justifies their suitability for use in advanced technologies [1–3,11–13]. Self-assembly is fast, easily scalable, and relatively cheap, and from a technological perspective, provides a pathway to synthesize materials from the nm to μm range [1,2]. Depending on the morphologies (shape, size, and periodicity), it can be suitable for numerous applications such as drug delivery [14], electronics [15], masks [2,16,17], nanopatterning [3,18–20], functional nanomaterials [3,21,22], and nanoporous membranes [3,23–26], among others. Block copolymers can also be combined with other materials, such as ceramic or metallic nanoparticles [3,27–32], leading to new physical, chemical, magnetic, or electric properties [3].

The basic key for self-assembly into ordered nanostructures is based on the blocks' thermodynamic incompatibility. Different morphologies can be obtained depending on the blocks' volume fraction, the variability of the Flory–Huggins interaction parameter (χ), the polymerization degree (N), and sample preparation parameters such as solvent or annealing temperature. The system will evolve to different morphologies to minimize free energy [27,33].

Selective chemical modification of one block by mild reactions is a beneficial technique to obtain new copolymers. A copolymer can be altered into a new material with different physical and chemical properties. It becomes essential when the polymerization of new monomers is complex, very difficult with low yields, or needs special conditions. By changing the chemical structure, the Flory–Huggins (χ) parameter is also altered, thus leading to nanostructures that can vary from the initial copolymer. Polydienes [especially poly(butadiene) and poly(isoprene)] are very suitable for chemical modification due to the existing double bond per monomeric unit [34]. Chemical modification can also be used for preparing amphiphilic copolymers, transforming one of the blocks into hydrophilic (PS modification with sulfonation, for example), which may result in interesting properties with a variety of applications, such as in micelles, polymeric surfactants, emulsifiers, foam, and stabilizers [35–38].

Atomic force microscopy (AFM) is a powerful technique for analyzing the surfaces of various samples and understanding their properties. It is a method which is continuously evolving, introducing new features and expanding its capabilities. AFM probes are becoming increasingly sophisticated, allowing for applications in the mechanical and electrical characterization of samples, as well as measurements in liquid environments and even within living cells [39]. AFM offers exciting potentials for cutting-edge applications, including the measurement of surface tension [40], adhesion [41], and the mechanical properties of soft tissues [42,43]. It can also be used for direct measurement of nanoindentation [44], investigation of magnetic properties [45], assessment of electrical conductivity [46], and analysis of the morphology and micromechanical properties of the membranes of specific human cells [47]. It is far from certain that AFM can provide micromechanical and physical characterization of various objects, such as red blood cells, with force-sensing capabilities to measure properties like surface topography and stiffness at the nanoscale [48].

This manuscript aims to analyze the surface morphology of a high molecular weight poly(styrene-*b*-isoprene) (PS-*b*-PI or SI) diblock copolymer before and after chemical modification exclusively of the PI segments by hydrogenation leading to poly(styrene-*b*-(ethylene-alt-propylene)) (PS-*b*-P(E-alt-P)) or SEP) and by sulfonation obtaining poly(styrene-*b*-sulfonated isoprene) (PS-*b*-(PI-r-PIs)) or SI/sulf). In Scheme 1 the chemical structures of the pristine diblock copolymer and the two modified cases are evident. The latter chemical modification does not yield to 100% modified monomeric units (42% is sulfonated PI). Besides the effect of chemical modification on the resulting morphologies, the sample preparation parameters are also discussed, such as solvent (toluene, tetrahydrofuran, and cyclohexane), film preparation technique (spin and drop casting), and annealing temperature (room temperature (R.T.), 80°, 100°, and 120 °C). Different chemical structures are sensitive to selective or nonselective solvents, while after modification, the nature of PI and the χ parameter between PI and PS changes, leading to different morphologies, also affected by annealing treatment and film preparation techniques. This manuscript is an add-on study to a previous work in which a PS-*b*-PI diblock copolymer was studied but with lower overall number average molecular weight [49]. The molecular weight parameter is important when it is considered in the context of microphase separation (degree of polymerization, N, is directly related to molecular weight and is an important variable for microphase separation fully dependent on the χN value). As molecular weight increases, χN increases, driving the system further into the microphase separation regime. The same copolymer and its lower-molecular-weight analogous may not fully organize, whereas the higher-molecular-weight analogous does, due to the increased χN value.

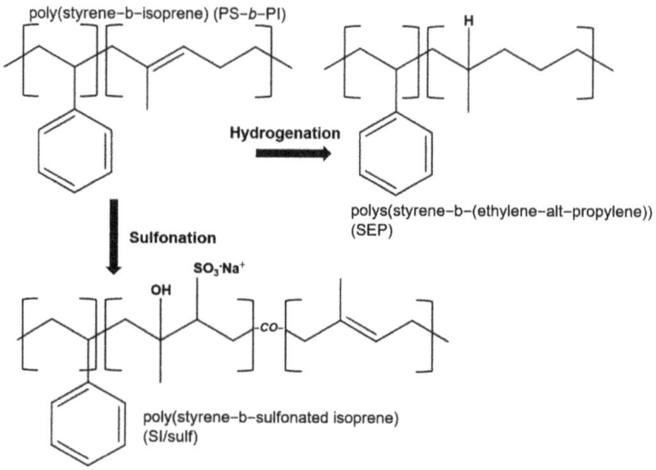

Scheme 1. Chemical structures of the poly(styrene-*b*-isoprene) (PS-*b*-PI or SI) diblock copolymer before and after chemical modification exclusively of the PI segments by hydrogenation leading to poly(styrene-*b*-(ethylene-alt-propylene)) (SEP) and by sulfonation obtaining poly(styrene-*b*-sulfonated isoprene) (SI/sulf).

In this study, the initial diblock copolymer has similar volume fractions for both blocks and a higher molecular weight than in our previous study. Finally, the molecular weight is also essential for determining the size of the microphase-separated domains, such as lamellae or cylinders. The longer chains must stretch farther to segregate into their respective domains, increasing the system's energy. To minimize this energy, the polymers form larger domains, which allows the chains to adopt less-stretched conformations.

2. Materials and Methods

2.1. Materials

The PS-*b*-PI (SI) diblock copolymer used in this work was synthesized by sequential anionic polymerization using high vacuum techniques. The number average molecular weight of the diblock copolymer was 187,000 g/mol, dispersity (Đ) lower than 1.1, with 53% being PS and 47% of PI. Purification of all chemical reagents (styrene and isoprene from Acros, Geel, Antwerpen Belgium and benzene from Lab-scan, Bangkok, Thailand) to the standards required for anionic polymerization has been described elsewhere [50]. Hydrogenation and sulfonation, both by mild reactions, were carried out as mentioned in the literature [49]. p-Toluenesulfonyl hydrazide (Sigma-Aldrich, St. Louis, MO, USA, 97%) was used for hydrogenation with p-xylene (Fluka, Pensacola, FL, USA, ≥99.5%) as solvent. Anhydrous sulfuric acid (Panreac, Barcelona, Spain, 96%) with 1,4-dioxane (Sigma-Aldrich, ≥99%) as solvent, methanol (Lab-scan, Puchheim, Germany, 99.8%), and 15 wt% aqueous solution of NaOH were used also in the sulfonation procedure. Finally, tetrahydrofuran (THF) (Lab-scan, 99.8%), toluene (Lab-scan, 99.8%), and cyclohexane (Panreac, 99.5%) were used as solvents for film preparation.

2.2. Characterization

Morphological characterization was conducted by atomic force microscopy (AFM). Thin films were prepared by casting or spin casting of 3 wt% solutions of copolymers in different solvents onto glass substrates. Spin-coating was conducted at 2000 rpm for 120 s using a Telstar Instrument P-6708D spin-coater(Terrassa, Barcelona, Spain). After complete evaporation of the solvent, films were then annealed in a high-vacuum oven (Binder, Tuttlingen, Germany) for 24 h at different temperatures. A Dimension 3100/Nanoscope IVA, Veeco, New York, NY, USA (from Digital Instruments Santa Barbara, CA, USA) was used. Tapping mode in air was employed using a TESP-V2 probe made from monolithic silicon which is highly doped to dissipate static charge, chemically inert, and offers a high mechanical Q-factor for high sensitivity. The AFM tip is uncoated and shaped like a polygon-based pyramid. The shape of the tip is standard with a radius of <8 nm (<12 nm guaranteed) and with cantilever shaped as beam with force constant 42 N/m, resonance frequency 320 kHz, length 125 μm, width 30 μm, and thickness 4 μm, at a scan rate of 1.0 Hz and a resonance frequency of ~300 kHz. Measurements were performed with 512 scan lines. The images presented here were scanned at a scale bar of 3 μm × 3 μm and a 3D representation was reconstructed from the same image at a scale bar of 1 μm × 1 μm. The profile of the mean value of the phase was also extracted at a diagonal of the initial image at 1.5 μm in the X axis with the Y axis representing the phase. For the measurements of the mean roughness S_a, three distinct areas of 3 μm × 0.5 μm of the initial image were selected and the mean value was presented. From the same areas, the mean value for S_k (skewness) was also presented. The skewness is presented only for the samples that had some type of morphology. The skewness will show a dominance of high peaks when it is positive or dominance of valleys when it is negative. In the case of 0 or close to 0, a symmetrical existence of both peaks and valleys is present. In the same images, the sizes of characteristic domains were also calculated by averaging 3 to 5 points. For the analysis of the AFM images, Gwyddion 2.69 (released 28 July 2025) (https://gwyddion.net/) was used.

Proton nuclear magnetic resonance spectroscopy (^{1}H-NMR) was used for the verification of the sulfonation. The sample was dissolved in deuterated chloroform ($CDCl_3$). The spectra were recorded at room temperature (R.T.) on an Avance Bruker 500 MHz (Rheinstetten, Germany) equipped with BBO z-gradient probe Bruker DSX NMR spectrometer (Billerica, MA, USA) using a rate of 5000 Hz, a frequency of 500 MHz, and a delay of 1 s.

For the Fourier-transform infrared spectroscopy (FTIR) and size exclusion chromatography (SEC) instruments, the experimental conditions can be found in the Supporting Information.

3. Results

3.1. Molecular Characterization

The molecular characterization instrumentation was thoroughly mentioned and analyzed for derivatives diblock copolymers in a previous work [49]. Size exclusion chromatography (SEC) was conducted to verify that no side reactions occurred during the synthesis of the diblock copolymer. Furthermore, Fourier transform infrared spectroscopy (FTIR), and proton nuclear magnetic resonances (^{1}H-NMR) were conducted to verify the success of the chemical modification reactions. Hydrogenation and sulfonation yields of 100% and 42% were obtained, respectively. Thus, in the SI/sulf copolymer, sulfonated and non-sulfonated PI blocks are present randomly located throughout the PI segments. The ^{1}H-NMR spectra for the result of the sulfonation reaction is shown in Figure S1 (Supporting Information).

3.2. Morphological Characterization

Solubility parameters for different solvents and blocks as well as polymer–polymer (χ_{AB}) and polymer–solvent (χ_{SP}) interaction parameters are given in Tables 1 and 2, respectively, together with the necessary equations needed for their calculation [51,52].

Table 1. Solubility parameter values for all solvents and polymeric blocks used.

Solvents Polymers	δ ($MPa^{1/2}$) [a] (Hansen)	δ (MPa) [b] (Van Krelen)	δ ($MPa^{1/2}$) [b] (Van Krelen)	δ (MPa) [b] (Van Krelen)
Solvents				
THF	19.46		19.50	
Toluene	18.16		18.20	
Cyclohexane	16.80		16.70	
Polymer blocks				
PS	19.07		19.10	
PI	17.94		17.40	
PEP	17.47 [c]		16.86 [d]	
PI/sulf	24.88 [c]		26.44 [d]	

[a] Overall solubility was calculated from $\delta^2 = {\delta_d}^2 + {\delta_p}^2 + {\delta_h}^2$ [51]. [b] Solubility values were found in the literature or calculated from Van Krelen [53]. [c] Solubility values were calculated from contributions of sulfonated PI groups by $\delta_d = \Sigma F_{di}/V$, $\delta_p = \sqrt{\Sigma {F_{pi}}^2}/V$ and $\delta_h = \sqrt{\Sigma E_{di}}/\sqrt{V}$ [53]. The rest were found in the literature [51]. Volume values were also calculated from the contributions of each group of the chemical structure [53]. [d] Solubility values were calculated from $\delta = (E_{coh}/V)^{1/2}$ by using contributions of each group to E_{coh} [53].

Table 2. χ_{AB} and χ_{SP} values for all solvents and blocks used for all samples.

χ_{AB} [a]	298 K (R.T.)	χ_{SP} [b]	THF	Toluene	Cyclohexane
$\chi_{PS/PI}$	0.05	χ_{PS}	0.09	0.17	0.36
$\chi_{PS/PEP}$	0.10	χ_{PI}	0.16	0.04	0.13
$\chi_{PS/PI/sulf}$	1.35	χ_{PEP}	0.20	0.12	0.16
$\chi_{PI/PI/sulf}$	1.93	$\chi_{PI/sulf}$	0.34	1.26	1.80

[a] χ_{AB} values were calculated from $\chi_{AB} \approx (\delta_A - \delta_B)^2 V_{reff}/RT$ equation, where V_{reff} = 100 cm^3/mol, R = 8.314 cm^3 MPa/K·mol and δ in MPa taken from Table 1 (Hansen). [b] χ_{SP} values were calculated from $\chi_{sp} = (aV/RT)[(\delta_{ds} - \delta_{dp})^2 + 0.25(\delta_{ps} - \delta_{pp})^2 + 0.25(\delta_{hs} - \delta_{hp})^2]$ equation, where V is in cm^3/mol, R = 8.314 cm^3 MPa/K·mol, a is constant equal to 0.6 [54], and δ_d, δ_p, and δ_h in $MPa^{1/2}$ were taken also from the literature [53].

The affinity between polymer–polymer and polymer–solvent is an essential parameter to explain the observed morphologies. The affinity of blocks with solvent molecules will affect their position and orientation onto the substrate after solvent evaporation. A solvent with high affinity will move the block towards the interface with air, while the block will be located at the substrate interface with a non-miscible solvent [55]. As evident by the calculated interaction parameter values, THF is a selective solvent for PS and PI/sulf (much lower affinity with non-polar toluene and cyclohexane) blocks. However, its affinity is higher with PS. On the other hand, toluene is a selective solvent for PI and PEP (lower affinity than PI); in this case, affinity with cyclohexane and THF is slightly lower but quite similar.

Regarding interaction parameters among the blocks, PS and PI present the highest affinity. PS and PI are supposed to present higher miscibility than PS and PEP so that a higher organized nanostructure can be expected for the SEP copolymer than for the initial diblock copolymer precursor. The SI/sulf case is more complicated since it contains three different blocks: PS (~50 wt%), PI (~30 wt%), and PI/sulf (~20 wt%). The overall morphology can be expected to be affected by more parameters than the other copolymers. As PI/sulf presents low affinity with PS and PI (1.35 and 1.93, respectively), immiscibility among the blocks is expected, leading to nanostructured morphologies due to chemical incompatibility among the segments [56]. Depending on the annealing treatment, different nanostructures are expected according to the literature [57].

The morphology of the initial SI copolymer is affected when chemical modification reactions occur. After complete hydrogenation of the PI, a new PEP block is generated (showcasing variable affinities with blocks and solvents), and therefore different morphologies are expected from those of the initial SI. In the same manner, after the sulfonation reaction, different topological structures are observed for the PI, PI/sulf, and PS blocks. Table 3 shows the morphologies obtained for all copolymers under different conditions. Figure 1 gives examples of all kinds of morphologies described in Table 3. In Figure S2 (Supporting Information), the phase images for all samples are presented with a 3D representation and a cut-off profile.

Table 3. Characteristic morphologies observed for all samples at different conditions.

	Copolymers					
	SI		**SEP**		**SI/sulf**	
	Casting Technique					
	Casting	**Spin Casting**	**Casting**	**Spin Casting**	**Casting**	**Spin Casting**
Morphologies	Samples		Samples		Samples	
Disorder	* 2C, 2F, 3C 3T, 3F, 4C 4T, 4F	2F, 3C, 3T 4C, 4T, 4F	3T	2C, 4T, 4F	2C, 2T, 4C 4T, 4F	1F, 4C, 4T 4F
Lamellae	-	3F	1T, 2T, 2F 4T	1T	2F	-
Lamellae (not fully organized)	1T, 1F, 2T	1T	1F	-	1T, 1F	-
Lamellae (not fully organized) with crystallites	-	2T	-	3T	-	1T, 2T, 2F
Network	-	-	4C, 4F	4C	-	3T
Spherical micelles	-	1C, 2C	1C, 2C, 3C	1C	3C 3T, 3F	-
Spherical micelles with crystallites	-	-	-	-	-	1C, 2C, 3F
Worm-like	1C	1F	3F	1F, 2T, 2F 3C	1C	3C
Worm-like with crystallites	-	-	-	3F	-	-

* Table is divided for different samples as long as for casting (c) and spin casting (sc) techniques. Abbreviations for each sample are referred as XY, where X is from 1–4 (R.T., 80, 100, 120 °C) and Y is for different solvents: C (cyclohexane), T (toluene) and F (THF).

In the phase images, it is evident that stiffer materials appear brighter, while the softer phases of the copolymers appear darker. This contrast highlights the distinct characteristics of each material, emphasizing their varying mechanical properties. Figure 1 clearly demonstrates that the polystyrene phase exhibits greater stiffness compared to the polydienic segments, as seen in Figure 1b and 1g, respectively. Furthermore, in the images featuring the hydrogenated sample (Figure 1c–e,h), the PS segment again stands out as brighter.

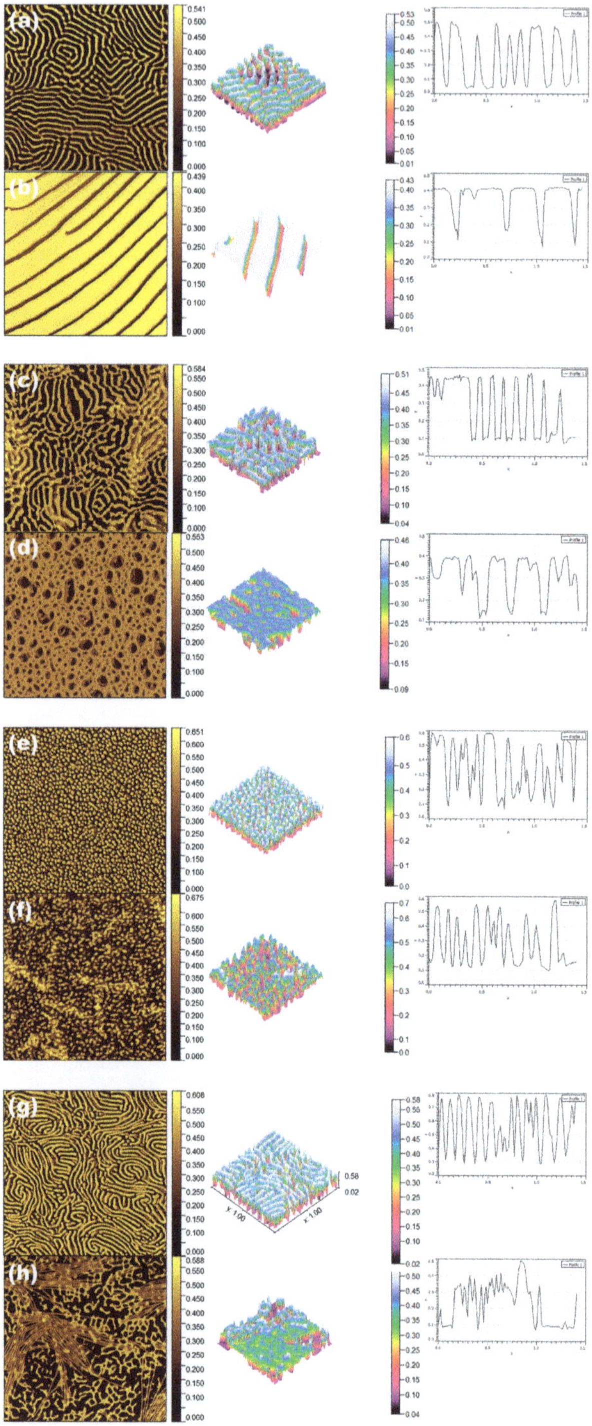

Figure 1. Examples of all morphologies observed by AFM phase images (3 μm × 3 μm) with a 3D representation of a 1 μm × 1 μm and a profile of the morphology of 1.5 μm diagonal: (**a**) lamellae

(SEP, spin casted from toluene annealed at R.T.), (**b**) not fully organized lamellae (SI, casted from toluene annealed at R.T.), (**c**) lamellae not fully organized lamellae with crystallites (SEP, spin casted from toluene annealed at 100 °C), (**d**) polymeric network (SEP, casted from cyclohexane annealed at 120 °C), (**e**) spherical micelles (SEP, casted from cyclohexane annealed at R.T.), (**f**) spherical micelles with crystallites (SI/sulf, spin casted from cyclohexane annealed at 80 °C), (**g**) worm-like (SI, casted from cyclohexane annealed at R.T.) and (**h**) worm-like with crystallites (SEP, spin casted from THF annealed at 100 °C).

Table 3 presents findings related to how various factors such as solvent type, annealing temperature, and casting technique influence the resulting morphologies of copolymer samples. Notably, disordered morphologies are predominantly observed in spin-casted samples that are subjected to high annealing temperatures of 120 °C across all solvent types used. This phenomenon is attributed to the rapid evaporation of the solvent, which does not allow adequate time for the molecules to organize themselves into distinct and well-defined nanostructures, as supported by prior research [58]. To achieve ordered or partially ordered morphologies, specific parameters must be optimized. The casting technique has a considerable impact, resulting in alternating lamellae structures, particularly when using toluene and THF at an annealing temperature of 80 °C. In contrast, the simple casting method tends to favor the formation of less ordered lamellae structures, especially for samples that are annealed at R.T. Furthermore, less organized lamellae containing crystallites are detected in samples that have been spin cast from toluene, suggesting that the solvent plays a role in determining the degree of order in the film structure. Network-type topologies are primarily observed in samples that have been spin cast from all solvent types and subsequently annealed at the high temperature of 120 °C.

In samples prepared with cyclohexane, spherical micelles become pronounced, particularly for those that are annealed from R.T. up to 100 °C. Notably, in spin-casted SI/sulf samples, crystallites are observed in conjunction with these spherical micelles. Worm-like structures are predominantly found in samples spin cast from THF, with the specific morphology depending on the copolymer under consideration. For example, the SEP sample spin cast with THF and annealed at 100 °C displays both crystallite formation and a worm-like structure. The fast evaporation rate of THF is a critical factor, restricting the time available for the samples to develop fully ordered nanostructures, a trend consistent with other solvents as well. When analyzing the effects of solvents, it becomes apparent that cyclohexane typically results in the formation of spherical micelles and disordered structures. Films derived from toluene tend to exhibit more organized domains, primarily in the form of lamellae. Conversely, samples made from THF exhibit a range of morphologies, which may include either lamellae or worm-like structures, depending on the specific conditions applied during processing. This variability leads to many observed disordered morphologies due to THF's high evaporation rate [59].

In terms of the effect of annealing temperature, the most well-ordered morphologies—primarily lamellae structures and spherical micelles—are most frequently identified at R.T. As the annealing temperature increases, a shift is noted towards more disordered structures alongside the simultaneous presence of ordered domains. This is primarily due to enhanced solubility at elevated temperatures, as indicated in previous studies [60]. The interplay between block miscibility and the thermal energy generated during annealing can inhibit the achievement of fully organized structures [61]. As the temperature rises, the increased thermal energy facilitates block mobility and rearrangement; however, the process of blocks migrating towards the surface complicates obtaining ideal morphologies.

The influence of the casting technique is particularly noteworthy, as crystallites have only been identified in samples that were spin cast, underscoring the importance of this method in structuring the final films [62].

For copolymers, microphase separation is strongly linked to the glass transitions of both blocks, enabling the blocks to move to overcome thermodynamic barriers and form stable nanostructures. In addition to the temperature barrier, annealing time plays a crucial role in forming nanostructures. The polyisoprene and polyethylene/polypropylene show glass transitions below R.T., meaning that these blocks have high mobility; in contrast, the sulfonated polyisoprene has a significantly higher glass transition temperature but is still close to R.T. The different temperatures used in this study are capable of high mobility for all blocks except PS. The PS is close to the glass transition at 80 °C and 100 °C and has high mobility at 120 °C. The 100 °C barrier is the key to achieving full mobility of all chains across all blocks. The higher temperature of 120 °C leads the copolymer to form a disordered, single-phase melt, as seen in the majority of samples, which do not show ordered structures at this temperature.

When examining the morphologies corresponding to various copolymers, it is evident that the SI copolymer precursor tends to yield less organized structures compared to its hydrogenated counterpart, SEP. This difference is likely due to variations in block miscibility, with SI displaying a lower interaction parameter [63] than SEP, thus primarily resulting in disordered morphologies such as lamellae structures, spherical micelles, and worm-like forms, though these phenomena are only observable under specific processing conditions. In contrast, the SEP copolymer typically exhibits more ordered structures like lamellae—predicted due to the molecular weight ratio of the constituent blocks—and also features some spherical micelles and worm-like morphologies due to its higher interaction parameter between blocks. The case involving the SI/sulf copolymer is more intricate, given the presence of multiple components such as PI, PI/sulf, and PS, which leads to a complex interaction landscape between block/block and block/solvent. This amphiphilic nature of the copolymer is a significant contributor to the prevalent formation of spherical micelles in a majority of cases. Exceptions to this trend include the lamellae and network morphologies, particularly when utilizing toluene and THF as solvents.

In terms of skewness (Table S1, Supporting Information), most of the samples have values less than 1 (negative or positive), indicating a slight dominance of peaks or valleys, but are relatively symmetrical in the organized or almost organized domains. Notably, some samples also showed values very close to 0, indicating a quite symmetrical distribution of peaks and valleys. Finally, the majority of more asymmetrical morphologies are those in which the lamellas have not formed, showing negative values and the majority of valleys. Some other morphologies with negative skewness are also the SEP samples, cast from cyclohexane annealed at 80 and 120 °C (micelles and network), and SI/sulf, cast from THF at 100 degrees (micelles). The only positive skewness value above 1 is the sample SI/sulf spic cast from cyclohexane annealed at 100 °C (worm-like). The nature and evaporation rate of cyclohexane are helping to create these asymmetries.

Regarding the roughness of the samples, the unorganized samples have a smoother surface, because no structure forms. If one looks more closely at the organized domains and the nature of each sample, as well as the individual parameters such as solvent, casting technique, and temperature, some interesting points can be extracted. Among the samples, the rougher surfaces are found in the SEP samples, with an average roughness of 0.148, while SI has 0.111 and SI/sulf has 0.087. In terms of the solvents, the difference is not so crucial, with cyclohexane and THF showing slightly higher roughness (0.121 and 0.122), probably due to their higher evaporation rates and the time structures have to evolve. In the casting technique, a pattern is observed in the spin-cast samples, showing higher roughness (0.127) than in the casted ones (0.112). Here, the influence of the substrate is evident, as evidenced by the formation of thinner films in the spin-cast samples. For the different annealing temperatures, all show similar values, with 120 °C slightly higher. Finally, the different morphologies show a mixed trend in roughness values. The structures with lamellae morphology have an average value of 0.104, but if we look closer, the organized lamellae and the non-organized ones show a significant difference, with values of 0.130 vs. 0.087. This is mainly because the organized lamellae have defined the two polymeric components, and the stiffer PS is contributing 100%. The micellar structures have a low value of 0.107, indicating relatively smooth surfaces. Also important are the network and worm-like structures, which can be interpreted as intermediate structures with the polymeric components not fully organized, and this peculiar mixing shows roughness values of 0.146 (worm-like) and 0.141 (network).

The sizes of the different domains within the organized structures are also important points that must be addressed. For the micelles found for SEP, the sizes are very close, ranging from 73 nm to 80 nm. A similar trend was also observed for the micelles found for the SI/sulf. For the network morphologies of the SEP, the casting technique plays a key role, as in simple casting, the sizes are similar for PS and PEP at 64 and 63 nm, respectively. By introducing the spin-casting technique for the same sample, the PS domain maintains the same size, while the PEP increases to 179 nm. Finally, the solvent plays a role: the SEP sample has a similar size when cast from cyclohexane, whereas casting from THF results in a PEP size of 115 nm.

The worm-like morphologies also show interesting domain sizes, with the SI case ranging from 16 nm (cast from cyclohexane annealed at RT) to 71 nm (spin-cast from cyclohexane annealed at 80 °C). In the case of SEP, the PS has quite similar sizes, ranging from 41 to 54 nm (smaller than in the SI), whereas the PEP has a higher value, ranging from 55 to 82 nm. The trend, as mentioned above, is directly related to the organization of the PEP during casting and annealing, compared to the PI block found in SI systems. The worm-like morphologies in the SI/sulf do not show significant size changes, with PS ranging from 39 to 52 nm.

Finally, for the lamellae morphology, the only example organized for the SI system shows PS domains of 56 nm and PI of 27 nm, spin-cast from the annealed THF at 100 °C. Lamellae morphology for the SEP samples shows an impressive trend: samples cast and spin-cast from toluene annealed at RT show differences. In casting, the PS is 63 nm, while the PEP is 90 nm; in spin casting, the PS is reduced to 46 nm and the PEP to 79 nm, resulting in larger PEP domains. The effect of the solvent can be seen for the SEP system casted and annealed at 80 °C, where from toluene has sizes of PS 59 nm and PEP 81 nm, but changing the solvent, and thus the affinity of the blocks, the sizes are increasing to 81 nm for PS and 130 nm for PEP; a 37% increase for the PS and 60% increase for the PEP. Finally, increasing the annealing temperature to 120 °C increases the domain size for PS to 138 nm (compared to 59 nm at 80 °C) and for PEP to 119 nm (compared to 81 nm at 80 °C), increases of 134% and 47%, respectively. The 120 °C annealing temperature is crucial because it is above the glass transition temperature of PS, allowing sufficient mobility for organization. In the case of SI/sulf, the only sample with lamellae morphology (cast from THF at 80 °C) shows similar domain sizes of 82 nm and 79 nm.

Figure 2 compares lamellae morphologies obtained for SI, SEP, and SI/sulf copolymers. Figure 2a corresponds to SI copolymer spin cast from THF and annealed at 100 °C. As can be seen, the nanostructure is not fully ordered. The fast evaporation rate of THF and spin casting technique gave lamellas little time for complete formation [64]. Figure 2b corresponds to SEP copolymer cast from toluene and annealed at R.T. Due to the lower evaporation rate of solvent (together with the casting technique), lamellas can be fully formed, and they also show higher immiscibility among blocks for SEP when compared to SI.

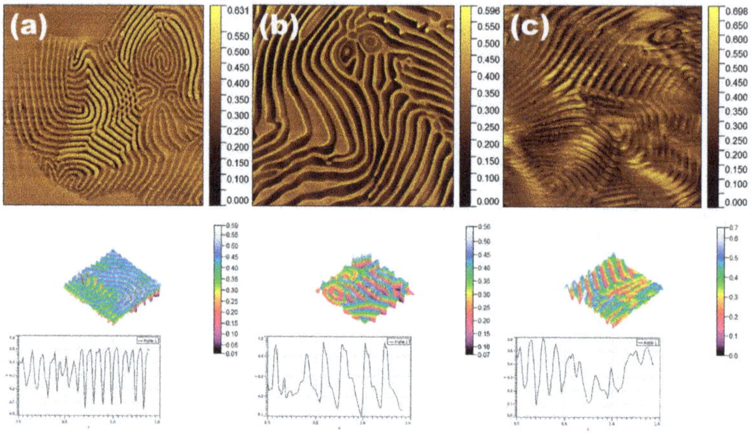

Figure 2. AFM phase images (3 μm × 3 μm) showing lamellae morphology observed for (**a**) SI, (**b**) SEP and (**c**) SI/sulf copolymers, showing phase and ordering differences. The 3D phase 1 μm × 1 μm and 1.5 diagonal profile sections are also shown.

Figure 2c shows lamellae morphology obtained for SI/sulf copolymer cast from THF and annealed at 80 °C. THF presents similar affinity with all blocks (χPS = 0.09, χPI = 0.16 and χPI/sulf = 0.34). With the casting technique, solvent evaporation is slower than by spin casting. Thus, llamelas have enough time to form fully.

The roughness of the different lamellae morphologies is obviously governed by the solvent, temperature, and casting technique, but it can be seen to be relatively smooth, with values of 0.064 to 0.094. The sizes of the PS domains are increasing from 56 nm at SI to 63 nm at SEP and 82 nm at SI/sulf. For the PI, the size is 27 nm; after modification, it increases to 90 nm at SEP and 79 nm for the sulfonated PI. The modification creates different balances among the PS chains, with the modified ones leading to changes in the sizes.

The solvent, temperature, casting technique, and chemical nature of blocks also affect topography and the phase difference of morphologies [65]. The size, roughness, and distribution of lamellae, spherical micelle, and worm-like domains will change by altering the abovementioned parameters. As an example, in Figure 3, a 3D representation of the morphologies obtained can be observed for SEP copolymer as cast from cyclohexane and annealed at R.T., 80, and 100 °C, respectively. The interaction parameters of PS and PEP with cyclohexane are 0.36 and 0.16, respectively (as evident in Table 2).

PEP presents a higher affinity with cyclohexane than PS, and the solvent evaporates from the surface; therefore, the PS chains are removing the solvent faster than the PEP blocks. The solvent is maintained longer with PEP chains giving an extra volume and PS blocks tend to aggregate for minimizing energy. In that manner, PEP blocks create spherical micelles on the surface. Domain size and aggregation are controlled by annealing temperature. An estimation of size, roughness, and aggregation has been carried out for topography and phase by AFM. Dimensions for PS are 53, 73, and 80 nm, by annealing at R.T., 80, and 100 °C, respectively. At 100 °C, PS blocks could have enough energy to rearrange and create more stable domains, something that can also be seen from roughness

(0.081), where at R.T. is rougher with a value of 0.157 and at 100 °C of 0.161. The 3D representation of the sample annealed at 80 °C has the most compact surface. The raise of the annealing temperature at 120 °C disrupts the morphology creating a network (Figure S2, Supporting Information).

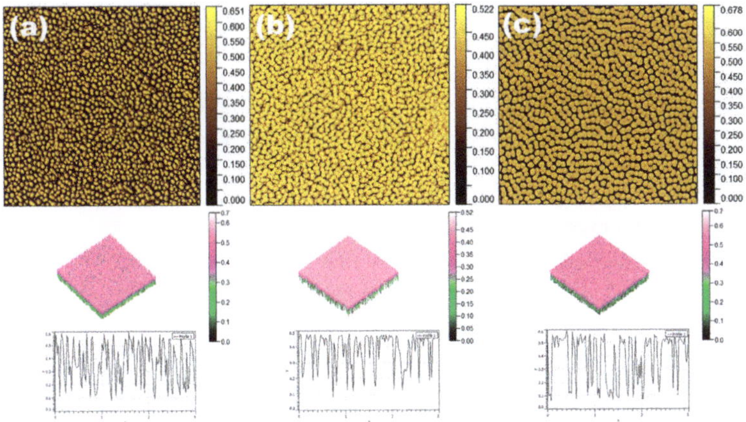

Figure 3. 2D and 3D representation of (3 μm × 3 μm) for SEP copolymer casted from cyclohexane and annealed at different temperatures: (**a**) R.T., (**b**) 80 °C, and (**c**) 100 °C, showing differences in aggregation and organization of spherical micelles. Lower images correspond to 3 μm horizontal profile sections, showing phase differences.

The effect of the casting technique on morphology is also crucial [66], as it can reveal information about the different phases. For example, Figure 4a,b shows a 3D representation of SEP copolymer (a) casted and (b) spin casted both from toluene and annealed at R.T. In Figure 4c,d, the 3D representation of the SI/sulf annealed at 80 °C and spin casted from (a) toluene and (b) THF can be observed to examine the differences in phase and topography.

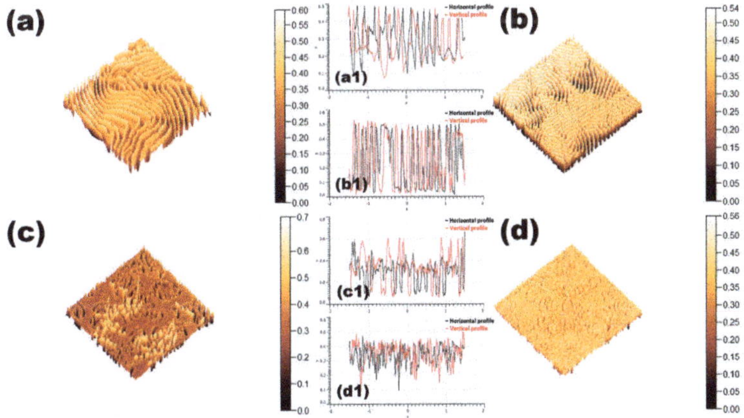

Figure 4. 3D representation of AFM images (3 μm × 3 μm) for SEP copolymer: (**a**) casted and (**b**) spin casted from toluene and annealed at R.T. In (**c**,**d**), the 3D representation of a SI/sulf annealed at 80 °C spin cast from (**c**) toluene and (**d**) THF. Inner images (**a1**–**d1**) show the distribution of roughness values found on the surface.

The effect of the casting technique is obvious in Figure 4a,b, where lamellae morphology was observed for both samples, but phase differences are evident. For the casted sample, the roughness showed value of 0.094, while in the spin casted the roughness is increased to 0.157 due to the effect of the substrate. Interestingly, the skewness is less than 1, with values of 0.10 and 0.42, respectively, indicating almost symmetrical distribution of peaks and valleys, with the peaks being slightly more. Moreover, the thickness of the lamellas is also affected by the casting technique, showing 63 nm for the PS and 90 nm for the PEP for the casted samples, whereas for the spin-casted ones, the values are decreasing to 46 nm (PS) and 79 nm (PEP). These differences are due to the slower solvent evaporation rate from the casted technique that results in longer times for lamellas to develop and create thicker domains. For the spin-cast sample, it is also observed that the faster evaporation rate creates a more uneven distribution of the different phase values observed from the histogram in Figure 4b compared to Figure 4a.

On the other hand, from Figure 4c,d, the effect of the solvent evaporation of toluene (c) vs. THF (d) can be seen. The sample from toluene led to a not-fully organized structure, but with potential to become fully organized. In contrast, the THF one showed a lamellae morphology but was not fully organized. The not fully organized domains lead to smooth surfaces with roughness of 0.099 and 0.052, where the size of the domains can be found at 72 for the modified PI (Figure 4c) and 35 for the PS (Figure 4d). In addition, the sample spin-cast from THF has evaporated faster, leading to the formation of regions with potential crystallites, as indicated by the high percentage of one phase, which is likely due to the crystallite phase.

4. Conclusions

This work evaluated the morphologies of three copolymer samples by modifying only the PI segments (hydrogenation and sulfonation) of an initial PS-*b*-PI diblock copolymer. By adjusting casting techniques, annealing temperatures, and solvents, distinct topological structures were observed through AFM characterization. The different interactions between polymeric blocks and solvents over the various experimental parameter led to four primary morphologies: (i) lamellae, (ii) spherical micelles, (iii) worm-like, and (iv) network. Most samples displayed organized domains, primarily consisting of lamellae.

The spin-cast samples predominantly exhibited disordered morphologies, especially at 120 °C across all solvents. Cast samples showed alternating lamellae structures using toluene and THF at 80 °C. Cyclohexane produced spherical micelles (annealed at R.T. to 100 °C), while THF showed mainly worm-like structures. The chemical nature of the polymeric blocks also influences morphology, with the pristine diblock showing a lower interaction parameter than SEP, resulting in more disordered morphologies compared to the ordered structures of SEP. The SI/sulf copolymer's interactions are complex due to multiple components—PI, PI/sulf, and PS—leading to intricate block/block and block/solvent interactions.

The sizes of organized domains depend on experimental parameters, with SEP micelles ranging from 73 nm to 80 nm and SI/sulf showing a similar trend. Using spin-casting keeps the PS domain size constant while increasing PEP up to 179 nm. The solvent also affected sizes.

The lamellae morphology of different copolymers revealed significant variations. PS domain sizes increased from 56 nm (SI) to 63 nm (SEP) and 82 nm (SI/sulf) respectively. In addition, PI's initial size was 27 nm, changing to 90 nm (SEP) and 79 nm (SI/sulf).

Overall, it should be noted that the observed morphologies are specific to certain AFM experimental conditions, which is a limitation. Although the AFM is a powerful technique for characterizing the self-assembly of nanostructures, it has to be used in parallel with other techniques to verify these structures. A future perspective is the preparation of copolymers with three blocks: fully hydrogenated, partially sulfonated, and modified with hydrophobic segments.

Supplementary Materials: The following supporting information can be downloaded at: https://www.mdpi.com/article/10.3390/polym17223047/s1, Figure S1: ^{1}H-NMR for the sulfonation of the sample SI; Figure S2: All morphologies observed by AFM phase images (3 μm × 3 μm) with a 3D representation of a 1 μm × 1 μm and a profile of the morphology of 1.5 μm diagonal; Table S1: Characteristic values of roughness Sa, skewness Sk and domain sizes of bright and dark areas for all samples. The abbreviation for the samples is sample-solvent-casting technique-annealing temperature.

Author Contributions: Conceptualization, N.P. and A.A. methodology, N.P. and G.K. investigation, N.P. resources, G.K. and A.A. data curation, N.P., A.A., and G.K. writing—original draft preparation, N.P. writing—review and editing, N.P., G.K., and A.A. supervision, G.K. and A.A. project administration, G.K. and A.A. funding acquisition, G.K. and A.A. All authors have read and agreed to the published version of the manuscript.

Funding: Funding from MICIU/AEI/10.13039/501100011033/ and FEDER (UE), is acknowledged for grant n° UEPID2021-126417NB-I00. N.P would like to acknowledge the European Union for the funding, HORIZON-WIDERA-2022-TALENTS-02, grant number Project 101090325—PuDeRegels.

Data Availability Statement: The original contributions presented in this study are included in the article/Supplementary Materials. Further inquiries can be directed to the corresponding author.

Acknowledgments: Technical and human support provided by SGIker (Advanced Research Facilities, created by the University of the Basque Country, San Sebastian, Spain) is also acknowledged.

Conflicts of Interest: The authors declare no conflicts of interest.

References

1. Rodríguez-Hernández, J.; Chécot, F.; Gnanou, Y.; Lecommandoux, S. Toward 'smart' nano-objects by self-assembly of block copolymers in solution. *Prog. Polym. Sci.* **2005**, *30*, 691–724. [CrossRef]
2. Darling, S.B. Directing the self-assembly of block copolymers. *Prog. Polym. Sci.* **2007**, *32*, 1152–1204. [CrossRef]
3. Kim, J.K.; Yang, S.Y.; Lee, Y.; Kim, Y. Functional nanomaterials based on block copolymer self-assembly. *Prog. Polym. Sci.* **2010**, *35*, 1325–1349. [CrossRef]
4. Mavroudis, A.; Avgeropoulos, A.; Hadjichristidis, N.; Thomas, E.L.; Lohse, D.J. Synthesis and Morphological Behavior of Model Linear and Miktoarm Star Copolymers of 2-Methyl-1,3-Pentadiene and Styrene. *Chem. Mater.* **2003**, *15*, 1976–1983. [CrossRef]
5. Winey, K.I.; Thomas, E.L.; Fetters, L.J. The ordered bicontinuous double-diamond morphology in diblock copolymer/homopolymer blends. *Macromolecules* **1992**, *25*, 422–428. [CrossRef]
6. Avgeropoulos, A.; Dair, B.J.; Hadjichristidis, N.; Thomas, E.L. Tricontinuous Double Gyroid Cubic Phase in Triblock Copolymers of the ABA Type. *Macromolecules* **1997**, *30*, 5634–5642. [CrossRef]
7. Breiner, U.; Krappe, U.; Thomas, E.L.; Stadler, R. Structural Characterization of the "Knitting Pattern" in Polystyrene-block-poly(ethylene-co-butylene)-block-poly(methyl methacrylate) Triblock Copolymers. *Macromolecules* **1998**, *31*, 135–141. [CrossRef]
8. Politakos, N.; Weinman, C.J.; Paik, M.Y.; Sundaram, H.S.; Ober, C.K.; Avgeropoulos, A. Synthesis, molecular, and morphological characterization of initial and modified diblock copolymers with organic acid chloride derivatives. *J. Polym. Sci. Part A Polym. Chem.* **2011**, *49*, 4292–4305. [CrossRef]
9. Karayianni, M.; Pispas, S. Block copolymer solution self-assembly: Recent advances, emerging trends, and applications. *J. Polym. Sci.* **2021**, *59*, 1874–1898. [CrossRef]
10. Doerk, G.S.; Stein, A.; Bae, S.; Noack, M.M.; Fukuto, M.; Yager, K.G. Autonomous discovery of emergent morphologies in directed self-assembly of block copolymer blends. *Sci. Adv.* **2023**, *9*, eadd3687. [CrossRef]
11. Kulkarni, A.A.; Doerk, G.S. Thin film block copolymer self-assembly for nanophotonics. *Nanotechnology* **2022**, *33*, 292001. [CrossRef]

12. Hu, X.-H.; Xiong, S. Fabrication of Nanodevices Through Block Copolymer Self-Assembly. *Front. Nanotechnol.* **2022**, *4*, 762996. [CrossRef]
13. Xiang, L.; Li, Q.; Li, C.; Yang, Q.; Xu, F.; Mai, Y. Block Copolymer Self-Assembly Directed Synthesis of Porous Materials with Ordered Bicontinuous Structures and Their Potential Applications. *Adv. Mater.* **2023**, *35*, 2207684. [CrossRef] [PubMed]
14. Kuperkar, K.; Tiwari, S.; Bahadur, P. Chapter 15—Self-Assembled Block Copolymer Nanoaggregates for Drug Delivery Applications. In *Applications of Polymers in Drug Delivery*, 2nd ed.; Misra, A., Shahiwala, A., Eds.; Elsevier: Amsterdam, The Netherlands, 2021; pp. 423–447. ISBN 9780128196595. [CrossRef]
15. Han, H.; Oh, J.W.; Park, J.; Lee, H.; Park, C.; Lee, S.W.; Lee, K.; Jeon, S.; Kim, S.; Park, Y.; et al. Hierarchically Ordered Perovskites with High Photo-Electronic and Environmental Stability via Nanoimprinting Guided Block Copolymer Self-Assembly. *Adv. Mater. Interfaces* **2022**, *9*, 2200082. [CrossRef]
16. Gabor, A.H.; Lehner, E.A.; Mao, G.; Schneggenburger, L.A.; Ober, C.K. Synthesis and lithographic characterization of block copolymer resists consisting of both poly (styrene) blocks and hydrosiloxane-modified poly (diene) blocks. *Chem. Mater.* **1994**, *6*, 927–934. [CrossRef]
17. Wan, L.; Ruiz, R.; Gao, H.; Albrecht, T.R. Self-Registered Self-Assembly of Block Copolymers. *ACS Nano* **2017**, *11*, 7666–7673. [CrossRef]
18. Park, M.; Harrison, C.; Chaikin, P.M.; Register, R.A.; Adamson, D.H. Block Copolymer Lithography: Periodic Arrays of ~10^{11} Holes in 1 Square Centimeter. *Science* **1997**, *276*, 1401–1404. [CrossRef]
19. Kim, D.H.; Suh, A.; Park, G.; Yoon, D.K.; Kim, S.Y. Nanoscratch-Directed Self-Assembly of Block Copolymer Thin Films. *ACS Appl. Mater. Interfaces* **2021**, *13*, 5772–5781. [CrossRef]
20. Yang, G.G.; Choi, H.J.; Han, K.H.; Kim, J.H.; Lee, C.W.; Jung, E.I.; Jin, H.M.; Kim, S.O. Block Copolymer Nanopatterning for Nonsemiconductor Device Applications. *ACS Appl. Mater. Interfaces* **2022**, *14*, 12011–12037. [CrossRef]
21. Bal, M.; Ursache, A.; Tuominen, M.T.; Goldbach, J.T.; Russell, T.P. Nanofabrication of integrated magnetoelectronic devices using patterned self-assembled copolymer templates. *Appl. Phys. Lett.* **2002**, *81*, 3479–3481. [CrossRef]
22. Huang, C.; Bai, K.; Zhu, Y.; Andelman, D.; Man, X. Design and Fabrication of Nano-Particles with Customized Properties using Self-Assembly of Block-Copolymers. *Adv. Funct. Mater.* **2024**, *34*, 2408311. [CrossRef]
23. Chen, L.; Phillip, W.A.; Cussler, E.L.; Hillmyer, M.A. Robust Nanoporous Membranes Templated by a Doubly Reactive Block Copolymer. *J. Am. Chem. Soc.* **2007**, *129*, 13786–13787. [CrossRef] [PubMed]
24. Hamta, A.; Ashtiani, F.Z.; Karimi, M.; Moayedfard, S. Asymmetric block copolymer membrane fabrication mechanism through self-assembly and non-solvent induced phase separation (SNIPS) process. *Sci. Rep.* **2022**, *12*, 771. [CrossRef] [PubMed]
25. Jackson, E.A.; Hillmyer, M.A. Nanoporous Membranes Derived from Block Copolymers: From Drug Delivery to Water Filtration. *ACS Nano* **2010**, *4*, 3548–3553. [CrossRef] [PubMed]
26. Hibi, Y.; Wiesner, U. Surface Segregation and Self-Assembly of Block-Copolymer Separation Layers on Top of Homopolymer Substructures in Asymmetric Ultrafiltration Membranes from a Single Casting Step. *Adv. Funct. Mater.* **2021**, *31*, 2009387. [CrossRef]
27. Listak, J.; Hakem, I.F.; Ryu, H.J.; Rangou, S.; Politakos, N.; Misichronis, K.; Avgeropoulos, A.; Bockstaller, M.R. Effect of Chain Architecture on the Compatibility of Block Copolymer/Nanoparticle Blends. *Macromolecules* **2009**, *42*, 5766–5773. [CrossRef]
28. Qi, L.; Colfen, H.; Antonietti, M. Synthesis and characterization of CdS nanoparticles stabilized by double-hydrophilic block copolymers. *Nano Lett.* **2001**, *1*, 61–65. [CrossRef]
29. Yang, G.G.; Choi, H.J.; Li, S.; Kim, J.H.; Kwon, K.; Jin, H.M.; Kim, B.H.; Kim, S.O. Intelligent block copolymer self-assembly towards IoT hardware components. *Nat. Rev. Electr. Eng.* **2024**, *1*, 124–138. [CrossRef]
30. Alexandridis, P.; Tsianou, M. Block copolymer-directed metal nanoparticle morphogenesis and organization. *Eur. Polym. J.* **2011**, *47*, 569–583. [CrossRef]
31. Kumar, L.; Singh, S.; Horechyy, A.; Fery, A.; Nandan, B. Block Copolymer Template-Directed Catalytic Systems: Recent Progress and Perspectives. *Membranes* **2021**, *11*, 318. [CrossRef]
32. Jeon, E.J.; Jean-Fulcrand, A.; Kwade, A.; Garnweitner, G. A room-temperature high performance all-solid-state lithium-sulfur battery enabled by a cross-linked copolymer@ceramic hybrid solid electrolyte. *Nano Energy* **2022**, *104 Pt A*, 107912. [CrossRef]
33. Angelopoulou, P.P.; Moutsios, I.; Manesi, G.-M.; Ivanov, D.A.; Sakellariou, G.; Avgeropoulos, A. Designing high χ copolymer materials for nanotechnology applications: A systematic bulk vs. thin films approach. *Prog. Polym. Sci.* **2022**, *135*, 101625. [CrossRef]
34. Escobar Barrios, V.A.; Herrera Nájera, R.; Petit, A.; Pla, F. Selective hydrogenation of butadiene–styrene copolymers using a Ziegler–Natta type catalyst: 1. Kinetic study. *Eur. Polym. J.* **2000**, *36*, 1817–1834. [CrossRef]
35. Kaditi, E.; Mountrichas, G.; Pispas, S. Amphiphilic block copolymers by a combination of anionic polymerization and selective post-polymerization functionalization. *Eur. Polym. J.* **2011**, *47*, 415–434. [CrossRef]

36. Perin, F.; Motta, A.; Maniglio, D. Amphiphilic copolymers in biomedical applications: Synthesis routes and property control. *Mater. Sci. Eng. C* **2021**, *123*, 111952. [CrossRef]
37. Wu, M.Q.; Wu, S.; Cai, Y.F.; Wang, R.Z.; Li, T.X. Form-stable phase change composites: Preparation, performance, and applications for thermal energy conversion, storage and management. *Energy Storage Mater.* **2021**, *42*, 380–417. [CrossRef]
38. Zuppolini, S.; Salama, A.; Cruz-Maya, I.; Guarino, V.; Borriello, A. Cellulose Amphiphilic Materials: Chemistry, Process and Applications. *Pharmaceutics* **2022**, *14*, 386. [CrossRef]
39. Xu, K.; Liu, Y. Studies of probe tip materials by atomic force microscopy: A review. *Beilstein J. Nanotechnol.* **2022**, *13*, 1256–1267. [CrossRef]
40. Sudersan, P.; Müller, M.; Hormozi, M.; Li, S.; Butt, H.-J.; Kappl, M. Method to Measure Surface Tension of Microdroplets Using Standard AFM Cantilever Tips. *Langmuir* **2023**, *39*, 10367–10374. [CrossRef]
41. Sychev, D.; Schubotz, S.; Besford, Q.A.; Fery, A.; Auernhammer, G.K. Critical analysis of adhesion work measurements from AFM-based techniques for soft contact. *J. Coll. Inter. Sci.* **2023**, *642*, 216–226. [CrossRef]
42. Navindaran, K.; Kang, J.S.; Moon, K. Techniques for characterizing mechanical properties of soft tissues. *J. Mech. Behav. Biomed. Mat.* **2023**, *138*, 105575. [CrossRef]
43. Norman, M.D.A.; Ferreira, S.A.; Jowett, G.M.; Bozec, L.; Gentleman, E. Measuring the elastic modulus of soft culture surfaces and three-dimensional hydrogels using atomic force microscopy. *Nat. Protoc.* **2021**, *16*, 2418–2449. [CrossRef] [PubMed]
44. Aldrich-Smith, G.; Jennett, N.M.; Hangen, U. Direct measurement of nanoindentation area function by metrological AFM. *Int. J. Mater. Res.* **2005**, *96*, 1267–1271. [CrossRef]
45. Gisbert, V.G.; Amo, C.A.; Jaafar, M.; Asenjo, A.; Garcia, R. Quantitative mapping of magnetic properties at the nanoscale with bimodal AFM. *Nanoscale* **2021**, *13*, 2026–2033. [CrossRef] [PubMed]
46. Lim, S.; Park, H.; Yamamoto, G.; Lee, C.; Suk, J.W. Measurements of the Electrical Conductivity of Monolayer Graphene Flakes Using Conductive Atomic Force Microscopy. *Nanomaterials* **2021**, *11*, 2575. [CrossRef]
47. Torkhov, N.A.; Buchelnikova, V.A.; Mosunov, A.A.; Ivonin, I.V. AFM methods for studying the morphology and micromechanical properties of the membrane of human buccal epithelium cell. *Sci. Rep.* **2023**, *13*, 10917. [CrossRef]
48. Demchenkov, E.L.; Nagdalian, A.A.; Budkevich, R.O.; Oboturova, N.P.; Okolelova, A.I. Usage of atomic force microscopy for detection of the damaging effect of $CdCl_2$ on red blood cells membrane. *Ecotoxicol. Environ. Saf.* **2021**, *208*, 111683. [CrossRef]
49. Politakos, N.; Moutsios, I.; Manesi, G.-M.; Artopoiadis, K.; Tsitoni, K.; Moschovas, D.; Piryazev, A.A.; Kotlyarskiy, D.S.; Kortaberria, G.; Ivanov, D.A.; et al. Molecular and Structure–Properties Comparison of an Anionically Synthesized Diblock Copolymer of the PS-*b*-PI Sequence and Its Hydrogenated or Sulfonated Derivatives. *Polymers* **2021**, *13*, 4167. [CrossRef]
50. Hadjichristidis, N.; Iatrou, H.; Pispas, S.; Pitsikalis, M. Anionic polymerization: High vacuum techniques. *J. Polym. Sci. Part A Polym. Chem.* **2000**, *38*, 3211–3234. [CrossRef]
51. Hansen, C.M. *Hansen Solubility Parameters a User's Handbook*, 2nd ed.; CRC Press: Boca Raton, FL, USA; Taylor & Francis Group: Abingdon, UK, 2007.
52. Miller-Chou, B.A.; Koenig, J.L. A review of polymer dissolution. *Prog. Polym. Sci.* **2003**, *28*, 1223–1270. [CrossRef]
53. Van Krevelen, D.W.; Te Nijenhuis, K. *Properties of Polymers: Their Correlation with Chemical Structure; Their Numerical Estimation and Prediction from Additive Group Contributions*, 4th ed.; Elsevier: Amsterdam, The Netherlands, 2009.
54. Lindvig, T.; Michelesen, M.L.; Kontogeorgis, G.M. A Flory–Huggins model based on the Hansen solubility parameters. *J. Fluid Phase Equilibria* **2002**, *203*, 247–260. [CrossRef]
55. Farrell, R.A.; Fitzgerald, T.G.; Borah, D.; Holmes, J.D.; Morris, M.A. Chemical Interactions and Their Role in the Microphase Separation of Block Copolymer Thin Films. *Int. J. Mol. Sci.* **2009**, *10*, 3671–3712. [CrossRef] [PubMed]
56. Jeong, J.W.; Park, W.I.; Kim, M.-J.; Ross, C.A.; Jung, Y.S. Highly Tunable Self-Assembled Nanostructures from a Poly(2-vinylpyridine-*b*-dimethylsiloxane) Block Copolymer. *Nano Lett.* **2011**, *11*, 4095–4101. [CrossRef] [PubMed]
57. Niu, S.; Saraf, R.F. Stability of Order in Solvent-Annealed Block Copolymer Thin Films. *Macromolecules* **2003**, *36*, 2428–2440. [CrossRef]
58. Shin, J.M.; Kim, Y.; Yun, H.; Yi, G.-R.; Kim, B.J. Morphological Evolution of Block Copolymer Particles: Effect of Solvent Evaporation Rate on Particle Shape and Morphology. *ACS Nano* **2017**, *11*, 2133–2142. [CrossRef]
59. Xue, L.; Zhang, J.; Han, Y. Phase separation induced ordered patterns in thin polymer blend films. *Prog. Polym. Sci.* **2012**, *37*, 564–594. [CrossRef]
60. Miwa, Y.; Usami, K.; Yamamoto, K.; Sakaguchi, M.; Sakai, M.; Shimada, S. Direct Detection of Effective Glass Transitions in Miscible Polymer Blends by Temperature-Modulated Differential Scanning Calorimetry. *Macromolecules* **2005**, *38*, 2355–2361. [CrossRef]
61. Almdal, K.; Koppi, K.A.; Bates, F.S.; Mortensen, K. Multiple ordered phases in a block copolymer melt. *Macromolecules* **1992**, *25*, 1743–1751. [CrossRef]

62. Nilsson, S.; Bernasik, A.; Budkowski, A.; Moons, E. Morphology and Phase Segregation of Spin-Casted Films of Polyfluorene/PCBM Blends. *Macromolecules* **2007**, *40*, 8291–8301. [CrossRef]
63. Lai, C.; Russel, W.B.; Register, R.A.; Marchand, G.R.; Adamson, D.H. Phase Behavior of Styrene–Isoprene Diblock Derivatives with Varying Conformational Asymmetry. *Macromolecules* **2000**, *33*, 3461–3466. [CrossRef]
64. Wu, Y.-H.; Lo, T.-Y.; She, M.-S.; Ho, R.-M. Morphological Evolution of Gyroid-Forming Block Copolymer Thin Films with Varying Solvent Evaporation Rate. *ACS Appl. Mater. Interfaces* **2015**, *7*, 16536–16547. [CrossRef]
65. Albert, J.N.L.; Epps, T.H., III. Self-assembly of block copolymer thin films. *Mater. Today* **2010**, *13*, 24–33. [CrossRef]
66. Luo, M.; Epps, T.H., III. Directed Block Copolymer Thin Film Self-Assembly: Emerging Trends in Nanopattern Fabrication. *Macromolecules* **2013**, *46*, 7567–7579. [CrossRef]

MDPI AG
Grosspeteranlage 5
4052 Basel
Switzerland
Tel.: +41 61 683 77 34

Polymers Editorial Office
E-mail: polymers@mdpi.com
www.mdpi.com/journal/polymers

www.ingramcontent.com/pod-product-compliance
Lightning Source LLC
LaVergne TN
LVHW070454061126
851858LV00011B/235
9783725880232